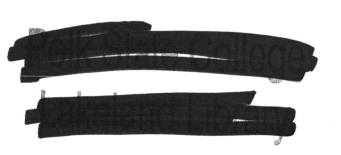

The Conversion to Sustainable Agriculture

Principles, Processes, and Practices

Advances in Agroecology
Series Editor: Clive A. Edwards

The Conversion to Sustainable Agriculture

Principles, Processes, and Practices

Edited by
Stephen R. Gliessman
Martha Rosemeyer

CRC Press
Taylor & Francis Group
Boca Raton   London   New York

CRC Press is an imprint of the
Taylor & Francis Group, an **informa** business

CRC Press
Taylor & Francis Group
6000 Broken Sound Parkway NW, Suite 300
Boca Raton, FL 33487-2742

© 2010 by Taylor and Francis Group, LLC
CRC Press is an imprint of Taylor & Francis Group, an Informa business

No claim to original U.S. Government works

Printed in the United States of America on acid-free paper
10 9 8 7 6 5 4 3 2 1

International Standard Book Number: 978-0-8493-1917-4 (Hardback)

Library of Congress Cataloging-in-Publication Data

The conversion to sustainable agriculture : principles, processes, and practices / editors: Stephen R. Gliessman, Martha Rosemeyer.
    p. cm. -- (Advances in agroecology)
  Includes bibliographical references and index.
  ISBN 978-0-8493-1917-4 (hardcover : alk. paper)
  1. Sustainable agriculture. 2. Sustainable agriculture--Case studies. I. Gliessman, Stephen R. II. Rosemeyer, Martha. III. Title. IV. Series: Advances in agroecology.

S494.5.S86C665 2010
630--dc22
                                       2009040253

Visit the Taylor & Francis Web site at
http://www.taylorandfrancis.com

and the CRC Press Web site at
http://www.crcpress.com

Contents

SECTION I   Basic Principles

SECTION II   Global Perspectives

SECTION III   The Way Forward

Preface

This book project began many years ago when the second editor, while still a graduate student, was asked by the first editor to carry out a literature search on the conversion process from conventional to alternative agroecosystems. During the course of this research, funded at that time by the Noyce Foundation, Martha Rosemeyer encountered Stuart Hill's three-level classification system for conversion. Using agroecology as a methodological tool for both researching and promoting the conversion process, and with growing awareness that any change in agriculture also implies social transformations, we eventually added a fourth level to Hill's taxonomy. We described the four levels of conversion in *Agroecology: The Ecology of Sustainable Food Systems* (CRC Press, 1997), but it remained to explore more deeply what conversion meant and to learn how it was actually proceeding around the world. With continuing support from the Ruth and Alfred Heller Chair in Agroecology at University of California–Santa Cruz (UCSC), we conceived of this project and pushed the book forward.

Eric Engles carried out his editing magic on all parts of the book, and ultimately was the person who really extracted the work from all of us. Master indexing was done by Michael Brackney. John Sulzycki, at CRC/Taylor & Francis, with all of his commitment to agroecology, created the space for this project in the first place. And finally, we sincerely appreciate and acknowledge the hard work of all the chapter authors in promoting the conversion process around the world, and thank them for their patience in bringing the book to completion.

Contributors

Antonio M. Alonso
Centro de Investigación y Formación de
   Agricultura Ecológica y Desarrollo
   Rural
Santa Fe (Granada), Spain

E. Ann Clark
Plant Agriculture
University of Guelph
Guelph, Ontario, Canada

David Dumaresq
The Fenner School of Environment and
   Society
The Australian National University
Canberra, ACT, Australia

Saan Ecker
The Fenner School of Environment and
   Society
The Australian National University
Canberra, ACT, Australia

Fernando R. Funes-Monzote
Estación Experimental "Indio Hatuey"
Universidad de Matanzas
Central España Republicana, Perico,
   Matanzas, Cuba

Stephen R. Gliessman
Department of Environmental Studies
University of California, Santa Cruz
Santa Cruz, California

Alba González Jácome
Universidad Iberoamericana AC
Santa Fe, Mexico City, Mexico

David Granatstein
Center for Sustaining Agriculture and
   Natural Resources
Washington State University
Wenatchee, Washington

Gloria I. Guzmán
Centro de Investigación y Formación de
   Agricultura Ecológica y Desarrollo
   Rural
Santa Fe (Granada), Spain

Kazumasa Hidaka
Agroecology Rural Community
   Management
College of Agriculture
Ehime University
Tarumi, Matsuyama, Japan

David Huggins
U.S. Department of Agriculture–
   Agricultural Research Service
   (USDA-ARS)
Washington State University
Pullman, Washington

Rachael J. Jamison
Washington State Department of
   Ecology
Lacey, Washington

Steve Jones
Mount Vernon Northwestern Research
   and Extension Center
Washington State University
Mount Vernon, Washington

Alireza Koocheki
Faculty of Agriculture
Ferdowsi University of Mashhad
Mashad, Iran

Carol Miles
Washington State University
Mount Vernon Northwestern
    Washington Research and Extension
    Center
Mount Vernon, Washington

Takuya Mineta
Laboratory of Environmental
    Evaluation
Department of Rural Environment
National Institute for Rural Engineering
    (NIRE)
Kannondai, Tsukuba, Japan

Joji Muramoto
Program in Community and
    Agroecology (PICA)
University of California, Santa Cruz
Santa Cruz, California

James Myers
Oregon State University
Department of Horticulture
Corvallis, Oregon

John H. Perkins
The Evergreen State College
Olympia, Washington

Paul Porter
Agronomy/Plant Genetics
University of Minnesota
St. Paul, Minnesota

María del Rocío Romero Lima
Programa de Agricultura Orgánica
Universidad Autonoma Chapingo
Chapingo, México

Martha E. Rosemeyer
The Evergreen State College
Olympia, Washington

Lori Scott
University of Minnesota
St. Paul, Minnesota

Steve Simmons
University of Minnesota
St. Paul, Minnesota

Jennifer Sumner
Adult Education and Community
    Development Program
OISE/University of Toronto
Toronto, Ontario, Canada

Section I

Basic Principles

1 The Framework for Conversion

Stephen R. Gliessman

CONTENTS

1.1 THE NEED FOR CONVERSION

As we near the end of the first decade of the twenty-first century, we are confronted with an increasing number of signs that our global food system is rapidly approaching, if not already in, a condition of crisis. Issues and problems that go beyond the litany of environmental degradation, pest and disease resistance, loss of genetic diversity, increasing dependence on fossil fuels, and others (Gliessman, 2007) now confront us, creating what is increasingly being called the food crisis. We now face a dramatic rise in food prices, increases in hunger and malnutrition, and even food riots in places in the world where people no longer have access to sufficient food. Making things worse, too many small traditional and family farmers have been forced off their land and out of agriculture due to a wide variety of reasons, including the neoliberalization of trade policy, the loss of support for local food production systems, the entrance of speculative financial capital into food markets, changes in diets and food preferences that accompany greater access to global markets, the agrofuel boom and resulting diversion of food energy to feed the global demand for energy, and the enormous spike in the cost of petroleum in 2008 that caused a rise in the cost for all fossil-fuel-based inputs to agriculture (Rosset, 2006, 2008).

On a global scale, agriculture was very successful in meeting a growing demand for food during the latter half of the twentieth century. Yields per hectare of basic crops such as corn, wheat, and rice increased dramatically, food prices declined, the rate of increase in food production was generally able to keep up with the rate of population growth, and chronic hunger diminished. This boost in food production was due mainly to scientific advances and technological innovations, including the development of new plant varieties, the use of fertilizers and pesticides, and the growth of extensive infrastructures for irrigation. But the elements of the food crisis noted above are signs that this era of ever-rising food production may be coming to an end. We may be approaching a limit in the amount of food that we can produce relatively inexpensively, given the limited amount of arable land left on the earth and the degraded condition of much that is already being cropped.

At the same time, we face a problem that in the long-term will be even more challenging to the global food system: the techniques, innovations, practices, and policies that have allowed increases in productivity have also undermined the basis for that productivity. They have overdrawn and degraded the natural resources upon which agriculture depends—soil, water resources, and natural genetic diversity. They have also created a dependence on nonrenewable fossil fuels and helped to forge a system that increasingly takes the responsibility for growing food out of the hands of farmers and farmworkers, who are in the best position to be stewards of agricultural land. In short, our system of agricultural production is unsustainable—it cannot continue to produce enough food for the growing global population over the long-term because it deteriorates the conditions that make agriculture possible.

Our global food system also faces threats not entirely of its own making, most notably the emergence of new agricultural diseases (such as mad cow and antibiotic-resistant salmonella), climate change, a growing demand for energy, and an approaching decline in the production of the fossil fuel energy that has subsidized agricultural growth.

Considering all these factors, it is clear that none of the strategies we have relied on in the past—creating higher-yielding varieties, increasing the area of irrigated land, applying more inorganic fertilizers, reducing pest damage with pesticides—can be counted on to come to the rescue. Indeed, it is becoming increasingly evident that these strategies, combined with the commoditization of food and the control of global food production by large transnational agribusiness interests, are a part of the problem, not its solution. The only way to avoid a deepening of the food crisis is to begin converting our unsustainable systems of food production into more sustainable ones. It is the goal of this book to establish a framework for how this conversion can be accomplished, and to provide examples from around the world where the conversion is under way.

1.2 GUIDING PRINCIPLES FOR CONVERSION

Farmers and ranchers have a reputation for being innovators and experimenters, constantly testing new seed, plants, breeds, inputs, and practices. They adopt new farming practices and marketing arrangements when they perceive that some benefit will be gained. The heavy emphasis on high yields and farm profits over the past 40 to 50 years has achieved remarkable results, but with an accompanying array of negative impacts that have restricted farmer-initiated innovation. After responding to this overriding economic focus in agriculture, many farmers are now choosing to make the transition to practices that not only are more environmentally sound in the short-term, but also have the potential for contributing to sustainability for agriculture in the long term (Gliessman, 2001). Several factors are driving the changes in our food systems that are facilitating this transition process. These include factors that range from on-farm issues to conditions well beyond farming communities:

- The uncertain cost of energy.
- The low profit margins of conventional practices.

- The development of new practices that are seen as viable options, especially in organic agriculture.
- Increasing environmental awareness among consumers, producers, and regulators.
- A better understanding of the close link between diet and the recent increases in health issues, such as obesity, diabetes, heart disease, and cancer.
- A growing appreciation for the need to integrate conservation and livelihoods in farming communities.
- New and stronger markets for organically and ecologically grown and processed farm products.

There are many factors that need to be dealt with in the process of converting to sustainable food systems. Many of these factors directly confront the farmer on the farm. As described in many of the chapters of this book, despite the fact that farmers often suffer both yield reduction and loss of profits in the first year or two after initiating conversion, most of those who persist eventually realize both economic and ecological benefits from having made the conversion. Obviously, a farmer's chances of making it through the transition process successfully depend in part on his or her ability to adjust the economics of the farm operation to the new relationships that come from farming with a different set of input and management costs. But as some chapters demonstrate, success in the conversion process is also dependent on factors beyond the farmer's control. These include the development of different marketing systems, pricing structures, policy incentives, and other changes that reach all aspects of the food system, from the grower on one end to the eater on the other.

While the economic goal of conversion is to maintain profitability, the ecological goal is to initiate a complex set of very profound changes. As the types of inputs change, and practices shift to ecologically based management, agroecosystem structure and function change as well. As some authors show in this volume, a range of ecological processes and relationships are altered, beginning with aspects of basic soil structure, organic matter content, and diversity and activity of soil biota. Eventually major changes also occur in the activity and relationships of weed, insect, and disease populations, especially the balance between beneficial and pest organisms. Ultimately, nutrient dynamics and cycling, energy use efficiency, and overall system productivity are impacted. Measuring and monitoring these changes during the conversion period can provide the foundations for developing practical guidelines and indicators of sustainability that will promote the changes that need to occur in the agriculture of the future.

The following principles serve as general guidelines for navigating the overall transformation that food systems undergo during the conversion process (Gliessman, 2007):

- Shift from extractive nutrient management to recycling of nutrients, with increased dependence on natural processes such as biological nitrogen fixation and mycorrhizal relationships.
- Use renewable sources of energy instead of nonrenewable sources.
- Eliminate the use of nonrenewable off-farm inputs that have the potential to harm the environment or the health of farmers, farmworkers, or consumers.

- When materials must be added to the system, use naturally occurring and local materials instead of synthetic, manufactured inputs.
- Manage pests, diseases, and weeds as part of the whole system instead of "controlling" them as individual organisms.
- Reestablish the biological relationships that can occur naturally on farms and ranches instead of reducing and simplifying them.
- Make more appropriate matches between cropping patterns and the productive potential and physical limitations of the agricultural landscape.
- Use a strategy of adapting the biological and genetic potential of agricultural plant and animal species to the ecological conditions of the farm rather than modifying the farm to meet the needs of the crops and animals.
- Value most highly the overall health of the agroecosystem rather than the outcome of a particular crop system or season.
- Emphasize the integrated conservation of soil, water, energy, and biological resources.
- Build food system change on local knowledge and experience.
- Carry out changes that promote justice and equity in all segments of the food system.
- Incorporate the idea of long-term sustainability into overall agroecosystem design and management.

To varying degrees, these principles are reflected in the conversion efforts described in the chapters of this book. The integration of these principles creates a synergism of interactions and relationships from the farm to the table that eventually leads to the development of the properties of sustainable food systems.

1.3 STEPS IN THE CONVERSION PROCESS

For many farmers and ranchers, rapid conversion to sustainable agroecosystem design and practice is neither possible nor practical. As a result, many conversion efforts proceed in slower steps toward the ultimate goal of sustainability, or are simply focused on developing food production systems that are somewhat more environmentally sound or slightly more economically viable or just. For the observed range of conversion efforts seen in this book, four distinct levels of conversion can be discerned. These levels—originally proposed by Hill as three steps (1985, 1998), and expanded to four levels in Gliessman (2007)—help us describe the steps that are actually taken in converting from modern conventional or industrial agroecosystems. They can serve as a map outlining a stepwise, evolutionary conversion process. They are also helpful for categorizing agricultural research as it relates to conversion.

- *Level 1: Increase the efficiency and effectiveness of conventional practices in order to reduce the use and consumption of costly, scarce, or environmentally damaging inputs.* The goal of this approach is to use inputs more efficiently so that fewer inputs will be needed and the negative impacts of their use will be reduced as well. This approach has been the primary emphasis of much of the agricultural research of the past four to five decades, through which numerous agricultural technologies and practices have been developed. Examples include optimal crop spacing and density, genomics, improved machinery, pest monitoring for improved pesticide application, improved timing of operations, and precision farming for optimal fertilizer and water placement. Although these kinds of efforts may reduce the negative impacts of conventional agriculture, they do not help break its dependence on external human inputs. While this may be a reason for arguing that they do not represent conversion at all, it must be recognized that in the real world of agriculture, level 1 efforts often represent a crucial foundation for initiating efforts at the other levels.

- *Level 2: Substitute conventional inputs and practices with alternative practices.* The goal at this level of conversion is to replace resource-intensive and environment-degrading products and practices with those that are more environmentally benign. The recent expansion in organic farming and ecological agriculture research has emphasized such an approach. Examples of alternative practices include the use of nitrogen-fixing cover crops and rotations to replace synthetic nitrogen fertilizers, the use of biological control agents rather than pesticides, and the shift to reduced or minimal tillage. At this level, the basic agroecosystem structure is not greatly altered; hence, many of the same problems that occur in conventional systems also occur in those with input substitution.
- *Level 3: Redesign the agroecosystem so that it functions on the basis of a new set of ecological processes and relationships.* At this level, overall system design eliminates or at least mitigates the root causes of many of the problems that still exist at levels 1 and 2. In other words, rather than finding sounder ways of solving problems, the problems are prevented from arising in the first place. Whole-system conversion studies allow for an understanding of yield-limiting factors in the context of agroecosystem structure and function. Problems are recognized, and thereby prevented, by internal site- and time-specific design and management approaches, instead of by the application of external inputs. An example is the diversification of farm structure and management through the use of rotations, multiple cropping, and agroforestry.
- *Level 4: Reestablish a more direct connection between those who grow the food and those who consume it, with a goal of reestablishing a culture of sustainability that takes into account the interactions between all components of the food system.* Conversion occurs within a social, cultural, and economic context, and that context must support conversion to more sustainable systems. At a local level, this means consumers value locally grown food and with their food purchasing, support the farmers who are striving to move through conversion level 1 to levels 2 and 3. In a sense, this means the development of a kind of "food citizenship," where everyone forms part of the system and both is able to influence change and be influenced by it. The more we move to this level of integration and action for change in food systems in communities around the world, the closer we move toward building a new culture and economy of sustainability (Hill, 1998; Gliessman, 2007).

In terms of research, agronomists and other agricultural researchers have done a good job of transitioning from level 1 to level 2. Research on the transition to level 3 has been very limited until recently, and work on level 4 is only just getting started. The chapters in this book describe work that is ongoing at several of these levels. The transition from level 1 to level 2 appears most commonly in the chapters of this book as the goal of reaching standards such as organic certification. As shown in Figure 1.1, we have seen considerable growth in the organic food industry in just the past decade, and this indicates that many farmers have reached level 2. The data presented here are from the sale of organic food in the United States, but are indicative of what is happening in other parts of the world as well.

But we must be sure that the movement toward sustainability does not stop at level 2. While the so-called mainstreaming of organic food availability signals a welcome shift in consumer consciousness, it also indicates that the most powerful elements of the conventional, industrialized food system are working to co-opt and contain change. We need to think beyond organic to all levels of the food system, with the idea of transcending product-focused thinking and maintaining a focus on achieving fully sustainable food systems.

In those chapters where agroecology provides the basis for researching level 3, we can see where the redesign and restructuring process is well under way. It is in those few examples in which all members of the food system value the principles of sustainability and relationship where we will we begin to find answers to larger, more abstract questions about the conversion process, such as what sustainability is

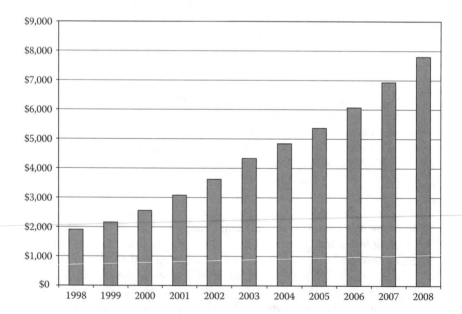

FIGURE 1.1 Sales of organic fruits and vegetables in the United States during the past decade. Sales are in the millions of U.S. dollars. (*Nutrition Business Journal* [http://nutritionbusinessjournal.com/natural-organic/news]; *Santa Cruz Sentinel*, March 18, 2009, pp. A1–A2.)

and how we will know we have achieved it. Ultimately, thinking about sustainability at level 4 can begin to guide the conversion process at all levels, promoting a more rapid transition to full food system sustainability for all parts, peoples, and scales of the global food system.

1.4   THE CHAPTERS IN THIS BOOK

The chapters that follow are highly diverse, each with a unique perspective shaped by the author's location, research, and central concerns. Some authors concentrate on explaining the challenges, while others look more closely at signs of progress and opportunities for change. Some choose a comprehensive overview approach, while others make use of more narrowly focused case studies and examples.

The second chapter provides a review of how researchers have attempted to apply the conversion framework, design experiments, and studies to evaluate conversion; carry out ecological, economic, and social analysis of results; and develop indicators that can tell us if particular conversion efforts are moving us toward sustainability. It is clear that we know how to study the pieces of agroecosystems separately, but we are still limited in our ability to work with the complexities of entire systems simultaneously. This is one of the reasons it is very easy to get stuck at level 2 in the conversion process.

The history of the conversion process as we have known it so far is essentially the history of the organic agriculture movement. This is the topic of Chapter 3. In this chapter, Jamison and Perkins trace the roots of the movement back to the early twentieth century and chronicle its development in the United States. They describe how, in its current phase of burgeoning popularity, the organic movement is in danger of getting stuck at level 2. Organic agriculture is increasingly being captured by market forces as production is concentrated in the hands of larger and vertically integrated growing, processing, shipping, and marketing operations. Knowing the history of the organic movement and the challenges it faces today provides the necessary context for understanding the conversion process as it is described in the chapters that follow.

Despite the fact that organic certification and expanding organic markets have motivated many farmers to convert to alternative production practices, it has also not been the only reason. As described by Porter, Scott, and Simmons in Chapter 4, there are many different constraints facing farmers in such places as the northwest Midwest of the United States. Farming in a difficult ecological transition zone with harsh winters and short growing seasons limits cropping options, and a combination of economic and social limitations limits choice and market access. But despite these limitations, farmers have been making the transition to more sustainable practices. The farmers themselves refer to an evolutionary or even "transformational" process they go through as they make the decision to change their farming systems, sharing in a set of revealing interviews how so much of the conversion process is determined by personal values, family needs, and even the degree of community support. Economics play an important role, but just deciding to farm differently, believing in the choice, and going through the learning process to make it happen shows how level 4 thinking is integral to driving the change process. As one farmer said, the conversion process is "not nearly complete, and we expect that it never will be. It's a biological system, alive and in need of observation, tending and rebalancing daily.... It takes time and a fair amount of courage and faith."

Every farming region faces different challenges to the sustainability of its food systems. Each region also takes different steps along the conversion pathway in confronting these challenges. As Miles et al. discuss in Chapter 5, the Pacific Northwest presents a spectrum of different challenges as one moves from the moister coastal areas through the interior valleys to the extensive dryland areas farther inland. The most obvious challenges are a combination of climatic, soil, and pest management issues. The most common approach to overcoming such challenges is to engage in level 1 conversion. Conventional farmers and researchers team up to develop new inputs and practices that increase efficiency, reduce environmental impacts, and improve economic return. But when we examine each region in particular, the value of different conversion approaches becomes apparent. The interior wheat-based dryland systems have tried to confront the overriding sustainability issue of soil erosion and degradation. At level 1, this has meant moving to conservation tillage to try to reduce soil exposure and loss to wind and water erosion. The description by Miles et al. of their ongoing research project that integrates direct seeding into crop residues with precision agriculture techniques, using multiple georeference points per acre that improve input use and yields, shows the strides that level 1 research can make. But the systems are still extremely intensive, and as the authors of this chapter state, the current systems are not sustainable. The conversion to level 2 and organic production still represents a very small proportion of the area farmed, and the farmers who have made this conversion face multiple challenges, including limited market access, a need for very intensive soil cultivation, and lack of appropriate seed and farming practices. Despite the fact that both these level 1 and 2 conversions generate beneficial ecosystem services, they are not rewarded at the marketplace. Some level 4 thinking is needed for this to happen.

In the intermediate orchard regions, Miles and coauthors show how conventional systems have been extremely innovative in developing new management technologies at level 1, and organic systems have developed very sophisticated management approaches at level 2, but neither one has advanced to the next two levels. Production remains intensive and single crop based, and market chains are long and distant from consumers. The authors' call for more direct market structures is a call for level 4 conversion, something that has become essential in the conversion process in the maritime horticultural zone. Under pressures from advancing urbanization, loss of agricultural land, and urban dwellers' concerns about noise, dust, smells, and pesticides, farmers in this region have moved ahead of their two interior counterparts with both level 3 and level 4 conversion steps. By moving to level 3 with crop rotations, diversification, and other redesign approaches, and to level 4 through direct marketing, farmers' markets, and community-supported agriculture (CSA), many farmers who were already at level 2 with organic certification are now moving beyond that toward sustainability.

In Chapter 6, Gliessman and Muramoto show how difficult it is for conventional strawberry growers on the central coast of California to take risks that threaten the economic viability of a crop that can cost in excess of $25,000 per acre to plant and maintain. Although some farmers in this area had begun to transition to level 2 organic production, it was not until a key input (methyl bromide) was banned that the pressure for conversion really took hold. The first step in the conversion process was the establishment of organic certification. But since there was so little knowledge about how to grow strawberries organically, the side-by-side comparison of conventional and organic plots during the transition process was a necessary first step. Once certification was achieved, however, continued monitoring and work with farmers was necessary in order to identify the limits to sustainability of the organic system. Once these limits could be identified, it became obvious that level 3 redesign needed to occur, and this type of work is ongoing, with many steps yet to be taken. Meanwhile, the push for alternative design and management strategies is being promoted by the level 4 connections that continue to develop between growers and consumers, with direct marketing, new relationships, and emerging understanding of food systems sustainability promoting deeper change.

The complexity of reaching levels 3 and 4 in the conversion process is highlighted by Clark and Sumner in Chapter 7. Their discussion of the contested nature of the term *sustainability* provides a useful way of thinking about how level 4 thinking can be generated *and* promoted. They refer to the work of McMurtry (2003) and his idea of "life capital" being "life-wealth that produces more wealth not just by sustaining it, but by 'value-adding' to it through providing more and better life goods." In other words, the values and beliefs inherent in the social, cultural, and environmental aspects of sustainability must be integrated with the economic aspects that so strongly guide most of conventional agriculture. The focus on enhancing the multiple and complex relationships that can occur in food systems becomes the focus of level 4 conversion. The lessons learned from farmers who have converted to organic production (mostly at level 2) give us ways of considering if organic is helping move food systems toward sustainability or diverting them from this goal. It is interesting to note that most farmers in Ontario would fall into what would be considered the small farmer category, and that off-farm income for these farmers was at least twice as much as on-farm income in the last census. As discussed above, combining these two modes of livelihood may be an important strategy for the economic side of farm system sustainability.

Clark and Sumner provide candid evidence from their analysis of Canadian organic agriculture that many farmers who convert do so for level 4 reasons—for "heartfelt concerns about personal and environmental health," coupled with an emerging alternative market structure with farmers' markets and CSA. The farmers who enter into conversion to organic are clear in their recognition of a range of ecological, agronomic, social, and economic constraints on the sustainability of the alternative systems. Since so many of the farmers who choose to farm differently mostly choose to "go it alone," a new set of economic and policy options are needed. Clark and Sumner show that no one event or issue pushes farmers in the direction of organic production, creating a "matrix of encouragement" that needs to be understood in order to move the conversion process forward more effectively. This can be aided greatly by further development of a range of ecological, social, and economic indicators. Clark and Sumner are especially effective in pointing out some of the social and ecological indicators that will help promote the conversion to level 4 in food systems, yet are realistic in the difficult challenges we face in ensuring that some of the same economic factors that are limiting options for alternative growers do not continue to threaten future food system sustainability.

In her review of the history of organic agriculture in Mexico (Chapter 8), Romero points out how the emergence of organic production systems has not been merely a matter of introducing technological change in agriculture, but is also a result of questioning the role of agriculture in society in general and of the kind of development model needed by the farming sector. She introduces arguments for level 4 thinking by raising the issue of food security and arguing for more equitable relationships between the rural and urban sectors, between agriculture and industry, and between energy and food policy. She calls for more participation of the peasant sector in the development of agricultural and food policy. She is clear in her claim that the elements of a new paradigm for food systems can be found in the peasant and indigenous communities that have spearheaded the conversion to organic agriculture in Mexico.

Interestingly, though, Romero also points out how—despite the fact that much of the conversion to organic is only taking place at level 2, with input substitution and policy support for the development of international export markets—there is the emergence of a "different kind of organic agriculture." This is one grounded in rural communities, and aims to build healthy soil, plants, animals, and human beings.

González Jácome goes even further back in history and describes the deep roots of traditional agriculture in Mexico, adding an even stronger cultural foundation to the argument that local knowledge, customs, and food system practices are crucial for sustainability. She is clear, however, about the seriousness of the threats to traditional agriculture, and how traditional farmers, despite having spearheaded much of the organic movement in Mexico, are also being threatened by globalization, lack of access to markets for their products, loss of local biodiversity, outmigration from rural communities, and a breakdown of the vital knowledge development and transfer processes so important for small-scale, rural cultures. There are many ways that traditional knowledge systems can help the conversion process, and most of them operate at levels 3 and 4. The human-directed selection and adaptation process that has gone on for eons must be allowed to continue, while it also provides local opportunity, farming system modification through the empirical process of trial and error, and the development of diverse sustainable livelihoods that protect local biodiversity and ecosystem services. But society must develop new priorities and policies to promote and protect traditional agriculture so that it can continue to serve as a foundation for the conversion to sustainability.

When faced with limited options or alternatives, a culture can also show remarkable resilience and ingenuity in the conversion process. As described by Funes-Monzote in Chapter 10, Cuba chose to make dramatic changes in its agricultural sector after the dissolution of socialist Eastern Europe and the USSR. With local agrarian production viewed as the key to food security for the country, Cuba developed a movement that has used input substitution focused at level 2 to transform a highly specialized, conventional, industrial, input-dependent food system into something far more sustainable. Due to the lack of access to external inputs and the persistence of local knowledge of how to manage diverse agricultural systems, the conversion was able to take place quickly and broadly (Funes et al., 2002). But as Funes-Monzote points out, it has become clear that neither the conventional model nor the input substitution model will be versatile enough to ensure the sustainability of an increasingly diverse and heterogeneous agriculture. It is time to move from level 2 to levels 3 and 4. The emerging mixed-farming systems that he reviews in Chapter 10 are excellent examples of the conversion to these next levels.

Guzmán and Alonso (Chapter 11) analyze the conversion process at yet another scale—that of the entire European Union. Here, a more unified political and economic structure has been called upon to promote the conversion to what is referred to as ecological agriculture. The conversion process most often begins at the individual farm level, with farmers entering into level 2 conversion in order to meet the requirements of alternative markets developing for ecological products, but this occurs within the context of shared agrarian legislation and common tools for the development of ecological agriculture that are variably applied in each EU country. The on-farm part of the conversion most often deals with farmers learning how to substitute conventional inputs and practices with accepted alternatives (level 2), but it also includes social and economic issues that go beyond this second level. The norms for ecological agriculture have been developed as part of EU policy, with subsidies often applied as an incentive for their adoption. But because the EU economic policy is more oriented toward intercountry commerce, and very little toward intracountry markets, economic barriers such as access to markets and credit complicate the conversion process considerably. Farmers generally need institutional support to successfully move to the next levels in the conversion process. Apart from direct economic subsidies, support can include funding for research and training in ecological agriculture, encouraging more local consumption of sustainable products, investment in alternative food chains, and even the development of legal structures that benefit alternative production systems. Each region or country has its own local character and set of conditions that can promote the transition to more sustainable food system levels, and as a result, each one needs its own unique set of programs, incentives, and structures.

We see another regional example of the conversion process in Chapter 12 (Muramoto et al.). Japan has a long history of small-scale, diverse, multifunction family farms. But as modernization has had its impacts, farms have begun to lose their biodiversity and closed nutrient cycles, and they have become more dependent on energy-intensive inputs. Coupled with the movement of people out of agriculture, the aging of those who are still in it, and the loss of food self-sufficiency, these trends have pushed Japan into a food system crisis. In spite of these trends and pressures, and in some cases in response to them, there is still a strong organic sector in the Japanese food system. Since rice is such an integral part of the Japanese diet, and because the Japanese have a strong preference for the taste and texture of rice grown in their own country, local rice systems have received considerable research attention, and this has helped them retain or reintegrate former sustainable practices and develop new, innovative ones. But it is the existing support for a culture of sustainability, most of it based in level 4 thinking, that forms the foundation of what could become a larger, more effective conversion movement in Japan.

In Chapter 13, Koocheki effectively integrates aspects of all levels of the conversion process as he describes the conversion to sustainability occurring in the Middle East, especially in Iran, and how the process might be accelerated. Considered to be one of the original centers of the origin of agriculture, the Middle East continues to harbor a rich heritage and culture of dryland farming, nomadic pastoralism, and sustainable water harvest and delivery systems. At the same time, population growth is putting pressure on food production, modernization has threatened the sustainability of water resources, excess fertilization is causing water pollution, and government policy promotes large-scale, simplified, high-input alternatives. Despite these challenges, progress is occurring as some farmers move through level 2 to level 3 in the conversion process. It will be level 4 thinking that will most likely move the process forward if, as Koocheki proposes, the region as a whole takes advantage of having a long tradition of small-scale, locally adapted, water-efficient, integrated agricultural practices from which it can draw for the conversion to sustainability.

Finally, Dumaresq and Ecker in Chapter 14 document how farmers in Australia have been working at level 3, especially when they design and implement more diverse cropping and grazing systems that integrate animals, crops, fallows, and pasture. The work of researchers—identifying indicators and monitoring the stages in the conversion process—is seen as an essential part of the conversion process. But for some of the more recent conversions, other issues and pressures are driving the conversion, such as nature conservation approaches, alternative food chains that better link farmers and consumers, and a growing environmental awareness and evolution of ethics and values grounded in the concept of sustainability. Level 4 appears to be gathering strength.

The success of all the movements toward sustainability documented in these chapters will depend in large part on how well farmer knowledge is combined with new agroecological principles and then linked out with the end users of the food system, the folks who gather around the table and give thanks for the sustainable systems that have brought them their food.

REFERENCES

Funes, F., García, L., Bourque, M., Pérez, N., and Rosset, P. 2002. *Sustainable agriculture and resistance. Transforming food production in Cuba.* Oakland, CA: Food First Books.

Gliessman, S.R. 2001. *Agroecosystem sustainability: Developing practical strategies.* Boca Raton, FL: CRC Press.

Gliessman, S.R. 2007. *Agroecology: The ecology of sustainable food systems.* Boca Raton, FL: CRC Press/Taylor & Francis Publishing Group.

Hill, S.B. 1985. Redesigning the food system for sustainability. *Alternatives* 12:32–36.

Hill, S.B. 1998. Redesigning agroecosystems for environmental sustainability: A deep systems approach. *Systems Research and Behavioral Science* 15:391–402.

McMurtry, J. 2003. The Life Capital Calculus. Paper presented at the Canadian Association for Ecological Economics Conference (CANSEE), Jasper, Alberta, Canada, October 18.

Rosset, P.M. 2006. *Food is different: Why we must get the WTO out of agriculture.* London: Zed Books.

Rosset, P.M. 2008. Food sovereignty and contemporary food crisis. *Development* 5:460–463.

2 What Do We Know about the Conversion Process?

Yields, Economics, Ecological Processes, and Social Issues

Martha E. Rosemeyer

CONTENTS

2.1   INTRODUCTION

As farmers reduce their dependence on externally produced agrochemical inputs for food production and convert to more sustainable agroecosystems, evaluating and documenting the success of these conversion efforts is paramount. Assessing the results of conversion using a variety of research methodologies and looking at the triple bottom line of ecological, economic, and social factors allows a more precise and complete picture to emerge. It also permits the identification of obstacles in conversion to organic and sustainable agriculture.

The most difficult period for farmers converting from agrochemical-intensive systems is the transition period. In developed countries (where agrochemically intensive production systems are the norm) this period is characterized by a reduction in yields compared to what was obtained in the former conventional system. Over time, productivity is recovered, but the depth of the yield decrease and the amount of time needed to complete the conversion process are crucial because they can spell the difference between success and failure. These important variables depend greatly on the type of crop or crops being farmed, the local ecological situation, the prior history of management and input use, and the particular weather conditions during the period of transition.

In developing countries, in contrast, adoption of organic techniques can mean higher yields. This may also be true when the initial system uses locally prevalent methods under field conditions (low-intensive or conventional with few agrochemical inputs). Badgley et al. (2007), in a survey of 293 publications of formal and informal literature, found that conversion to organic agriculture in developing countries resulted in a 20 to 90% increase in yields, whereas conversion in developed countries resulted in a 3 to 20% decrease in yields. Another study surveyed 208 conversion projects in 52 developing countries and determined that of the 98 projects with the most reliable yield data, intensification of cultural techniques enabled an average per-project increase in food production of 93% (Pretty et al., 2003).

Farmers in developing countries who were using relatively chemically intensive methods prior to conversion experience yield decreases, similar to their counterparts in developed countries. This suggests strongly that the effect on yield of the conversion depends on the intensity of the preconversion starting point (Parrott et al., 2006). In India, a study of seven farm pairs determined that the length of the transition period was positively correlated with the amount of mineral fertilizers previously used (de Jager and van der Werf, 1992). In Africa, conversion to organic fruit production for export was associated with increased yields in coffee, pineapples, and cacao, a result that the authors ascribe to the low amount of inputs used in conventional production (Gibbon and Bolwig, 2007).

The scope of the conversion—whether the farm is using input substitution to meet the current standards for certified organic production (level 2) or undergoing a full-scale system redesign (level 3)—is obviously a strong determinant of the length of the transition period. For some short-term annual crops, the time frame for a conversion to level 2 might be as short as three years, and for perennial crops and animal systems the time period is probably at least five years or longer. Level 3 conversions can be even more lengthy, due to the large-scale biological and infrastructural changes that are involved. Data are scant on level 3 transitions, so most of this discussion will focus on level 2 conversions. These primarily involve conversions from conventional to organic systems, but also included are conversions to bioorganic, biodynamic, and more sustainable production systems (as defined by the respective studies).

One the most difficult aspects of conversion from conventional to organic may be the effort involved in rethinking one's farm or agroecosystem, especially with the increased complexity of a level 3 system. The current agricultural focus on the production of only one or two cash crops may make the substitution of organic inputs for agrochemical (level 2) easier and more attractive than more complete redesign of the system. In the current conventional paradigm, the main crop is incorporated into a limited rotation. In a level 2 conversion there is often little change in the rotation sequence, and organically acceptable fertilizers and pesticides are substituted for conventional. Level 3 change, in contrast, demands the incorporation of diverse crops, the use of rotations, and the planting of perennials—which under certain circumstances of land ownership may be precluded. In addition to land tenure issues, transitioning to an organic system with rotation and perennials presents other challenges, such as new infrastructure needs and the need for diverse marketing strategies. The integration of livestock found on a diversified farm, though common in the past, challenges people without livestock experience. Additionally, sourcing organic stock and feed during certain periods may be a challenge to a more diverse level 3 system.

Why is the transition period important to study? For farmers converting from systems dependent on high agrochemical inputs, this is the critical period when the learning curve is steep and the farmer is not necessarily rewarded with profitability. If the transition period can be shortened or eliminated, or its challenges mitigated, this would allow more farmers to overcome a major barrier of conversion to organic or more sustainable production systems. In developing countries, understanding the yield increase that usually comes with conversion to organic practices can allow targeting of critical resources to small farmers in order to increase both food sovereignty and security for parts of the population experiencing the current food and economic crisis.

2.2 FARMER PARTICIPATION

A farmer's involvement in developing the parameters of the study (defining the question or hypothesis, determining the type of research, etc.) is an important factor affecting both the applicability of the results of a study and its success in encouraging farmers to make changes in their agricultural systems (Selener, 1997). Exactly how and where the study is conducted (on an experiment station or on a farm) can affect the engagement of farmers. Even if the study is not conducted on a commercial farm, the representativeness of the land chosen will determine credibility of the study in the eyes of potential adopters (Petersen et al., 1999). An optimal research situation is a functioning commercial crop production unit whose owner-operator wishes to convert to a recognized alternative type of management, such as certified organic agriculture, and wants to participate in the redesign and management of the farm system during the conversion process (Swezey et al., 1994; Gliessman et al., 1996). Such a "farmer first" approach is considered essential in developing viable farming practices that have a realistic chance of adoption.

There are various levels of farmer participation in on-farm studies; they range from trials in which researchers make the management decisions to those in which farmers are the decision makers. Although this forms a continuum, Selener roughly breaks the possibilities down to four categories: (1) researcher-managed on-farm trials, (2) consultative researcher-managed on-farm trials, (3) collaborative farmer-researcher on-farm trials, and (4) farmer-managed participatory research. To establish collaborative and farmer-managed studies, it is necessary to establish open, respectful, equal communication between farmers and researchers (Selener, 1997). Historically, communication between farmers and scientists has been difficult, especially cross-culturally (Dusseldorp and Box, 1993). However, whatever form is taken, the more farmer collaboration there is, the greater the chances of acceptance by farmers (Selener, 1997; Rosemeyer, unpublished manuscript).

2.3 TYPES OF CONVERSION RESEARCH

There are many possible approaches to the study of the conversion process and the transition period: case studies, surveys, on-farm comparisons, systems experiments, and single-factor experiments. Examples of each approach and discussions of what data each can yield are discussed below. Observations on experimental design, including control treatments and the lengths of experiments, follow those discussions.

2.3.1 CASE STUDIES

A case study approach on an individual farm that describes the system and the economic and social factors of conversion in depth may be a compelling choice for farmers. For example, a study of the Krusenbaum dairy farm's conversion to organic yielded important insights into critical social and economic factors (Posner et al., 1998). A case study of a converting vegetable farm in Cornwall, United Kingdom, has not only documented yields, weed control, pests and diseases, and soil fertility, but also examples of effective marketing and innovation (Sumption et al., 2004).

2.3.2 Surveys

Surveys can provide important data on such issues as what motivates farmers to initiate conversion and what social and economic factors are crucial in successfully negotiating the transition process. As a good example, Jamison (2003) surveyed conventional and organic farmers to study the values that caused them to undergo the transition process. Surveys can also be used comparatively to study different types of systems, approaches, or strategies. For example, Lockeretz (1995) compared equal numbers of conventional and organic farms. Additionally, survey data from a representative sample of conventional and organic farms in transition can be essential in choosing an appropriate site for an experimental study of conversion processes.

2.3.3 Farm Comparisons

A common type of study compares transitioning farms with conventional counterparts (Swezey et al., 1998; de Jager and van der Werf, 1992). Pairing transitioning farms and conventional farms with respect to physical factors can provide some control by eliminating confounding factors such as differing soil types and other environmental factors. As an example of this type of study, in South India each of six transitioning farms was compared to a conventional equivalent with respect to agronomic and economic factors, including labor, and these farm pairs studied over a six-year rotation (de Jager and van der Werf, 1992).

2.3.4 Systems Experiments

Large-scale experiments comparing the performance of systems of production have been found to be of specific interest to farmers considering conversion. This type of experiment compares complex systems side by side in order to understand how they function as a whole (Drinkwater, 2002). The emphasis of study is usually on the interrelationships among the components of the agricultural system, such as those between plants and animals and between plants and elements of the physical environment, such as soil and water. Both the components and their relationships vary when comparing one system with another (e.g., a corn–soybean rotation versus grazing), making a cause-and-effect relationship more difficult to determine. However, comparing an entire system itself to another entire system is more realistic, and thus more credible to farmers, since the farmer is essentially choosing one system over another in considering what type of operation to adopt. Systems experiments can take place on either an experiment station or a farm, and farmers are usually at least consulted on the management, if not included as a part of the decision-making group. This type of experimentation may provide insights into ecologically significant parameters, as well as interactions between fertilization, pests and disease, and the environment (Bellows, 2002). The Rodale Farming Systems Trial (FST) study in Pennsylvania, initiated in 1981, was one of the first systems studies in the United States. It compares grain rotations with and without animal inputs, that is, with manures or green manure crops, respectively (Peters, 1991; Petersen et al., 1999). The Wisconsin Integrated Cropping Systems Trials (WICST), begun in 1989, is another grains system study designed so that all rotations of each system are present in all years, facilitating comparison between systems (Posner et al., 1995).

The more similar the experiment is to real conditions, the greater the farmer confidence in the results. In systems experiments that involve transition to organic management, appropriately sized buffer strips between management treatments should be implemented; however, this has not always been the case (Lipson, 1997). Additionally, if all experimental treatments are present in all years, the effect of the year's weather can be separated out as a variable (this is the case, for example, in the WICST) (Posner et al., 1995). It is possible to maintain the same rotational sequence, eliminating as a variable the difference in the initial crops undergoing conversion, if the study is given a "staggered" start. This means that for a four-year rotation, it will take four years for all the treatments to be present in all years. This increases the size of the experiment because there are more treatments; however, it is warranted since the type of crop that begins the conversion affects both yield and economic profitability. For example, in livestockless conversions in the United Kingdom, even after three years following the two-year transition period, the nature of the specific crop treatments planted during the transition period was reflected in different soil mineral nitrogen levels, yields, and gross margins (Rollett et al., 2007). This type of experiment provides useful information for the farmer on how to avoid the economic hardship of the transition period.

2.3.5 SINGLE-FACTOR EXPERIMENTS

Factorial experiments can complement systems experiments. Factorial design can be used to isolate specific components and identify cause-and-effect relationships (Drinkwater, 2002). Additionally, laboratory experiments can be critical for determining the mechanism of an observed interaction in a systems experiment (Bellows, 2002). For example, at the two sites of the WICST plots, side experiments on weed control, rotational grazing, and cover crop selection, designed to isolate the specific factors responsible for observed system results, complemented the main systems experiments (Posner et al., 1995). More recently, experiments on certified organic land have been designed to help understand rotational effects on organic grain production (Hedtcke and Posner, 2005). Bulluck et al. (2002) compared alternative and synthetic fertilizer treatments in replicated field trials on three conventional and organic farms to see if beneficial microbes, including the fungus *Trichoderma*, were antagonistic to disease. On conventional farms, the use of alternative fertilizers, such as those used on organic farms, increased *Trichoderma* and decreased the plant pathogens *Pythium* and *Phythopthora*. These studies elucidate the mechanisms that explain the lower incidence of plant disease observed with alternative fertilizer use (Bulluck and Ristaino, 2002); as such, they contribute to an understanding of the transition process.

2.3.6 STUDY DESIGN CONSIDERATIONS

In addition to comparing transitional farms with conventional control plots or farms, ecological studies may benefit from comparison with a nearby site supporting relatively undisturbed habitat such as prairie or a forest (Leite et al., 2007) or a previous native or traditional agricultural system, such as a 1,000-year-old grass ley (Blakemore, 2000). Less overall energy (human, animal, or fossil fuel) is expended if the agroecosystem mimics the original ecosystem (Gliessman, 2000), and thus a native ecosystem may serve as an important baseline.

A number of authors have emphasized the extended amount of time needed to compare systems posttransition (Clark et al., 1998; Petersen et al., 1999; Reganold, 1988; Campos et al., 2000; de Jager and van der Werf, 1992), with the value of the experiment increasing over time. In a similar vein, a critique of methodologies for the comparison of organic and conventional farming systems emphasizes the importance of long-term case studies (Lee and Fowler, 2002). The differences between the organic and conventional treatments postconversion highlight critical indicators that might be followed during transition.

The idea of farmer and researcher knowledge being complementary has been proposed by a number of researchers (Kloppenburg, 1991; Lyon, 1996; Bigelaar, 1997; Dusseldorp and Box, 1993). This complementarity is reflected in the types of experiments that are most compelling to each group. Case studies of farms undergoing transition provide a deeper understanding of the interactions of social, ecological, and economic factors, prioritizing the most important interactions and identifying important indicators that more hypothesis-driven experiments can subsequently explore. Surveys set the context for the question and frame its relevance. Systems experiments (especially those most useful to a farmer) and factorial studies that test single parameters (which are typically employed by researchers) support and validate each other, overcoming the specificity of a particular set of results at one or two locations and providing a more precise mechanism or explanation for the results in question.

2.4   PARAMETERS FOR EVALUATING THE CONVERSION PROCESS

In addition to monitoring the yield during the transition process as is most common, it is important to consider a variety of other parameters. These can be divided into the categories of ecological, economic, and social; the specific factors to study depend on the nature of the question. Economic factors may include profitability and gross margin. Ecological factors may include soil microfauna, soil chemistry, populations of pests, the incidence of diseases, nutrient cycling, and energy use. Social factors may include labor, health of farmers and farmworkers, general farm family well-being, and whether the next generation takes over the farm. Expectations for how these factors may change both during transition and as a result of conversion may play an important role in motivating a farmer or farming family to make the conversion (Padel, 2008; Padel and Foster, 2006; Jamison, 2003).

2.4.1   THE YIELD PARAMETER

Of critical importance to the farmer, the yield parameter is the result of many factors. Yields during conversion can either decrease or increase, depending on the chemical intensity of the initial system, as mentioned above. When yields decline as a result of conversion, the length of the transition period is often defined as the number of years it takes for the yield of the converted system to return to the yield observed before conversion (or to reach the level of an equivalent conventional system).

Yields are reported in a number of ways; however, usually it is in terms of main cash crop productivity per unit of land. During the conversion to organic, which usually means moving to a longer rotation, yields from the entire rotational sequence should be assessed as more accurately indicative of the farmer's profit or loss. In addition to yields, determining management effort per unit of production may be especially instructive (Lee and Fowler, 2002), as is calculating the bottom line, or income less expenses. Since most published conversion studies are input substitution focused and concern one cash crop, the following review is organized according to type of crops grown: grains, vegetables and small fruits, perennials, and animals.

2.4.1.1 Grains

The transition time for temperate grain-based systems is usually reported to be about three to five years. In the Rodale Farming Systems Trial, the number of years it took for yields of organic maize and soybeans to equal those of conventional maize and soybeans was determined to be about four years (Petersen et al., 1999). In a study of the transition of grains to certified organic in Iowa, Delate and Cambardella (2004) concluded that the transition time was three to four years, with organic corn yield reaching conventional in its fourth year and soybeans in the third year. In Ohio, organic corn appeared to gain parity with conventional in year 2, but organic soybeans did not do so until year 3 (Stinner et al., 2004). In a 10-year study of grain production systems in Japan that started in 1976, the conversion from chemically-intensive to nature farming techniques required a three- to five-year transition period, based on the point at which soil and agroecosystem stabilization and equilibration had occurred (Nakamura et al., 2000).

2.4.1.2 Vegetables and Small Fruits

Vegetable crops may take a longer time to transition than small grains. In a number of studies, perhaps too short in length, organic vegetable yields did not reach conventional production levels. A three-year study of conversion from pasture to vegetables found that although sweet pepper yields were similar between organic and conventional during the three years of transition, conventional cucumbers had higher yields than organic under the first two years, and yields of conventional sweet corn surpassed organic in all three years (Russo and Taylor, 2006). Another study found that yields of transitional organic strawberries in their third year were still significantly lower than yields for conventional strawberries (Gliessman et al., 1996). In a study that averaged 21 years of production, including the transition period, Mäder et al. (2002) reported that organic potato yields were 58 to 66% of conventional yields, not yet reaching parity with conventional.

2.4.1.3 Perennials

For perennial crops, there appears to be great variability in the length of the transition period. On one hand, there is evidence that organic production for some crops has difficulty ever achieving the same yield as conventional; on the other hand, some studies have reported higher yields for organic. At three sites in Costa Rica, organic coffee yields four to six years after conversion from agrochemically intensive production were 44 to 72% those of conventional plots (Campos et al., 2000). During transition in California, apple yield was higher in the organic treatments in two of three years due to the efficacy of hand thinning in organic systems versus chemical thinning in conventional. Additionally, the number of fruits per tree and the total fruit weight per tree were higher in the organic than in the conventional, although average fruit size was smaller (Swezey et al., 1998). In Washington State, organic apples yielded similarly to conventional apples by the third year after conversion (Reganold et al., 2001). In California, cotton yields during a six-year transition did not seem to demonstrate a recuperation of yield, and the average yield reduction over this period was 34% (Swezey et al., 2007). In India, cotton yields were reduced during transition from conventional cotton, but in a survey of organic and conventional farms it was determined that postconversion parity with conventional farms was achieved (Eyhorn et al., 2007).

2.4.1.4   Animals

Dairy farmers generally experience lower milk production when converting to an organic, pasture-based system. Despite this lower production, most organic dairy farmers have transitioned successfully because of lower costs and the organic price premium (Friedman, 2003). In addition, if conventionally raised animals are first transitioned to a pasture-based system, the transition to organic is less difficult (Friedman, 2003). It is interesting to note that in transitional herds of milking sheep in Romania, reproductive indices, milk production, and milk quality increased relative to conventional (Man et al., 2007).

2.4.2   ECONOMIC ANALYSIS: PROFITABILITY

The economic return to an agricultural enterprise, often measured as gross margin,* is determined by a number of factors, including yields, costs of production, prices, and other market variables. Farmers who convert to organic production are often in a favorable position with respect to at least a few of these factors: yield can increase (especially for small holders in developing countries), input expenses may be lower, a price premium may exist for the organic products, there may be access to more lucrative markets, prices may be more stable, and there may be less competition with others (since supply chains are often more direct and organic is less of a commodity) (Parrott et al., 2006). Since so many factors determine the economic bottom line, overall profitability may be achieved even when one or two factors are not favorable compared to conventional production. Dairy farms in Pennsylvania that converted to organic production, for example, had higher costs for seed, forage, animal certification, and infrastructure, but were profitable due to the price premium

---

* "A 'gross margin' is the gross income from an enterprise less the variable costs incurred in achieving it. It does not include fixed or overhead costs such as depreciation, interest payments, rates, or permanent labor. The gross margin budgets are intended to provide a guide to the relative profitability of similar enterprises and an indication of management operations involved in different enterprises" (http://www.dpi.nsw.gov.au/agriculture/farm-business/budgets/about/intro).

for organic milk (Rotz et al., 2008). In India, farmers who converted from conventional to organic cotton reported increased profitability with no changes in yield, mainly because input expenses to organic cotton were 10 to 20% lower (Eyhorn et al., 2007).

Various modeling studies show positive economic effects as a result of converting to organic farming (Rost et al., 2007; Acs et al., 2007; Kerselaers et al., 2007). A Belgian linear programming model simulating an individual farm-level conversion to organic farming shows positive economic benefit; however, the economic potential of conversion depends on farm type and characteristics. The authors also identify the transition period as a time when there are higher risks and liquidity problems. A model based on data from the Netherlands shows that conversion to organic farming is more profitable than staying conventional, but passing through a two-year transition period is economically difficult. However, conversion can be less profitable if depreciating machinery is made superfluous by conversion or if price premiums for organic products are not sufficiently high (Acs et al., 2007).

2.4.2.1 Grains

The Rodale Farming Systems Trial (FST) reported long-term data on the economic factor with respect to grain production. In these trials, net returns (revenue minus explicit costs representing the return to transitional investment costs, unpaid family labor, and management, expressed in terms of dollars/acre) were a little more than half those of conventional during the years of the transition period, but were greater than conventional for each of the next two five-year periods after transition. While the biological transition in these systems (defined as the time it took for yields of all crops to reach parity with conventional) took four years, it was estimated that it might take more than a decade to recover the income lost during the transition period. However, these estimates were made without considering price premiums. Production costs in the organic plots of the Rodale FST were overall 26% less in the legume system, despite higher machinery costs due to expanded rotations (Petersen et al., 1999). Later analysis showed that the higher prices that organic foods command in the marketplace still make the net economic return per acre either equal to or higher than that of conventionally produced crops, even though cash crops cannot be grown as frequently over time on organic farms because of the dependence on rotations and other cultural practices to supply nutrients and control pests, and despite the fact that labor costs average about 15% higher in organic farming systems (Pimentel et al., 2005). In the WICST plots, postconversion gross margin analyses show that even in difficult years, organic production premiums allow commensurate economic profitability (Hedckte and Posner, 2005).

2.4.2.2 Vegetables and Small Fruits

Economic analyses of the conversion of vegetable and small fruit production systems are scant. A study of five farms in the United Kingdom undergoing conversion of vegetable production showed that net farm income declined by an average of 66% during transition due to a fall in output and higher labor costs, but eventually recovered to within 36% of preconversion levels. The authors conclude that the key factors influencing profitability in organic vegetable production are the starting financial position of the farm prior to conversion, the rate at which the farm converts, and the price of organic vegetables received once conversion is completed (Firth et al., 2004). In strawberry systems converted from conventional to organic, slower plant growth, lower yields, and increased labor requirements were observed, but price premiums for organic fruits permitted favorable economic returns (Gliessman et al., 1996). In vegetable and small fruit production in particular, it appears that price premiums are often the key to profitability.

2.4.2.3 Perennials

In a six-year study of conversion of California cotton, costs of production per bale were on average 37% higher for organic than for conventional cotton, primarily due to significantly lower yields in organic cotton compared with conventional cotton (Swezey et al., 2007). These costs would need to be absorbed by a price premium in order for organic production to be economically viable. In Switzerland, a study determined that even though labor costs in organic tree fruit production exceed those of integrated (diversified) fruit production by 7% due to blossom thinning by hand, manual weed management, and mouse control, the economic benefit of organic orchards was 16% higher than that of integrated fruit production (Weibel et al., 2004).

2.4.2.4 Summary

In summary, the studies, experiments, and models cited above reveal two important points. First, the economic indicator most useful for farmers is gross margin over time in comparison with a similar conventional system. Gross margins with and without price premiums, calculated over longer-term rotations, reflect most accurately the economic reality of the process of conversion. Second, the farmer is financially vulnerable during the transition period. The risk is that the lower yields seen during this period are not completely compensated for by price premiums (Kerselaers et al., 2007). Financial support during the transition period, therefore, may be critical to farm survival.

However, it is necessary to put the financial vulnerability of transitioning farmers in perspective. If the external environmental costs (externalities) of conventional agriculture—which are currently borne by society or postponed into the future—were quantified and included in the cost-benefit calculus, organic farming would be much more attractive economically (Vandermeer and Perfecto, 2005).

2.4.3 Ecological Analysis

Changes in yield during the transition period (both the decrease and the recovery) are due to the combined effects of a number of ecological factors, both environmental and biological. Ecological analysis can determine the critical mechanisms that control changes in yield.

During the transition from agrochemical-intensive to organic methods, changes in biological and ecological processes contribute to a reduction in yields or increase the costs of maintaining yield (Dabbert and Madden, 1986). The conversion from agrochemically intensive to organic production centers on the substitution of biological interactions for the application of chemical agents, and the desired biological interactions may take time to reestablish. In other words, the yield dip appears to be due to the lag time in the establishment of populations of beneficial organisms.

Biological interactions break down organic material to free plant nutrients; maintain checks on weeds, pests, and diseases; fix atmospheric nitrogen in the soil; affect soil texture and chemistry; favorably modify the crop environment; and so on. The presence of extensive biological interactions in organic systems is one of the key differences between conventional and organic farms (Lipson, 1997; Bellows, 2002; Petersen et al., 1999; Drinkwater et al., 1995).

Since biological interactions are key to the functioning of organic systems, the extent of these interactions in a particular system should be indicative of the system's progress in transitioning from a chemically dependent system to one based on biological processes. We might call this progress the degree of conversion. But since there are so many complex biological interactions in an organic system, how does one measure their extent? Practically speaking, there are three main methods for measuring biological activity and interactions:

1. Measure the specific *activity* involved in the interaction or its physical results (e.g., nitrogen fixation or pest damage).
2. Measure the *abundance* of the organisms responsible for the activity or interaction (e.g., the number of nitrogen-fixing bacteria or the size of pest populations).
3. Measure the *diversity* of the organisms responsible for the activity or interaction (e.g., the diversity of natural enemies of a pest).

Different cropping systems appear to rely on different types of biological interactions to achieve maximal function. This means that the biological interactions most crucial to reestablish during the transition process will vary by type of system. For example, the key relationships in grain-bean crop rotations appear to be those associated with weed control (Posner et al., 2008); in certain fruit and vegetable systems they are the interactions affecting nitrogen availability and control of insect pests and diseases (e.g., control of late blight in potato) (Pimentel et al., 2005; Piorr and Hindorff, 1984). Below, we review research in two categories of interaction:

1. Interactions affecting nutrient dynamics—specifically the activity, abundance, and diversity of soil microbial and macrofauna; and
2. Interactions affecting weed, insect, and disease activity and abundance and the diversity of their "natural enemies."

Energy use as a result of ecological changes during conversion and the effect at the ecosystem level will be reviewed at the end of this section.

2.4.3.1 Nutrient Dynamics

In the temperate zone, nitrogen is generally the most limiting nutrient, and a number of studies report that nitrogen deficiency is a major and widespread problem during conversion (e.g., Dabbert, 1986; Peters, 1991). Evidence that nitrogen limitation contributes to the yield dip during conversion is supported by reports of a smaller negative effect of conversion where legumes are the initial conversion crop.

Treatments during conversion in which N was supplied through legumes or was not needed as extensively (low-N-use crops) did not exhibit as great a reduction in yield during conversion. Legume crops generally maintain their yields during the conversion process (Dabbert and Madden, 1986), and legumes are used to build fertility during this critical phase. Researchers of the Rodale FST concluded that one should avoid starting a conversion with corn or other N-demanding crops (Peters, 1991). The above-mentioned UK study found that starting the conversion process with a soil-building phase that includes legumes (with small grains) resulted in higher average annual gross margins (Rollett et al., 2007). Where no nutrient-building phase took place in conversion, high nitrogen-demanding vegetables (e.g., sweet corn) did not reach yield parity over three transition years, although parity was reached after one year for bell pepper and two years for cucumber crops (Russo and Taylor, 2006).

2.4.3.2 Soil Nutrient Availability

There is widespread evidence that nitrogen limits yield during conversion. In a UK study of conversion to organic-field vegetable production, nitrogen availability was low in organic conversions, especially where there was a lack of investment in soil fertility (Sumption et al., 2004). Another UK study of stockless and compostless transition found that yield variation in oats was sensitive to fall mineral soil nitrogen levels and spring weed abundance (Sparkes et al., 2006).

During the transition period of an organic apple system in California, potentially mineralizable nitrogen and microbial biomass carbon were more sensitive indicators of system change than total nitrogen or organic carbon (Swezey et al., 1998). In the Rodale FST, potentially mineralizable nitrogen, labile carbon, and microbial biomass were significantly higher in organic systems (Drinkwater et al., 1995). In these experiments, soil nitrate leaching decreased in crop rotations with either legumes or animals, and these rotations retained significantly more nitrogen and carbon than conventional over a 15-year period, with implications for carbon sequestration in the soil (Drinkwater et al., 1998).

On Spanish calcareous soils, although similar amounts of N were applied, microbial biomass N was significantly greater in animal compost treatment than that of conventional with applied mineral fertilizer in two of five years. Interestingly, although applied P and K levels were lower in the organic animal compost treatment than the conventional, available P and K in the soil over five years was higher. The authors conclude that in the case of P, this is probably due to release from the calcareous soil or increased rapidity of soil P cycling due to higher microbial activity (Melero et al., 2008). This again supports the fundamental difference between organic and conventional agriculture with respect to the role of biota in system function.

2.4.3.3 Microbial Biomass

Breakdown of organic matter is conducted by soil microorganisms and macrofauna. Low availability of nitrogen in an organically managed system implies either lack of biological fixation or insufficient release of N from decomposition during the growth period. If not due to cold weather, both are symptoms of a low population or activity of appropriate organisms. A number of studies have found that microbial biomass increased with organic farming during the transition period and beyond (Smukler et al., 2008; Petersen et al., 1999; Mäder et al., 2002; Swezey et al., 1998). In a Washington State apple orchard, four years after conversion from conventional, microbial biomass increased and other biological properties were found to be enhanced (Glover et al., 2000). In the Rodale FST, soil respiration (which correlates with microbial biomass) was found to be higher in postconversion organic plots than in conventional plots (Petersen et al., 1999). Over 21 years of study, the flux of phosphorus through microbial biomass was faster in organic soils, and more P was retained in the microbial biomass than in conventional plots (Mäder et al., 2002).

Mycorrhizal fungus infection of plant roots in organic vegetable crops increased during the transition period (Smukler et al., 2008) and after (Petersen et al., 1999; Mäder et al., 2002). After 22 years, arbuscular mycorrhizal fungi (AMF) spore abundance and species diversity were significantly higher in the organic than in the conventional systems. Furthermore, the AMF community differed in the conventional and organic systems in that certain species were only found in organic systems. The authors conclude that some AMF species present in natural ecosystems are conserved under organic farming in comparison to conventional; this might severely impact the agroecosystem function of conventional agriculture (Oehl et al., 2004).

2.4.3.4   Soil Biodiversity

A number of studies show that higher diversity of soil biota is clearly correlated with increased ecosystem function (for a review see Balvanera et al., 2006). However, not all biodiversity (species richness) acts upon function in the same way; that is, not all biodiversity is created equal. For example, adding a nitrogen-fixing microorganism will have a different effect from adding a decomposer, yet both are adding diversity as measured by species richness. Additionally, there is much apparent redundancy in function among species of organisms, so functional diversity may actually be the critical question. Consequently, an increased diversity of functional groups appears to be more important than an increased number of species (Balvanera et al., 2006; Swift et al., 2004). In laboratory experiments, net N mineralization increased with a higher diversity of nematode life strategy groups. Likewise, the effect of earthworm species on nitrogen mineralization was dependent on their ecological traits (Postma-Blaauw, 2008).

In studies postconversion, soil biodiversity has been shown to increase over time in organically managed soil. In the Rodale FST, the long-term history of management (organic versus conventional) had more effect on microbial community composition than did residues applied, as determined by the methods of both fatty acid methyl ester analysis (FAME) and substrate utilization. Not only was the decomposer biomass greater in the organic treatments, but also species composition changed (Petersen et al., 1999). In a long-term study of bioorganic and biodynamic treatments versus conventional systems in Switzerland, the more diverse soil microbial community found in the organic treatments decomposed more ^{14}C-labeled plant material and light fraction particulate organic matter (POM) than did the community in conventional soils. These results support the hypothesis that a more diverse soil microbe and macrofauna community as found commonly in organic systems is more efficient in resource utilization (Mäder et al., 2002). The evidence as a whole indicates that management history is reflected in changes in soil microbe and macrofauna populations over the long-term, and the status of these soil communities at any one time is a major determinant of soil function and thus the stage of transition.

2.4.3.5 Nematode Community Composition

Nematodes are particularly important in making nutrients available to plants, since bactivorous nematodes excrete mineral N through their metabolism of bacteria. In one study, changing management from conventional to organic over five years both increased the diversity and changed the composition of nematode communities (Tsiafouli et al., 2007). A gradual decline of plant parasites and an increase in bactivorous and fungivorous nematodes were reported. Changing management regime caused a greater impact on genera-level changes in nematode populations than seasonal agricultural practices, and in conventional cultivation the latter were almost entirely masked. In another study, bactivorous nematodes became more common in the organic treatment over a four-year transition period, while plant parasitic root lesion nematode, *Pratylenchus crenatus,* became more predominant in the conventional. The structure of the nematode community was similar in the two treatments, which was attributed to excessive tillage in both treatments preventing population increases of the more tillage-sensitive groups, mainly the omnivorous and predatory nematodes (Briar et al., 2007). The WICST experiments also show a difference in nematode community composition between organic and conventional treatments; however, continuous conventional corn was found to have significantly fewer plant parasitic nematodes than organic after three years of transition (MacGuidwin, 1993). Differences in previous experiment management history were still reflected in nematode populations years after conversion (MacGuidwin, personal communication, 2001).

In two laboratory experiments complementing field experiments on conversion, the relationship between soil biodiversity and nitrogen cycling was studied. The difference in life strategies between nematode species *within* the same trophic group was found to be of importance for their communal effect on N mineralization (Postma-Blaauw, 2008). With respect to the nematode community, changes in community composition may take longer than the transition period to be detected, and to understand the effects of nematodes on nutrient cycling, we may need to refine our assessment of functional diversity so that it involves more than just diversity of and within trophic groups.

2.4.3.6 Internal Nutrient Cycling

Greater biodiversity and abundance of soil organisms suggests a greater on-farm capture of nutrients, and thus an increase in internal nutrient cycling, one of the central goals of sustainable agriculture. In a comparison of organic and conventional grain production systems over a transitional period of four years, nitrate nitrogen was found to be higher in the conventional system, whereas the organic system had higher N in the microbial biomass, indicating differences in nitrogen pools between the two systems (Briar et al., 2007). In an irrigated California tomato–corn rotation the organic system treatments, which received high inputs of organic matter (OM), "released NH^{4+} in a gradual manner and compared with the low OM input conventional system, supported a more active microbial biomass with greater N demand that was met mainly by NO^{-3} immobilization" (Smuckler et al., 2008, p. 185). The authors of this study found that during the transition period after conversion, the ratio of ammonia to nitrate was higher in the system converted to organic; however, lower nitrate levels were adequate for plant growth, indicating a higher rate of internal nutrient cycling (Smuckler et al., 2008). In systems with reduced tillage or reduced external inputs there was an increase in the abundance of certain soil biota associated with higher nutrient use efficiency (Postma-Blaauw, 2008). A challenge of transition may be in the synchronization of crop needs with C and N inputs so that plant needs are met without excessive nitrate leaching or microbial immobilization (Burger and Jackson, 2003).

Including animals in cropping systems can provide for a faster cycling of nutrients. Oberson et al. (1996) determined that in years 13 and 14 after conversion, bioorganic and biodynamic treatments that used manure had higher levels of phosphatase, higher mineralization of organic C, and consequently more organic P than treatments that used mineral fertilizer. A different P signature in the organic treatment was due to faster turnover of P through biota, especially in biodynamic and bioorganic. Oehl et al. (2001) found that 20 years after conversion, biodynamic systems using farmyard manure and slurry amended with biodynamic preparations showed faster microbiological cycling of P (due to higher phosphatase and higher turnover of organic substrates) than organic systems using slightly composted farmyard manure and slurry, and both performed better in this respect than conventional systems using agrochemical fertilizer.

2.4.3.7 Agroecosystem Diversity

One of the pillars of organic farming system practice is to increase total system biodiversity (Pimentel et al., 2005). This may be achieved by increasing the species and varieties or breeds of crops and livestock raised in a system (diversity in space), increasing crop rotation (diversity over time), removing biocides of all types, and maintaining wild areas. In paired studies of conventional and transitioning organic farms, trees and livestock were found to be more common on organic farms (de Jager and van der Werf, 1992). In a comparison to conventional farms, organic farms that had converted an average of 7.3 years earlier had larger areas of seminatural habitat and more diversity in arable fields (Gibson et al., 2007). Literature reviews examining the wildlife conservation value of organic farms find significantly greater abundance and species diversity of plants, birds, and bats than for conventional farms (Hole et al., 2005; Fuller et al., 2005). Another review of 66 papers found that organic farms, on average, had 50% greater abundance of wildlife and 30% more species than conventional farms (Bengtsson et al., 2005).

2.4.3.8 Summary

An important point that emerges from the research reviewed above is that deliberate, planned increases in the biodiversity of agroecosystems tend to result in increases in associated biodiversity as well (Swift et al., 2004). In other words, increasing the forms of agroecosystem diversity over which the farmer has direct control (crops, livestock, wild areas) results in increases in forms of diversity over which the farmer does not have direct control (soil microbe and macrofauna, terrestrial and aquatic wildlife, etc.). Additionally, since there appears to be a redundancy of function in the biological world, the diversity of functional groups (and perhaps diversity within certain trophic groups of nematodes), as opposed to species diversity, appears to be a key indicator of agroecosystem function.

The most useful parameters for ecological assessment of the conversion process may be the following: the amount of potentially mineralizable N and the ratio of ammonium to nitrate (as activity measures); the amount of microbial biomass, the extent of mycorrhizal fungus colonization, and counts of macrofauna (as abundance measures); and the taxonomic or trophic diversity of soil organisms (as a proxy for functional diversity).

2.4.4 PESTS

In the category of pests we include weeds, herbivorous insects, and diseases. All three types of pests can cause significant problems during the transition process, when the various biocides used in conventional systems have been withdrawn and biological control functions have not yet developed to their full potential. Therefore, studies involving measurement of pest damage and how it changes over time during the conversion process are important assessments of conversion.

2.4.4.1 Weeds

Although a few studies mention that weeds were not a problem during transition (e.g., Smukler et al., 2008), in most experiences during (and after) transition, weed competition appears to be a key factor reducing yield. In a study of cotton systems converted to organic, it was found that yield was significantly lower than conventional in all six years of transition studied; the authors conclude that weed pressure may best explain the yield difference (Swezey et al., 2007). The number of weeds and their biomass increased over a six-year transition period in a system converted to organic in Germany (Belde et al., 2000); in another study, Stinner et al. (2004) found that weeds impaired planting during wet springs and decreased organic yields during transition.

To assess the effect of weeds on yield in low-input systems compared to conventional, Posner et al. (2008) combined WICST data with that from other published reports on low-input systems (Table 2.1). Across the studies, the degree of weed control in the low-input systems correlated with yield. Closer examination of the data revealed an interaction between treatments and weather: in the 34% of site-years with low yields, wet weather made mechanical tillage difficult, preventing adequate weed control, and this resulted in yields averaging only 74% of conventional systems. In the other 66% of the cases, where mechanical weed control was successful, the yield of the low-input crops was 99% of conventional systems.

The weed seed bank has generally been found to increase during and after the transition period (Riemens et al., 2007; Turner and Bond, 2004; Kummel et al., 2005). Including cereals in rotations, as is common, appears to increase viable weed seed numbers (Turner and Bond, 2004). In an 11-year study to determine the effect

TABLE 2.1

Low-Input versus Conventional Cropping System Yields as Influenced by Weed Control, from Field Trials for Row Crops and Nonrow Crops

Study Citation	State (U.S.)	Sites	Site-Years	Weed Control[a]	Corn	Soybean	Small Grain[b]	Forage
Liebhardt et al., 1989[c]	PA	1	1	Poor	84			
		1	1	Good	112			
		1	2	Unrated		103	90w	
Porter et al., 2003	MN	2	6	Poor		64		
		2	8	Good		98		
		2	14	Unrated	92		100o	96
Delate and Cambardella, 2004	IA	1	1	Good	114	111		
Smith and Gross, 2006	MI	1	4	Poor	72			
WICST	WI	2	6, 9[d]	Poor	75	79		
			10, 13[d]	Good	98	94		
			15, 16[e]	Unrated			93w	100

The header "Low-Input Yield as a Percent of Conventional System Yield" spans the Corn, Soybean, Small Grain, and Forage columns.

Source: Data from Posner et al. 2008. *Agronomy Journal* 100:253–260. (With permission.)

[a] Weed control in the low-input system determined by visual ratings or biomass.

[b] Small grain: w = wheat and o = oat.

[c] Results given here are after three transition years. Authors presented forage yields for low-input systems but without comparison to conventional yields.

[d] Number of site-years for corn and soybean, respectively.

[e] Number of site-years for wheat and forage crops, respectively.

of an increased weed seed bank during transition, a one-time pulse of wild oat seeds was found to have relatively few long-term agronomic effects (Maxwell et al., 2007), suggesting that increases in the weed seed bank during conversion can potentially be controlled over the long-term.

With respect to control, the extent of the increase in the weed seed bank can be limited by vigilance in preventing weeds from forming seeds (Riemans et al., 2007). Additionally, a level 3 conversion to a diversified farming system that includes a range of crop species of different heights that provide cover is likely to provide the greatest opportunity for weed seed destruction by seed-eating animals (Heggenstaller et al., 2006).

Not all weeds compete effectively with the crop, and type of fertilizer may affect weed competition. Organic versus agrochemical fertilization can affect the competitive ability of the weed to reduce the biomass of the crop (Davis and Liebman, 2001). Although the abundances of various weeds increased with time during conversion to organic in a Finnish study, the abundance of only two of the species (*Elymus repens* and *Circium arvense*) was negatively correlated with crop dry weight (Riesinger and Hyvönen, 2006).

Weeds are not always detrimental to a system in their overall effect, as they can serve various positive functions ecologically (as insect "traps" or repellents, as habitat and food for beneficials, as components of higher biodiversity). In this context, it is important to note that weed diversity often increases during transition (de Jager and van der Werf, 1992; Turner and Bond, 2004). In a study to determine whether organic farming can restore weed diversity to preintensification levels, it was found that herbicide-sensitive nitrophilous species immediately increase and that perennials and species less responsive to high nitrogen apparently take longer to become established (Hyvönen, 2007).

2.4.4.2 Insects

Pest pressure during the conversion process may be more important in fruits and vegetables than in grains. Insect pressure from corn borer (*Ostrinia nubilalis*) and bean leaf beetle (*Cerotoma trifurcata*) was below economic threshold levels and did not appear to be significant in a Midwestern grain system in the process of conversion (Delate et al., 2002). However, demonstrating the vulnerability of fruit crops, it was found in a three-year apple conversion in California that secondary lepidopteran pests (apple leafroller and orange tortrix, *Argyrotaenia citrana*) caused greater fruit scarring in the converted fields than in the conventional fields in the last year of conversion; similarly, apple leafhopper (*Typhlocyba pomaria*) had denser populations and caused more leaf damage in the converted fields than in the conventional fields in the second and third years of transition (Sweezey et al., 1998).

Like weeds, insects do not always function as pests. Many insects (and other arthropods) play important roles in biological control as parasitoids and predators of herbivorous insects. Increased populations of these beneficials during and after conversion have been consistently reported (Drinkwater et al., 1995). In a study of strawberry conversion in California, there was little economically important pest damage in the organic system over the three-year study period, while at the same time an increase in naturally occurring insect predators was observed (Gliessman et al., 1996). In all but one year of a six-year study, cotton fields converted to organic production in California had significantly greater insect predators than conventional fields (Swezey et al., 2007). When nine conventional and organic farms in California were compared using canonical discriminant analysis, it was determined that pest abundance did not differ significantly between the two types of farms, but the organic farms had higher natural enemy abundance and greater species richness of all functional groups of arthropods (herbivores, predators, parasitoids, and others); the authors concluded that natural enemies appeared to substitute for pesticides (Letourneau and Goldstein, 2001).

An increase in beneficial insects can be facilitated by raising and releasing them as needed. In Cuba, during the large-scale conversion of the country to organic-style management, the monitoring systems and biological control became important tools in the management of pests. Local production of biological control agents replaced many of the former imported insecticides (see Funes-Monzote, Chapter 10, this volume).

Increased populations of birds may be significant with respect to control of insect, rodent, and weed populations. Rachmann et al. (2006) found that aerial hunters and raptors significantly preferred converted organic farms over conventional; they also found that the organic farms supported significantly higher densities of raptors during autumn and winter and more seed-eating and insect-eating birds in autumn.

2.4.4.3 Diseases

A lack of disease can indicate that ecological interactions are keeping pathogenic organisms in check and maintaining a level of nutritional health that increases crop plants' resistance to disease (Vaarst et al., 2004; Parrott et al., 2006); conversely, problems with disease during the conversion process may be the result of ecological interactions having not been fully reestablished. Crop rotations (a form of plant community change over time) have long been known to suppress disease (Curl, 1963), and organic farming usually involves more complex and lengthy rotations than conventional, particularly when a soil-building phase is included.

Benítez et al. (2007) determined that a transition management strategy involving the planting of mixed hay increased levels of bacterial populations associated with disease-suppressive bacteria more than tilled fallowing or open-field vegetable treatments. Soil from the hay treatment consistently had the lowest incidence of damping off in greenhouse tests, and was correlated with specific gene sequences presumably indicative of certain microorganisms involved with disease suppression. In another study, higher propagule densities of the disease-suppressive fungus *Trichoderma* were found in the soil of organic farms than were found in conventional farm soil. In addition, greater propagule densities of *Trichoderma,* thermophilic bacterial species, and enteric bacteria were detected in plots amended with organic matter than in plots fertilized with agrochemicals, and these higher densities were associated with lower densities of the propagules of the plant pathogens *Pythium* and *Phytopthora* (Bulluck et al., 2002).

In some cases, diseases can impact established, postconversion organic vegetable production. Mäder et al. (2002) found that yields in organic potato plots 21 years after conversion were 58 to 66% those of conventional equivalents due to potassium deficiency and late blight, *Phytophtora infestans.* With respect to fruit production in Europe, three diseases have been identified as key challenges to conversion of apple and pear due to lack of natural control: apple scab (*Venturia* sp.), sooty blotch (*Glosodes pomigena),* and fire blight (*Erwinia amylovora*) (Weibel et al., 2004). On the other hand, during large-scale, input substitution conversion of salad greens in California, 87 to 90% of visually inspected samples were without leaf or root disease (Smukler et al., 2008).

2.4.4.4 Summary

If problems with pests (weeds, insects, or diseases) increase during the transition period, they usually decrease over time as biological interactions develop and the ecological robustness of the system is restored; however, pest problems may not always disappear entirely. Also, certain environmental conditions favorable to disease or unfavorable to pest control measures—particularly unusual periods of cold and wet—may cause pest problems to increase temporarily.

Judging from the research reviewed above, the pest-related parameters that are most useful for assessing conversion are the following: the number of interventions required over time for control of weeds, insects, or diseases; the degree to which the use of toxics is reduced (even those agents allowed by the Organic Materials Review Institute and consistent with U.S. organic standards); and changes in populations of biological control organisms, including parasites, parasitoids, and predators of potential pests, as well as soil organisms known to suppress disease. When measuring parasitoid and predator populations, it is important to consider the data in light of pest population levels, since populations of parasitoids may be dependent on the number of hosts present, which may change as the system transitions (BaoYu et al., 2007).

2.4.5 ENERGY

About 72% of energy use on conventional farms is due to the energy embodied in fertilizers and pesticides (USEPA, 2008). Fossil-fuel–based nitrogen fertilizers, in particular, use much energy in their production via the Haber process. In addition, as agrochemical fertilizers applied to soil partially denitrify, they contribute to greenhouse gases through the release of methane (CH_4) and nitrous oxide (N_2O), which are 21 and 310 times more powerful, respectively, at causing radiative forcing than CO_2 (USEPA, 2008). A decrease in the use of nonrenewable energy and a corresponding increase in the use of renewable energy are characteristic of conversion to organic production. Much of this reduction comes from abandoning fossil-fuel-based nitrogen fertilizers.

When assessing changes in energy use during conversion, it is important to quantify both the energy use per hectare and the energy use per unit of production. For example, during the first three years of the WICST study, the conventional corn and corn-soy rotations used more energy than the organic, although the energy output to energy input ratios were lower in the organic plots due to lower yields (Posner et al., 1993). However, the WICST study found that the energy output to energy input ratio of the organic grain treatment postconversion was twice that of the conventional treatments (excluding human labor and the sun's energy) (Posner et al., 1995). In a long-term (21-year) study in Switzerland, Mäder et al. (2002) found that the energy required to produce a crop dry matter unit in the organic treatment was 20 to 56% lower than in the conventional treatment, which corresponded to a 34 to 53% lower use of energy per unit of land area. During the nine-year postconversion, the Rodale study found that the animal- and legume-based grain systems required about 30% less fossil fuel energy input per hectare than the conventional system, while grain yields in the two were statistically similar (except in one year) (Pimentel et al., 2005).

Similar differences in energy use are seen in perennial crops. In a study that compared energy use in organic and conventional apricot production in Turkey, it was found that the total energy requirement under conventional apricot farming was 38% higher than organic on a per-hectare basis; in these systems the ratio of energy output to energy input was 2.22 for organic and 1.45 for conventional (Gündoğmus and Bayramoglu, 2006). In another study, organic apple systems were found to be more energy efficient than conventional, with overall lower energy inputs and a 7% greater output-to-input ratio (Reganold et al., 2001). In a study in Costa Rica that used cluster analysis to compare 39 coffee farms grouped into three models of small coffee production, it was determined that the organic model achieved the best results from the point of view of energy efficiency; in this system 0.51 MJ kg^{-1} was invested to produce each 1 kg of coffee as cherry-like fruit, which was half of the energy required to produce 1 kg of coffee as cherry-like fruit in the conventional or mixed models of coffee production (Mora-Delgado et al., 2006).

A global-level analysis comparing the energy use of organic and conventional systems found that organic farming uses about 30% less energy than nonorganic per ton of cereal or vegetables produced and about 25% less for meat and dairy products (Azeez, 2007). In the United Kingdom, the Department of Environment, Food and Rural Affairs (DEFRA) carried out studies that found that the energy involved in organic production was significantly less per unit of production than conventional in 11 of the 15 crop or livestock operations examined. Only organic poultry, egg, potato, and long-season greenhouse tomato production systems used more energy than their conventional equivalents per unit of production, mainly due to lower yields. The study concluded that conversion to an organic diet would decrease energy use by 29% in comparison to conventional (DEFRA, 2008).

Often analyses of energy use appropriately place energy use calculations within a larger context of overall environmental impact, and include factors such as water use and greenhouse gas emissions. The authors of an Australian study that used life-cycle assessment* and included direct and indirect effects found that organic farming can reduce energy use, greenhouse gas emission, and the total water use involved with food production (Wood et al., 2006). A nine-year Michigan State University study concluded that the global warming potential of organic systems is only 43% that of conventional on a per-unit yield basis (Robertson et al., 2000). Another Australian study concluded that organic agriculture produces about half of the greenhouse gas intensity per unit that conventional farming produces (Wood et al., 2006).

In organic and conventional sugar cane production, calculated energy use and greenhouse gas emissions were similar per dry unit weight, due to the lower yield of the organic. The fossil fuel energy use avoided by not using synthetic fertilizers and chemicals is offset by the more intensive use of machinery and the transport of low-density nutrient sources in organic systems. However, growing organic cane may provide greenhouse gas benefits if the expected lower levels of denitrification with organic fertilization are taken into consideration. Additionally, the system enhances water and soil quality by eliminating inorganic fertilizers, herbicides, and pesticides, and it would be useful to quantify these benefits in economic terms (Renouf et al., 2005).

---

* "Life-Cycle Assessment is a technique for assessing the environmental aspects and potential impacts associated with a product, service or process. It compiles an inventory of a system's inputs and outputs evaluating the potential environmental impacts associated with these and interpreting the results in order to determine relative performance and scope for improvement where appropriate. It is now being applied to the analysis of agricultural systems and technologies with special reference to farming" (DEFRA, 2008).

In summary, although there are few studies of energy use during the conversion period per se, there are a number of postconversion energy analyses showing that organic farming uses 30 to 50% less energy than conventional, and that conversion to organic results in a 40 to 50% reduction in greenhouse gas emissions. Even though the energy use in organic systems is in general favorable compared to conventional, there is still in organic systems a great deal of energy use based on fossil fuels that has not been seriously addressed. Modeling studies in Denmark show that it is possible, through on-farm production of biofuels and biogas production from grass and clover, to reduce this use significantly (Halberg et al., 2008).

Energy use indicators to follow during the transition period are fossil fuel use per unit of production; the ratio of output of the crop in kcal to total energy inputs in kcal, both per hectare and per crop unit; ratio of energy output to fossil fuel input; greenhouse gas radiative forcing equivalents per unit of production; and life-cycle assessment of the product using both fossil fuel energy use and greenhouse gas emissions.

2.4.6 SOCIAL FACTORS ANALYSIS

A number of parameters have been evaluated in attempting to determine the social effects of converting to organic. These parameters include availability of employment, well-being of farm families, life satisfaction of farmers, and whether or not the farm is maintained by the next generation. Here, labor will be included as a social factor as well, although it also enters into economic analysis.

2.4.6.1 Labor

From the point of view of the owner, the higher labor intensity of organic farming can be a challenge. Both limited access to labor and the cost of labor have been identified as barriers to conversion (Strochlic and Sierra, 2007). Based on results from the Rodale FST, total labor needs are about 35% higher in organic grain systems; however, because the extra labor needs are distributed over the entire year, the farm family can perform most of the additional labor without hiring any workers, which means that the amount of nonfamily labor needed is about the same as that of conventional systems. Pimentel et al. (2005) conclude that organic systems need about 15% more labor than conventional on average, with the differential in individual systems ranging from 7 to 75%. In a survey of Washington State organic farmers, 57% said high labor costs were a considerable or moderate problem and 40% said the inability to find adequate labor was a considerable or moderate problem in making the organic farm successful (Goldberger, 2008). It has been difficult for small farmers to pay union wages for agricultural labor (there was only one unionized organic farm in California as of 2009, for example); however, with the rise of large organic operations (Halberg et al., 2006), there might be more of an opportunity for companies to respond in a financially appropriate manner, whether their workers are union or nonunion.

From the perspective of the rural economy and farmworkers, the higher labor requirement of organic farming is positive, since increasing employment benefits agricultural communities. Agricultural labor can enhance community well-being and rural social capital (Pearson, 2007). In the United Kingdom, organic farms provide 32% more jobs than conventional farms (Green and Maynard, 2006). A survey of 23% of all organic farms (1,144) in the United Kingdom and Republic of Ireland (IE) found that the farm-size-weighted full-time equivalent (FTE) per area for organic farms (4.33 FTE per 100 ha) was almost twice that of conventional farms. The authors predict that there would be 19% more farming jobs in the United Kingdom and 6% more farming jobs in the IE if 20% of the farms of both countries were to become organic (this compares to 1 to 2% of farms at present) (Morison et al., 2005).

2.4.6.2 Satisfaction and Motivation

Why do people choose to farm organically? When organic farmers express what has motivated them to convert to organic production, they most often mention improved food quality, improved health and environmental protection, professional challenge, greater fairness in the food chain, and maintenance of farm income (Padel, 2008). In general, U.S. organic farmers are college educated and choose to enter farming for "the joy of it" (Lockeretz, 1995). In Brazil, organic producers are characterized by a relatively high level of education and professional experience, and typically have enough land available to integrate farming, livestock production, and forest activities (Mazzoleni and Nogueira, 2006).

Supporting the theory of conventionalism* (Padel, 2008), there is currently data to suggest that the newer wave of farmers converting to organic in developed countries are more economically motivated than their predecessors (Jamar et al., 2007 [Belgium]; Tranter et al., 2007 [Great Britain]; Flaten et al., 2006 [Norway]; Strochlic and Sierra, 2007 [United States]). However, the data gathered in a focus group study targeting the values of organic producers entering the sector at different times in Austria, Italy, the Netherlands, the United Kingdom, and Switzerland do not support the idea that recently converting producers are less committed to core organic values than previously established organic farmers (Padel, 2008). In addition, it was determined that farmers in Poland are converting for reasons related to family health, safety of agricultural products, and soil condition (Padel and Foster, 2006).

In Belgium, converting beef farmers were grouped into four categories according to their motivations and perspectives: environmentalist, market, opportunist, and holist. Those motivated by the market achieved higher levels of production in their systems than did environmentalists and holists. The market-motivated group increased livestock density through feedstuff importation, which is a common technique of input substitution. Consequently, however, they applied more farm manure to fodder crops, leading to a higher nitrate load and increased leaching risk, compromising some environmental goals (Jamar et al., 2007). Not only does this support the conventionalist theory, but it also highlights the varying effects of the input substitution versus systems-level approaches with respect to environmental goals, and it

---

* Organic agriculture has moved from a loosely coordinated and local network of producers with a local certifier to a globalized system of trade linking spatially distant producers and consumers with a national certifying organization. Specialization and enlargement of farms, decreasing product prices, and increasing debt loads have been characteristics. This has been called "conventionalism" and is the description of the changes that conventional agriculture has passed through on the path to current globalization and specialization (Halberg et al., 2006).

challenges current organic standards and regulations based mainly on level 2 input substitution conversions.

In developed countries, an assessment of the likelihood of the next generation taking over the farm indicates the attractiveness of the farm system. In the United Kingdom, a survey of organic farmers shows that organic farmers are younger, more optimistic, and more entrepreneurial than their counterparts in conventional farming. The average age of organic farmers was 49, whereas conventional farmers averaged 56 years old. Some 64% of organic farmers versus 51% of conventional farmers expect their children to continue to farm. Three times more organic farms are involved in a direct marketing system or farmers' market than are conventional farms, indicating that organic farmers have a substantially different type of operation (DEFRA, 2008), which may contribute to farmer well-being.

2.4.6.3 Health and Well-Being

Health, including mental health, is an important measure of the social sustainability of a farming operation. As an extreme measure of the lack of well-being, the suicide rate among conventional farmers has been documented in developed and developing countries. Suicide has been mainly correlated with a high level of indebtedness and inability to pay off debts due to high interest rates, international competition, droughts, and the high cost of seeds, fertilizer, and pesticides. In India, the Independent Human Rights Law Network believes that more than 10,000 farmers have committed suicide over the last five years in five important farming states. According to the Indian National Crime Records Bureau, at least 87,567 farmers committed suicide in the country between 2002 and 2006 (Motlagh, 2008), although other estimates go as high 200,000 farmer suicides (Shiva, personal communication, 2008). The situation in India has been so severe that in 2006 the Indian government took over farm debt for the first time, assuming $840 million in loans. Most of this debt was due to farmers taking out loans to buy costly farm inputs. More recently state banks cancelled $15 billion of debt to small and marginal farmers (Motlagh, 2008).

In India, organic farming has been proposed as an antidote to the recent spike in the suicide rate associated with farmer indebtedness (Dogra, 2006). Where an organic farming project (Integrated Natural Sustainable Agriculture Programme) was introduced in a cotton-growing area in India, farmer suicides were significantly reduced and 88% of farmers reported a boost in confidence and self-reliance (Dogra, 2009). More extensive information and further documentation are needed. Additionally, in developing countries there is evidence that the growth of organic production has offset or reversed rural urban migration (e.g., in Niger [Hassane et al., 2002]). In a number of developing countries, organic farming has been associated with increased capacity to solve problems, increased self-esteem, and increased children's health and nutrition (Scialabba and Hattam, 2002).

In general there is a dearth of research on the social effects of conversion from conventional to organic production. Future studies might document how the suicide rate and health status of farmers change with conversion to organic production in developed and developing countries. More data concerning employment on organic farms in the United States is critical as well, as are surveys of social parameters, pre- and postconversion.

2.5 CONCLUSIONS: RESEARCH AND EXTENSION NEEDS

The rate and success of farmer conversions to organic production can be increased if more attention and funding are given to research and extension activities. The U.S. public or governmental sector is currently providing less technical support to organic agriculture than is justified by its economic and environmental benefits (Marshall, 1991; Lipson, 1997; Scooby et al., 2007). The level of funding of research in organic systems does not reflect organics' 4% market share, even with recent funding increases in the 2008 U.S. Farm Bill (Scowcroft, 2008). An investment in research on certified organic research areas and on-farm trials might help farmers negotiate the conversion process with less stress and more success (Karlen et al., 2007).

Of use to policymakers would be research using life-cycle analysis to compare conventional and organic options, as well as research that calculates the full costs (including externalities) to society of both types of production (Pearson, 2007). Research into both direct and indirect aspects of life-cycle assessments may be key in being able to assess total environmental impact, and might fill a critical information gap for policymakers and regulators in promoting the conversion to sustainable forms of agriculture (Pearson, 2007). For example, although Australian life-cycle assessments of organic farming show that direct energy use and greenhouse gas emissions are higher in organic production than in conventional, when other environmental effects and the indirect, externalized effects of energy use and greenhouse gases are taken into account, organic farming's impacts on the environment are significantly lower than those of conventional (Wood et al., 2007; Renouf et al., 2005).

Life-cycle assessments may also identify research topics and areas that need further regulation as part of organic standards. For example, a life-cycle assessment of dairy farming in the Netherlands reported better energy use and lower eutrophication potential per kilogram of milk for organic systems than conventional, but found that the organic system had higher global warming potential than conventional, implying that higher ammonia, methane, and nitrous oxide emissions occur on organic farms per kilogram of milk. This study helped identify environmental problems in organic dairying, that is, imported concentrated feed and roughage (Thomassen et al., 2008). A complete life-cycle assessment will also allow a more complete ecological appraisal of local versus nonlocal production, in addition to organic versus conventional.

Real improvement in agricultural sustainability requires that its assessment involve more than monitoring changes for research and policy evaluation purposes. Producers need practical information and guidance on how to change their systems to affect the triple bottom line of economic, ecological, and social factors (Lampkin et al., 2006). The lack of farmer-ready extension information on conversion, including availability of and access to production and market information as well as training in organic management systems, is considered a major barrier (Lohr and Salomonsson, 2000). In the EU, practical tools to evaluate conversion have been developed with stakeholder participation: these include farm income and financial monitoring and goal making, animal welfare assessments, socioeconomic assessments, and attempts to evaluate ecological, environmental, and social factors together. The experiences from these projects provide a basis for developing coherent, integrated approaches to sustainability assessment at the farm level, with the aim of developing practical management tools (Lampkin et al., 2006). In the United States, the lack of government support and funding has provided an opportunity for the nongovernmental organization, the Organic Trade Association, to initiate a Web site to facilitate conversion for both processors and growers (http://www.howtogoorganic.com/).

The conversion process is information-intensive relative to conventional production, and requires well-trained individuals who are proactive and holistic in their management strategies (Melone, 2006). Government extension agents can play a catalytic role in ensuring that adequate information is made available to those who need it. For example, in the Netherlands a network of transition advisors translate appropriate research for farmers and help them develop a farm and business plan. Integrated in a team with the other agriculture advisors, they assist the farmers to make the conversion (Zimmerman, 1997). In the United States, the training of extension agents to help farmers with conversion has had good but variable results (Park and Lohr, 2007).

The health of the entire system—including soil, plants, animals, and people—is the ultimate evaluator of the conversion process; however, evaluations of total system health during transition are few. Future research is needed at various levels: interdisciplinary research to develop, test, and monitor practical tools to aid farmers in conversion and maintain them in the organic program; life-cycle assessments for organic and sustainable production that can become tools for policymakers; and long-term interdisciplinary research to make a more apparent link between soil, plant, animal, and human health. Furthermore, it is important to determine the influence of international trade in organic products on local production and subsequent conversion.

ACKNOWLEDGMENTS

The author thanks the Jesse Smith Noyes Foundation for the graduate student stipend that supported the original literature search on conversion. Additionally, thanks go to Eric Engles for helpful editorial comments, and to Liza Rognas of the Evergreen State Library for persistence in procuring references. Special thanks to Mario Gadea-Rivas for patience and support during the writing of this chapter.

REFERENCES

Acs, S., Berentsen, P.B.M., Wolf, M. de, and Huirne, R.B.M. 2007. Comparison of conventional and organic arable farming systems in the Netherlands by means of bio-economic modelling. *Biological Agriculture and Horticulture* 24:341–61.

Azeez, G. 2007. Organic energy. *New Scientist* 2611:21.

Badgley, C., Moghtader, J., Quintero, E., Zakem, E., Chappell, M.J., Avilés-Vázquez, K., Samulon, A., and Perfecto, I. 2007. Organic agriculture and the global food supply. *Renewable Agriculture and Food Systems* 22:86–108.

Balvanera, P., Pfisterer, A.B., Buchmann, N., He, J.S., Nakashizuka, T., Raffaelli, D., and Schmid, B. 2006. Quantifying the evidence for biodiversity effects on ecosystem functioning and services. *Ecological Letters* 9:1146–56.

BaoYu, T., Yang, J., and Zhang, K. 2007. Bacteria used in the biological control of plant-parasitic nematodes: populations, mechanisms of action, *Microbial Ecology* 61:197–213.

Belde, M., Mattheis, A., Sprenger, B., and Albrecht, H. 2000. Long-term development of yield affecting weeds after the change from conventional to integrated and organic farming. *Zeitschrift für Pflanzenkrankheiten und Pflanzenschutz Sonderh* 17:291–301.

Bellows, B. 2002. An introduction to organic research. In *Southern organic resource guide: A reference handbook of organic resources in the South including Arkansas, Kentucky, Louisiana, Mississippi, and Tennessee*, 123–128. Available online at http://attra.ncat.org/sorg/research/.

Bengtsson, J., Ahnström, J., and Weibull, A.C. 2005. The effects of organic agriculture on biodiversity and abundance: A meta-analysis. *Journal of Applied Ecology* 42:261–69.

Benítez, M.S., Tustas, F.B., Rotenberg, D., Kleinhenz, M.D., Cardina, J., Stinner, D., Miller, S.A., and Gardener, B.B.M. 2007. Multiple statistical approaches of community fingerprint data reveal bacterial populations associated with general disease suppression arising from the application of different organic field management strategies. *Soil Biology and Biochemistry* 39:2289–301.

Bigelaar, C. den. 1997. *A synthesis of the FTPP farmer-initiated research and extension practices initiative in East Africa*. Network Paper 21f, ODI—Rural Development Forestry Network, pp. 9–18.

Blakemore, R.J. 2000. Ecology of earthworms under the 'Haughley experiment' of organic and conventional management regimes. *Biological Agriculture and Horticulture* 18:141–59.

Briar, S.S., Grewal, P.S., Somasekhar, N., Stinner, D., and Miller, S.A. 2007. Soil nematode community, organic matter, microbial biomass and nitrogen dynamics in field plots transitioning from conventional to organic management. *Applied Soil Ecology* 37:256–66.

Bulluck, L.R., Brosius, M., Evanylo, G.K., and Ristaino, J.B. 2002. Organic and synthetic fertility amendments influence soil microbial, physical and chemical properties on organic and conventional farms. *Applied Soil Ecology* 19:147–60.

Bulluck, L.R., III, and Ristaino, J.B. 2002. Effect of synthetic and organic soil fertility amendments on southern blight, soil microbial communities, and yield of processing tomatoes. *Phytopathology* 92:181–89.

Burger, M., and Jackson, L.E. 2003. Microbial immobilization of ammonium and nitrate in relation to ammonification and nitrification rates in organic and conventional cropping systems. *Soil Biology and Biochemistry* 35:29–36

Campos, E., Ramírez, G., Fonseca, C., and Obando, J.J. 2000. Programme for the production of organic coffee in Costa Rica. *Boletin PROMECAFE* 87/88:13–16.

Clark, M.S., Ferris, H., Klonsky, K., Lanini, W.T., Bruggen, A.H.C. van, and Zalom, F.G. 1998. Agronomic, economic, and environmental comparison of pest management in conventional and alternative tomato and corn systems in northern California. *Agriculture Ecosystems and Environment* 68:51–71.

Curl, E.A. 1963. Control of plant diseases by crop rotation. *Botanical Review* 29:413–79.

Dabbert, S. 1986. A dynamic simulation model of the transition from conventional to organic farming. MS thesis, Pennsylvania State University.

Dabbert, S., and Madden, P. 1986. The transition to organic agriculture: A multi-year simulation model of a Pennsylvania farm. *American Journal of Alternative Agriculture* 1:99–107.

Davis, A., and Liebman, M. 2001. Nitrogen source influences wild mustard growth and competitive effect on sweet corn. *Weed Science* 49:558–66.

Delate, K., and Cambardella, C.A. 2004. Agroecosystem performance during transition to certified organic grain production. *Agronomy Journal* 96:1288–98.

Delate, K., Cambardella, C.A., and Karlen, D.L. 2002. Transition strategies for post-CRP certified organic grain production. *Crop Management*, August, pp. 1–9.

Department of Environment, Food and Rural Affairs of the UK (DEFRA). 2008. The contribution organic farming makes in supplying public goods. Available online at http://www.defra.gov.uk/farm/organic/policy/actionplan/pdf/org-245.pdf.

Dogra, B. 2006. Organic farming, answer to farmers' suicides? Inter Press Service, July 18. Available online at http://www.commondreams.org/headlines06/0718-05.htm.

Dogra, B. 2009. Organic farming brings hope to Vidarbha's distressed farmers. Infochange: Agriculture. Available online at http://infochangeindia.org/200812247549/Agriculture/Stories-of-change/Organic-farming-brings-hope-to-Vidarbha-s-distressed-farmers.html.

Drinkwater, L. 2002. Cropping systems research: Reconsidering agricultural experiment approaches. *HortTechnology* 12:326–515. Available online at http://horttech.ashpublications.org.

Drinkwater, L.E., Letourneau, D.K., Workneh, F., van Bruggen, A.H.C., and Shennan, C. 1995. Fundamental differences between conventional and organic tomato agroecosystems in California. *Ecological Applications* 5:1098–112.

Drinkwater, L.E., Wagoner, P., and Sarrantonio, M. 1998. Legume-based cropping systems have reduced carbon and nitrogen losses. *Nature* 396:262–65.

Dusseldorp, D. van, and Box, L. 1993. Local and specific knowledge: Developing a dialogue. In *Cultivating knowledge: Genetic diversity, farmer experimentation and crop research*, ed. W. de Boef, K. Amanor, and K. Welland with A. Blebbington. London: Intermediate Technology Publications, pp. 20–26.

Eyhorn, F., Ramakrishnan, M., and Mäder, P. 2007. The viability of cotton-based organic farming systems in India. *International Journal of Agricultural Sustainability* 5:25–38.

Firth, C., Schmutz, U., Hamilton, R., and Sumption, P. 2004. The economics of conversion to organic field vegetable production. Organic farming: Science and practice for profitable livestock and cropping. In *Proceedings of the BGS/AAB/COR Conference*, Newport, Shropshire, UK, April 20–22, 2004, p. 19-2.

Flaten, O., Lien, G., Ebbesvik, M., Koesling, M., and Valle, P.S. 2006. Do the new organic producers differ from the 'old guard'? Empirical results from Norwegian dairy farming. *Renewable Agriculture and Food Systems* 21:174–82.

Friedman, D. 2003. *Transitioning to organic production*. Sustainable Agriculture Network, USDA-SARE, Waldorf, MD: Sustainable Agriculture Productions.

Fuller, R.J., Norton, L.R., Feber, R.E., Johnson, P.J., Chamberlain, D.E., Joys, A.C., Mathews, F., Stuart, R.C., Townsend, M.C., Manley, W.J., Wolfe, M.S., Macdonald, D.W., and Firbank, L.G. 2005. Benefits of organic farming to biodiversity vary among taxa. *Biology Letters* 1:431–34.

Gibbon, P., and Bolwig, S. 2007. *The economics of certified organic farming in tropical Africa: A preliminary assessment*. Danish Institute for International Studies Working Paper 2007, Copenhagen, Denmark: Danish Institute for International Studies.

Gliessman, S. 2000. *Agroecosystem sustainability: Developing practical strategies*. Boca Raton, FL: CRC Press.

Gliessman, S., Werner, M., Allison, J., and Cochran, J. 1996. A comparison of strawberry plant development and yield under organic and conventional management on the central California coast. *Biological Agriculture and Horticulture* 12:327–38.

Glover, J.D., Reganold, J.P., and Andrews, P.K. 2000. Systematic method for rating soil quality of conventional, organic, and integrated apple orchards in Washington State. *Agriculture, Ecosystems and Environment* 80:29–43.

Goldberger, J. 2008. Certified organic production: The experiences and perspectives of Washington farmers. Washington State University. Available online at http://www.crs.wsu.edu/facstaff/goldberger/organicsurvey/goals.html.

Green, M., and Maynard, R. 2006. The employment benefits of organic farming. *Aspects of Applied Biology* 79:51–55.

Gündoğmus, E., and Bayramoglu, Z. 2006. Energy input use on organic farming: A comparative analysis on organic versus conventional farms in Turkey. *Journal of Agronomy* 5:16–22.

Halberg, N., Alroe, H.F., and Kristensen, E.S. 2006. Synthesis: Prospects for organic agriculture in a global context. In *Global development of organic agriculture: Challenges and prospects*, ed. N. Halberg, H.F. Alroe, M.T. Knudsen, and E.S. Kristensen, pp. 343–367, Cambridge, MA: CABI.

Halberg, N., Dalgaard, R., Olesen, J.E., and Dalgaard, T. 2008. Energy self-reliance, net-energy production and GHG emissions in Danish organic cash crop farms. *Renewable Agriculture and Food Systems.* 23:30–37.

Hassane, A., Martin, P., and Reij C. 2002. *Water harvesting, land rehabilitation and household food security in Niger.* Amsterdam: IFAD/Vrije Universteit.

Hedtcke, J., and Posner, J., eds. 2005. *The Wisconsin Integrated Cropping Systems Trials 10th Annual Technical Report—2003–2004.* University of Wisconsin, Mimeo, Madison, WI.

Heggenstaller, A.H., Menalled, F.D., Liebman, M., and Westerman, P.R. 2006. Seasonal patterns in post-dispersal seed predation of *Abutilon theophrasti* and *Setaria faberi* in three cropping systems. *Journal of Applied Ecology* 43:999–1010.

Hole, D.G., Perkins, A.J., Wilson, J.D., Alexander, I.H., Grice, P.V., and Evans, A.D. 2005. Does organic farming benefit biodiversity? *Biological Conservation* 122:113–30.

Hyvönen, T. 2007. Can conversion to organic farming restore the species composition of arable weed communities? *Biological Conservation* 137:382–90.

Jager, A. de, and van der Werf, E. 1992. Ecological agriculture in South-India: An agro-economic comparison and study of transition. *Mededeling—Landbouw-Economisch Instituut* 459:80.

Jamar, D., Stilmant, D., and Baret, P.V. 2007. Fodder productions in organic suckling beef farming systems: Impact of the driving forces that have led to system conversion to organic rules. Permanent and temporary grassland: Plant, environment and economy. In *Proceedings of the 14th Symposium of the European Grassland Federation*, Ghent, Belgium, September 3–5, 2007, pp. 560–63.

Jamison, R. 2003. Certified organic farming in Washington State: A feminist empiricist examination. Thesis (Masters of Environmental Studies), The Evergreen State College, Olympia, WA.

Kerselaers, E., Cock, L. de, Lauwers, L., and Huylenbroeck, G. van. 2007. Modeling farm-level economic potential for conversion to organic farming. *Agricultural Systems* 94:671–82.

Kloppenburg, J. 1991. Social theory and the de/reconstruction of agricultural science: Local knowledge for an alternative agriculture. *Rural Sociology* 56:519–48.

Kummel, H., Doll, J., Hedtcke, J., and Cook, A. 2005. Weed seedbank community changes on WICST: 1992–2004. In *The Wisconsin Integrated Cropping Systems Trials —2003–2004*, ed. J. Hedtcke and J. Posner, 95–118, Mimeo, Dept. of Agronomy, Madison, WI: University of Wisconsin.

Lampkin, N., Fowler, S.M., Jackson, A., Jeffreys, I., Lobley, M., Measures, M., Padel, S., Reed, M., Roderick, S., and Woodward, L. 2006. Sustainability assessment for organic farming—Integrating financial, environmental, social and animal welfare benchmarking. *Aspects of Applied Biology* 79:9–13.

Lee, H., and Fowler, S. 2002. A critique of methodologies for the comparison of organic and conventional farming systems. In *Proceedings of the UK Organic Research 2002 Conference*, March 26–28, 2002, pp. 281–84. Available online at http://orgprints.org/8234.

Leite, L.F.C., Mendonça, E.S., and Machado, P.L.O.A. 2007. Influence of organic and mineral fertilisation on organic matter fractions of a Brazilian Acrisol under maize/common bean intercrop. *Australian Journal of Soil Research* 45:25–32.

Letourneau, D.K., and Goldstein, B. 2001. Pest damage and arthropod community structure in organic vs. conventional tomato production in California. *Journal of Applied Ecology* 38:557–70.

Liebhardt, W.C., Andrews, R.W., Culik, M.N., Harwood, R.R., Janke, R.R., Radke, J.K., and Rieger-Schwartz, S.L. 1989. Crop production during conversion from conventional to low-input methods. *Agronomy Journal* 81:150–59.

Lipson, M. 1997. *Searching for the "O-word": Analyzing the USDA Current Research Information System for pertinence to organic farming.* Organic Farming Research Foundation, Santa Cruz, CA.

Lockeretz, W. 1995. Organic farming in Massachusetts: An alternative approach to agriculture in an urbanized state. *Journal of Soil and Water Conservation* 50:663–67.

Lohr, L., and Salomonsson, L. 2000. Conversion subsidies for organic production: Results from Sweden and lessons for the United States. *Agricultural Economics* 22:133–46.

Lyon, F. 1996. How farmers research and learn: The case of arable farmers of East Anglia, UK. *Agriculture and Human Values* 13:39–47.

MacGuidwin, A. 1993. Nematode communities in the Wisconsin Integrated Cropping Systems Trials (WICST). In *WICST Second Report 1992*, 8–9.

MacGuidwin, A. 2001. Personal communication. Department of Plant Pathology, University of Wisconsin.

Mäder, P., Fliessbach, A., Dubois, D., Gunst, L., Fried, P., and Niggli, U. 2002. Soil fertility and biodiversity in organic farming. *Science* (Washington) 296:1694–97.

Man, C., Ivan, I., Maerescu, C., and Ciupe, M. 2007. Studies concerning the evolution of some reproduction indices in sheep from conversion and organic farms. Bulletin of University of Agricultural Sciences and Veterinary Medicine Cluj-Napoca. *Animal Science and Biotechnologies* 63/64:576.

Marshall, G. 1991. Organic farming: Should government give it more technical support? *Review of Marketing and Agricultural Economics* 59:283–96.

Maxwell, B.D., Smith, R.G., and Brelsford, M. 2007. Wild oat (*Avena fatua*) seed bank dynamics in transition to organic wheat production systems. *Weed Science* 55:212–17.

Mazzoleni, E.M., and Nogueira, J.M. 2006. Organic agriculture: Basic characteristics of producers. *Revista de Economia e Sociologia Rural* 44:263–93.

Melero Sánchez, S., Madejón, E., Herencia, J.F., and Ruiz-Porras, J.C. 2008. Long-term study of properties of a xerofluvent of the Guadalquivir River Valley under organic fertilization. *Agronomy Journal* 100:611–18.

Melone, B. 2006. Broadening the education infrastructure in organic agriculture for farmers. *Crop Management*, September, pp. 1–8.

Morison, J., Hine, R., and Pretty, J. 2005. Survey and analysis of labour on organic farms in the UK and Republic of Ireland. *International Journal of Agricultural Sustainability* 3:24–43.

Motlagh, J. 2008. India's debt-ridden farmers committing suicide. *San Francisco Chronicle Foreign Service*, March 22.

Nakamura, Y., Fujikawa, T., and Fujita, M. 2000. Long-term changes in the soil properties and the soil macrofauna and mesofauna of an agricultural field in northern Japan during transition from chemical-intensive farming to nature farming. *Journal of Crop Production* 3:63–75.

Oberson, A., Besson, J.M., Maire, N., and Sticher, H. 1996. Microbiological processes in soil organic phosphorus transformations in conventional and biological cropping systems. *Biology and Fertility of Soils* 21:138–48.

Oehl, F., Sieverding, E., Mäder, P., Dubois, D., Ineichen, K., Boller, T., and Wiemken, A. 2004. Impact of long-term conventional and organic farming on the diversity of arbuscular mycorrhizal fungi. *Oecologia* 138:574–83.

Padel, S. 2008. Values of organic producers converting at different times: Results of a focus group study in five European countries. *International Journal of Agricultural Resources, Governance and Ecology* 7:63–77.

Padel, S., and Foster, C. 2006. Local and/or organic: A balancing of values for producers and consumers. *Aspects of Applied Biology* 79:153–57.

Park, T.A., and Lohr, L. 2007. Meeting the needs of organic farmers: Benchmarking Organizational performance of university extension. *Review of Agricultural Economics* 29:141–55.

Parrott, N., Olesen, J.E., and Hogh-Jensen, H. 2006. Certified and non-certified organic farming in the developing world. In *Global development of organic agriculture: Challenges and prospects*, ed. N. Halberg, H.F. Alroe, M.T Knudsen, and E.S. Kristensen, pp. 153–180, Cambridge, MA: CABI.

Pearson, C. 2007. Regenerative, semiclosed systems: A priority for twenty-first-century agriculture. *BioScience* 57:409–18.

Peters, G.A. 1991. Azolla and other plant-cyanobacteria symbioses: Aspects of form and function. *Plant and Soil* 137:25–36.

Petersen, C., Drinkwater, L., and Wagoner, P. 1999. *The Rodale Institute Farming Systems Trial: The first fifteen years*. Kutztown, PA: Rodale Press.

Pimentel, D., Hepperly, P., Hanson, J., Douds, D., and Seidel, R. 2005. Environmental, energetic, and economic comparisons of organic and conventional farming systems. *BioScience* 55:573–82.

Piorr, H.P., and Hindorff, H. 1984. The implication for plant diseases and pests during the conversion from conventional to biological agriculture. In *Proceedings of the Fifth IFOAM International Scientific Conference: The Importance of Biological Agriculture in a World of Diminishing Resources*, Verlagsgruppe Witzenhausen, pp. 421–35.

Porter, P.M., Huggins, D.R., Perillo, C.A., Quiring, S.R., and Crookston, R.K. 2003. Organic and other management strategies with two and four year crop rotations in Minnesota. *Agronomy Journal* 95:233–44.

Posner, J., Frank, G., Nordlund, K., and Schuler, R. 1998. Constant goal, changing tactics: A Wisconsin dairy farm start-up. *American Journal of Alternative Agriculture* 13:50–60.

Posner, J., Mulder, T., and Tangeldine, H. 1993. *Energy use and output/input ratios on the Wisconsin Integrated Cropping Systems Trials*. Integrated Cropping Systems Trials Second Report 1992, University of Wisconsin, pp. 118–19.

Posner, J.L., Baldock, J.O., and Hedtcke, J.L. 2008. Organic and conventional production systems in the Wisconsin Integrated Cropping Systems Trials. I. Productivity 1990–2002. *Agronomy Journal* 100:253–60.

Posner, J.L., Casler, M.D., and Baldock, J.O. 1995. Wisconsin Integrated Cropping Systems Trials: Combining agroecology with production agronomy. *American Journal of Alternative Agriculture* 10:98–107.

Postma-Blaauw, M.B. 2008. Soil biodiversity and nitrogen cycling under agricultural (de-) intensification. Thesis, Wageningen University, Wageningen, Netherlands.

Pretty, J.N., Morison, J.I.L., and Hine, R.E. 2003. Reducing food poverty by increasing agricultural sustainability in developing countries. *Agriculture, Ecosystems and Environment* 95:217–34.

Reganold, J., Glover, J., Andrews, P., and Hinman, H. 2001. Sustainability of three apple production systems. *Nature* 410:926–30.

Reganold, J.P. 1988. Comparison of soil properties as influenced by organic and conventional farming systems. *American Journal of Alternative Agriculture* 3:144–55.

Renouf, M.A., Antony, G., and Wegener, M. 2005. Comparative environmental life cycle assessment (LCA) of organic and conventional sugarcane growing in Queensland. In *Proceedings of the 2005 Conference of the Australian Society of Sugar Cane Technologists*, Bundaberg, Queensland, Australia, May 3–6, pp. 312–23.

Riemens, M.M., Groeneveld, R.M.W., Lotz, L.A.P., and Kropff, M.J. 2007. Effects of three management strategies on the seedbank, emergence and the need for hand weeding in an organic arable cropping system. *Weed Research* (Oxford) 47:442–51.

Riesinger, P., and Hyvönen, T. 2006. Impact of management on weed species composition in organically cropped spring cereals. *Biological Agriculture and Horticulture* 24:257–74.

Robertson, G., Paul, E., and Harwood, R. 2000. Energy use in intensive agriculture: Contributions of individual gases to radiative forcing of the atmosphere. *Science* 289:1922–25.

Rollett, A., Wilson, P., and Sparkes, D. 2007. The economic legacy of stockless organic conversion strategies. *Biological Agriculture and Horticulture* 25:103–22.

Rotz, C.A., Karsten, H.D., and Weaver, R.D. 2008. Grass-based dairy production provides a viable option for producing organic milk in Pennsylvania. *Forage and Grazinglands*, March, p. FG-2008-0212-01-RS.

Russo, V.M., and Taylor, M. 2006. Soil amendments in transition to organic vegetable production with comparison to conventional methods: Yields and economics. *HortScience* 41:1576–83.

Scialabba, N., and Hattam, C. 2002. *Organic agriculture, environment and food security.* Environment and Natural Resources Series 4. Rome, Italy: UNFAO. www.fao.org/DOCREP/005/Y4137E/Y4137E00.htm.

Scooby, J., Landeck, J., and Lipson, M. 2007. *2007 national organic research agenda: Soil, pests, livestock, genetics: Outcomes from the Scientific Congress on Organic Agricultural research (SCOAR).* Santa Cruz, CA: Organic Farming Research Foundation.

Scowcroft, B. 2008. Organic farming research foundation applauds farm bill victories for organic farmers. Available online at http://www.reuters.com/article/press.

Selener, D. 1997. Farmer participatory research. In *Participatory action research and social change*, ed. D. Selener, pp. 149–188. Ithaca, NY: Cornell Participatory Action Research Network, Cornell University.

Smith, R.G., and Gross, K.L. 2006. Weed community and corn yield variability in diverse management systems. *Weed Science* 54:106–13.

Smukler, S.M., Jackson, L.E., Murphree, L., Yokota, R., Koike, S.T., and Smith, R.F. 2008. Transition to large-scale organic vegetable production in the Salinas Valley, California. *Agriculture, Ecosystems and Environment* 126:168–88.

Sparkes, D.L., Rollett, A.J., and Wilson, P. 2006. The legacy of stockless organic conversion strategies. *Aspects of Applied Biology* 79:139–43.

Stinner, D., Stinner, B., Grewal, P., Klompen, H., McCartney, D., Phelan, L., Cardina, J., Alexnadrou, A., Batte, M., Rigot, J., and Michel, F. 2004. *Field crop transition experiment—The first four years.* Organic Food and Farming Education and Research Program Field Day, Wooster, OH: Ohio State University.

Strochlic, R., and Sierra, L. 2007. Conventional, mixed and "deregistered" organic farmers: Entry barriers and reasons for exiting organic production in California. California Institute of Rural Studies. Available online at http://www.cirsinc.org/Documents/Pub0207.1.PDF.

Sumption, P., Firth, C., and Davies, G. 2004. Observations on agronomic challenges during conversion to organic field vegetable production. In *Proceedings of the BGS/AAB/COR Conference: Organic Farming: Science and Practice for Profitable Livestock and Cropping*, Newport, Shropshire, UK, April 20–22, 2004, pp. 176–79.

Swezey, S., Goldman, P., Bryer, J., and Nieto, D. 2007. Six-year comparison between organic, IPM and conventional cotton production systems in the Northern San Joaquin Valley, California. *Renewable Agriculture and Food Systems* 22:30–40.

Swezey, S., Werner, M., Buchanan, M., and Allison, J. 1998. Comparison of conventional and organic apple production systems during three years of conversion to organic management in coastal California. *American Journal of Alternative Agriculture* 13:162–80.

Swezey, S.L., Werner, M., Buchanan, M., and Allison, J. 1994. Granny Smith conversions to organic show early success in Santa Cruz County. *California Agriculture* 48:36–44.

Swift, M.J., Izac, A.-M.N., and van Noordwijk, M. 2004. Biodiversity and ecosystem services in agricultural landscapes—Are we asking the right questions? *Agriculture, Ecosystems and Environment* 104:113–34.

Thomassen, M.A., Calker, K.J. van, Smits, M.C.J., Iepema, G.L., and Boer, I.J.M. de 2008. Life cycle assessment of conventional and organic milk production in the Netherlands. *Agricultural Systems* 96:95–107.

Tranter, R.B., Holt, G.C., and Grey, P.T. 2007. Budgetary implications of, and motives for, converting to organic farming: Case study farm business evidence from Great Britain. *Biological Agriculture and Horticulture* 25:133–51.

Tsiafouli, M.A., Argyropoulou, M.D., Stamou, G.P., and Sgardelis, S.P. 2007. Is duration of organic management reflected on nematode communities of cultivated soils? *Belgian Journal of Zoology* 137:165–75.

Turner, R.J., and Bond, W. 2004. Participatory organic weed management: *Rumex* spp. control—A farmer perspective. In *Proceedings of the 6th EWRS Workshop on Physical and Cultural Weed Control*, Lillehammer, Norway, p. 185.

USEPA. 2008. Methane: http://www.epa.gov/methane/scientific.html; Nitrous oxide: http://www.epa.gov/nitrousoxide/scientific.html.

Vaarst, M., Roderick, S., Lund, V., Lockeretz, W., and Hovi, M. 2004. Organic principles and values: The framework for organic animal husbandry. In *Animal health and welfare in organic agriculture*, ed. M. Vaarst, S. Roderick, V. Lund, and W. Lockeretz, pp. 1–13, Cambridge, MA: CABI.

Vandermeer, J., and Perfecto, I. 2005. The future of farming and conservation. *Science Now*, May 27, pp. 39–40.

Weibel, F.P., Häseli, A., Schmid, O., and Willer, H. 2004. Present status of organic fruit growing in Europe. *Acta Horticulturae* 638:375–85.

Wood, R., Lenzen, M., Dey, C., and Lundie, S. 2006. A comparative study of some environmental effects of non-organic and organic farming in Australia. *Agricultural Systems* 89:324–48.

Zimmerman, K.L. 1997. Organic farming research in the Netherlands and comments on the report of Els Wynen. *REU* (Agricultural Economics Research Institute, Netherlands) *Technical Series* 54:70–73.

3 The History of Organic Agriculture*

Rachael J. Jamison and John H. Perkins

CONTENTS

3.1 INTRODUCTION

Organic agriculture has become the fastest-growing sector in U.S. agriculture (Jamison, 2003; Heckman, 2006). Internationally, it is a multibillion-dollar industry fully integrated into the global food system. Despite this rapid growth, significant barriers still hinder the future course of the organic sector, and the history of organic agriculture can help us to understand the nature of these impediments. This chapter traces the historical origins of organic agriculture with emphasis on the United States.

3.2 DEVELOPMENT OF U.S. AGRICULTURE

The factors affecting the development of agriculture in the United States have been many and varied, ranging from the climatic and geophysical to the institutional and social. From the late 1800s through the 1900s, these factors have driven rapid agricultural change and created a highly differentiated, heavily capitalized commercial enterprise.

---

* A portion of this chapter is adapted from Jamison (2003).

3.2.1 REGIONAL DIFFERENTIATION

Cropping patterns in different parts of the country developed in unique ways, resulting in regional agricultural specialization. These regional differences, taking the form
of crop "belts," such as the corn–soybean belt of the Midwest and the cotton belt of
the South, characterized U.S. farming by the beginning of the 1900s (Haystead and
Fite, 1955; Cochrane, 1979). This differentiation arose in part because differences
in soil, climate, and water availability made it economically advantageous to grow
certain crops in a region and not others. This had to do primarily with production
costs: the crops that became dominant in an area were those with the lowest costs of
production. Railroads, and now trucks and air freight, enabled rapid transportation
of farm products nationwide, which promoted and still favors specialization. It was
not that other crops would totally fail when grown outside their preferred belt, but
the competitive pressures of agriculture dealt harsh penalties to farmers seeking to
grow crops in areas with other than the lowest possible production costs.

As powerful as the physical and biological factors were in shaping what crops
were grown where, economic, social, institutional, and political factors also contributed to the development of specialized cropping regions and reinforced the tendency
toward regional specialization. Of particular importance were patterns in land tenure
and labor, the potential for export markets, the creation of educational and research
facilities, and the history of subsistence production in the area.

Cotton, for example, developed in the Southeast in conditions that favored the
power of large landowners, who farmed first with slave labor and later in share-crop-
ping schemes. In the upper Midwest, south-central Canada, and New England, in
contrast, the smallholder agrarian tradition favored the development of dairy farms
that could be smaller in scale and run primarily with the labor and management
of owner-operators and their families. Vegetable and fruit production in the West
relied on government provision of water and large, migratory workforces to support
large-scale holdings and export of most of the production to other parts of the United
States or overseas.

3.2.2 MECHANIZATION AND INTENSIFICATION

Independent of regional specialization, farmers across the United States faced a
seemingly inexorable force: the pressure to reduce production costs and increase
yields with new and more efficient technology and farming practices, even though
existing practices produced ample yields nationwide. According to the theory of
the treadmill (Cochrane, 1979), this pressure was the inevitable result of a competitive environment coupled with alienable land and constant technological innovation. Poverty and possibly bankruptcy were the bleak outcomes of farming without
aggressively adopting innovative production technologies and the crops best suited
to the individual farmer's region. As farmers responded to the pressure of the treadmill, they increased production, which tended to make supply outpace demand—and
this condition only reinforced the dynamic of the treadmill.

The treadmill hypothesis provides an overarching explanation for why farmers
have historically sought higher yields even in the face of inadequate demand for their

crops and low prices. This framework also explains why farm units tended to grow larger and the number of farmers steadily smaller: technologically progressive growers bought out their less progressive neighbors as the latter left agriculture, either through choice, foreclosure, retirement, or death.

Mechanization, agricultural chemicals, and breeding programs fueled the operation of the treadmill during the twentieth century. Farmers either adopted the new practices or could not produce yields at costs enabling them to compete in the market. Tractors and other farm equipment allowed sharp reduction in labor needed. Mechanization, more than any other practice, drove the exodus of people from rural areas and fostered the growth of larger farms.

The development of agricultural chemicals worked harmoniously with the growing mechanization of American agriculture. Immediate production costs declined, while crop yields increased drastically. Fertilizers, pesticides, and growth regulators allowed significant reduction in pressures from perennial weed and insect pests. By 1945, nitrogen fertilizers became a necessity for the progressive farmer. Chemical programs on the increasingly larger farming operations permanently launched American agriculture into the industrial age.

Plant and animal breeding, especially after 1940, increasingly aided the progressive farmer. New varieties of crops and livestock yielded more of what the markets demanded. Farmers continued specializing and many farms elected to raise either animals or plants, but not both. Steady turning of the treadmill led most farmers either to stay on by adopting new practices or to get off and get out of agriculture. A nonprogressive farmer who tried to stay in the game was usually dumped off the end of the treadmill in a heap.

By the late twentieth century the earlier revolutions in mechanization and chemicals were fully established in the American agricultural economy. Land prices had adjusted upward to reflect the higher yields and lower production costs. Labor had left the rural areas, and city dwellers no longer had the knowledge or skills needed to be productive farmworkers.

3.2.3 Appearance of Negative Consequences

The intensification and industrialization of agriculture led some to conclude that the new practices were ultimately self-defeating. Fertilizers and pesticides caused pollution as they escaped from their place of application. Pesticides were especially damaging, as they were by design toxic to living organisms. Machinery, too, despite the miracle of freeing people from hard drudgery, left its scars in depopulated rural towns and villages. People were not needed anymore, and they became redundant in the countryside.

Increasingly, studies showed that heavy reliance on pesticides created instability through pollution, development of resistance in target species, and destruction of nontarget organisms. Fertilizer runoff created intense algal blooms in fresh and marine waters, killing desirable species. Chemical changes leading to nitrous oxide created a greenhouse gas that threatened climate. All these problems left many wondering if the benefits of modern, intensive agriculture were really worth it once all the costs were totaled. That the whole enterprise might have the stability of a house of cards led critics to charge that modern agriculture was, in a word, unsustainable.

The technological transformation of American agriculture had enormous social consequences as well (Lobao and Meyer, 2001). Where more than one-third of the population lived on farms in 1900, by 2000 the proportion was less than 2%. Farming for most Americans had ceased to be a "household livelihood strategy." At the national or macro level, the changes were swiftest after 1945. Many white farmers left, and almost all African American farmers departed. Large, well-capitalized farm and nonfarm firms ended up dominating the food system. These macro-level changes had counterparts at the community and individual levels. Today one can drive through the Midwest and see small towns devoid of the vitality they once held.

Multiple reform efforts blossomed in contentious disputes about agricultural technology and the rapid changes affecting communities and individuals. Organic agriculture was one of the more prominent agricultural reform movements, and it asserted the ability to provide high-quality food without many of the pollution problems of the fertilizers and pesticides of modern agriculture. The early forms of the "organic movement" were also deeply agrarian: advocates cared about the fate of rural communities, social justice, and alienation of consumers from farm producers. Put in the context of the current volume, organic farming came to represent the most important form of conversion from conventional agriculture to more sustainable alternatives.

3.3 BEGINNINGS OF ORGANIC FARMING

To understand organic agriculture as a social movement—a phenomenon with technical, scientific, social, economic, political, and philosophical components—it must be examined in light of the history narrated above. In this context, organic agriculture arose as an alternative to what had become the taken-for-granted method of growing food. It had origins, early advocates and promoters, and a period of growth during which it penetrated into the mainstream.

The organic food movement united those concerned with environmental health with those focused on human health; it challenged corporations dependent on sales of chemical pesticides and fertilizers; it contested the idea that humans could control natural systems; and it instigated one of the fastest-growing markets of the late twentieth century, the market for organic food products.

3.3.1 PHILOSOPHICAL ROOTS

The roots of U.S. organic agriculture lie in the advocacy of a small group of people, mostly from the United Kingdom, who in the late 1930s and early 1940s began to recognize the connections among farming practices and the health of soil, people, and the environment. The British organic movement was part of the larger social turmoil that rocked the United Kingdom during the years between World War I and World War II. British advocates of organic practices fell across a wide spectrum of political thought, from socialist to fascist to conservative nationalist to deeply Christian (Conford, 2001).

Although he did not use the word *organic* to describe the farming methods he explored, Englishman Sir Albert Howard is often credited with the first articulation of the biological principles underlying organic production. Howard attended Cambridge University, where he earned bachelor's and master's degrees in natural science. In 1903, after time with the British Colonial Service in the Caribbean, he attended the Wye College of Agriculture in England (Perkins, 1997). On assignment in Pusa, India, as an economic botanist with the Imperial Agricultural Research Institute of Britain, Howard conducted research on methods of obtaining higher-yielding wheat varieties. After retirement from Imperial service, he moved to the Institute for Plant Industry at Indore in central India.

Before he went to India, Howard saw science as an effort to bring the perspectives of genetic and taxonomic theory from the laboratory to the field. Truth lay in performing studies that were interpretable within the basic science framework he had learned at Cambridge. During his time in India, and especially after he went to Indore, Howard reoriented his science to accommodate the knowledge of peasant farmers, time-tested field results, and the constraints of markets. He realized the importance of producing results that real farmers could use under their conditions and for their markets (Gieryn, 1999).

In 1943 Howard authored *An Agricultural Testament*, which formally introduced the Indore process to Western audiences. In time, the Indore process became known as composting, which today is one of the fundamental practices allowing organic farmers to return organic matter and its attendant nutrients to the soil.*

Howard was wary of the long-term effects of monocrop, chemical-intensive agriculture on soil fertility in industrialized countries. "The capital of nations which is real, permanent, and independent of everything except a market for the products of farming, is soil" (Howard, 1943, p. 219). After studying Indian methods of farming, Howard suggested that in order to maintain the integrity of the soil, farmers must engage in composting, cover cropping, and crop rotations—elements central to contemporary organic agriculture. He formulated the law of return, which codified his view that farming should be seen as an integrated system that generated no waste. Soil required the decaying parts of plants and animals to generate humus, without which neither plants nor animals nor people could hope to enjoy health (Heckman, 2006).

The United States was not without its proponents of biologically based agriculture at this time. Concurrent with Howard's work, William Albrecht, professor of soils at the University of Missouri, argued that soil was one of America's most valuable natural resources. He encouraged farmers "to restore fertility by the use of lime and fertilizer ... to put some lands permanently into sod crops ... and to use sod more regularly in rotations on tillable cropped lands ... and use ... such farm wastes as crop residues and manures" (Albrecht, 1938, p. 21). His recommendations would ultimately become central to organic farming practices. Albrecht also began to raise questions related to the connection between agricultural practices and human health, an issue that would later become a major driver in the growth of the organic food movement.

---

* Composting is "the product of a managed process through which microorganisms break down plant and animal materials into more available forms suitable for application to the soil" (Electronic Code of Federal Regulations, 2008, 7 CFR 205.2).

U.S. farmers were offered a "road out of this impasse" by entrepreneur and publisher J. I. Rodale. Rodale began his career as an electric equipment manager and founder of numerous how-to and health care magazines, including *Prevention*. He lived in rural Pennsylvania in the late 1930s and became interested in agricultural practices that retained soil health.

Rodale, like Howard and Albrecht, contended that soil health was paramount in long-term agricultural success and supported farming practices that maintained and built soil fertility. He, too, perceived human health as intrinsically tied to agricultural practices. His interest in farming practices and human and environmental health prompted him to publish the magazine *Organic Gardening and Farming,* which provided instruction to farmers on how to integrate biological principles into food production on all scales. The publication also served as a bridge between the agricultural community and the American consumer, planting the seed for the concept that the manner in which food was produced can have significant impacts on individual health.

Interest in his work grew and he founded the Soil and Health Foundation, now known as the Rodale Institute. The institute, today, continues to publish books, educate farmers, promote organic agriculture, and conduct research on organic and other chemical-free farming practices.

The term *organic farming* was coined by Lord Northbourne (born Walter Earnest Christopher James) in his 1940 book *Look to the Land.* The term *organic* came from his conception of "the farm as organism." He described a holistic, ecologically balanced approach to farming. The term *organic,* as a description of the practices advocated by Northbourne, Rodale, Albrecht, and Howard, took hold and today continues to describe a way of farming that supports the health of soil, plants, and people (Jamison, 2003; Heckman, 2006).

Howard, Northbourne, Albrecht, and Rodale provided organic agriculture with a solid scientific and philosophical foundation. They saw the imperative for respecting nature with science, not simply dominating and controlling it instrumentally. To make this point forcefully, Howard used the metaphor of war in his most outspoken work, *The War in the Soil* (1946). Here he railed against what he saw as a simplistic mindset derived from the German chemist Otto Von Liebig: just put the right mineral nutrients near a plant, and the plant will thrive. Howard and the others saw instead that soil was a kind of living entity and that using the soil for human purposes required a subtle partnership with nature. The law of return summarized his deeply held belief that waste must return to the soil for health. This co-equal status of science and nature philosophy continues to drive much of the critical thinking about organic agriculture to this day.

3.3.2 Concern about the Health Effects of Pesticides

As the work of these men began to reach deeper into the agricultural community, farmers interested in maintaining soil health and wary of the impacts of agricultural chemicals on human health began to incorporate the farming methods they described. The benefits afforded to consumers of purchasing organically produced food were still relatively unknown, however, and the market for organic food was small through the 1940s and 1950s; governments collected no statistics and there was no legal definition of *organic.* The eventual blossoming of consumer interest in organic food can be largely credited to the work of two women: Lady Eve Balfour and Rachel Carson. In addition to supporting the ecological benefits of organic practices, Balfour and Carson focused their work on connecting agricultural practices to human health. In many ways, it was they who brought organic food to the tables of the masses.

Largely inspired by Howard's work, Lady Eve Balfour devoted her life to the study of the environmental benefits of organic production methods and worked to connect these benefits to human health. "My subject is food, which concerns everyone; it is health, which concerns everyone; it is the soil, which concerns everyone— though they may not realize it" (Balfour, 1943).

Balfour's book *The Living Soil,* published in 1943, compiled her research on the connections between food and health. Response to her work was so great that, in 1945, a meeting was called to bring together others interested in these topics. The result of this meeting was the founding of the Soil Association, an organization that has grown into a leading international organization in research on alternative agriculture and is now the largest organic certification agency in the world.

Balfour, one of the first women to study agriculture at Reading University, initiated the Haughly experiment, the first and longest study of the microbial benefits of organic compared to conventional farming. Started in 1939 and not completed until 1972, the Haughly experiment intended to

> observe and study nutrition cycles, functioning as a whole, under contrasting methods of land use, but on the same soil and under the same management, the purpose being to assess what effect, if any, the different soil treatments had on the biological quality of the produce grown thereon, including its nutritive value as revealed through its animal consumers. (Balfour, 1976, p. 14)

This study was the first to suggest increased microbial activity and absorbable nutrient levels on organically managed plots compared to conventionally managed land. It also made clear that pest pressures were not more significant on organic sites than they were on conventional ones. By bringing to the public's attention information that confirmed the benefits of organic farming for both the environment and the individual, Balfour initiated the discussion about the inextricable connection between human health and food production.

Although Balfour's work had fairly wide influence, it was not until 1962 that the dominant view of chemical-intensive agriculture as benign was deeply challenged. This occurred with the publication of Rachel Carson's *Silent Spring,* which unequivocally pointed to the immediate threats to individual health posed by the application of synthetic agricultural chemicals. Carson changed the way the public viewed synthetic pesticides by bringing together research that identified specific consequences of exposure, from cancer to infertility. These revelations would impact consumers' food choices into our present time.

Carson described neighborhoods where birds' songs were silenced by death from pesticide exposure. She spoke of lakes and rivers filled with fish too toxic for human consumption. Carson described unborn babies developing cancers *in utero* or after birth as a result of the mother's exposure to carcinogenic chemicals during pregnancy. *Silent Spring* made it impossible to ignore the immediate risks of pesticide exposure.

> Through all these new, imaginative, and creative approaches to the problem of sharing our earth with other creatures there runs a constant theme, the awareness that we are dealing with life, with living populations and all their pressures and counter-pressures, their surges and recessions. Only by taking account of such life forces and by cautiously seeking to guide them into channels favorable to ourselves can we hope to achieve a reasonable accommodation between the insect hordes and ourselves. (Carson, 1962, p. 12)

Although her research focused on the acute symptoms of chemical exposure such as cancer and birth defects, Carson alluded to what would later be known as endocrine disruption, a topic more deeply explored in later years by Theo Colborn and colleagues (1996). Many credit Carson's work with the banning of DDT, the establishment of the Environmental Protection Agency, and the birth of the organic food industry.

By raising the public's awareness of the connections between farming practices and the health of soils and people, Balfour and then Carson planted the seeds for what would grow into an international political and environmental movement.

3.3.3 ORGANIC FARMING TAKES ROOT IN THE UNITED STATES

Concurrent with consumers beginning to ask for pesticide-free food products, farmers who had begun to experience the repercussions of conventional agriculture—a decline in soil and environmental health, the loss of profits due to the expanding global market for food, and the loss of rural culture—were drawn to organic practices. There was also a growing community of formerly urban residents who were drawn "back to the land" and, desiring to rebuild the human-environment relationship, were naturally called to organic practices. These two groups constituted the first wave of American organic farmers. The meeting point between the increasing awareness of detrimental farming practices and farmers looking for a way back to the right ways of farming gave rise to the seed of organic agriculture as an industry. The organic farm of the 1960s and 1970s epitomized many of the values that underlie the contemporary mythos of organic: small in scale, focused on direct markets, aimed at revitalizing rural communities.

The 1970s and 1980s saw tremendous growth in consumer demand for organic food in the United States. This growth was largely a result of the heightened awareness between food and human health that Carson initiated. Core consumers tended to seek organic products directly from farmers, farmers' markets, and local food co-ops. It was not until major food safety issues began to surface, however, that organic moved into the mainstream in America.

The Alar incident of the late 1980s demonstrated consumers' willingness to respond immediately to perceived threats from agricultural chemicals. Alar is a material used on apples postharvest to preserve crispness and appearance from field to market. In 1989, the CBS television series *60 Minutes* reported that residues of the chemical on apples posed health threats to children, increasing their risks of contracting certain kinds of cancer by up to 100 times.

Apple sales immediately plummeted. Apple growers in Washington State, the country's leading apple producer, estimated losses upwards of $200 million (Smith, 1994). What the incident demonstrated was that consumers were growing more responsive to issues of food safety and were willing to pay higher prices for food to avoid risks. Where conventional products symbolized risk, organic products began to symbolize safety.

Although the market was growing, there existed no standards governing how organic products had to be grown in order to be labeled organic. Product integrity was based largely on the trust inherent in the consumer–farmer relationship. This worked well when a consumer had access to a farm directly. However, when buying a product from a store, there was no assurance that the food product was, in fact, organic. As consumer interest grew, so did the demand for a way of verifying that a product had indeed been grown in accordance with organic principles.

3.4   GROWTH OF AN INTERNATIONAL INDUSTRY

Increased awareness of the benefits of organic farming to human and environmental health coupled with a deeper understanding of the detrimental environmental impacts of conventional farming practices resulted in increased demand for organic food worldwide. As demand was growing in other parts of the world and international trade of organic products was increasing, questions were being raised about the products' integrity.

In 1972, in an effort to both establish an organic standard and provide consumer assurance, the International Federation of Organic Agriculture Movements (IFOAM) was formed. In the beginning, IFOAM consisted of members in numerous European countries and India and Canada. Norms were developed that provided a baseline standard under which a product must be produced if it was to carry the "organic" label. The IFOAM norms served as a means of ensuring that internationally traded organic products were produced under a similar standard.

Based in Germany, IFOAM worked closely with the Soil Association in the United Kingdom, which, at this point, was the premiere organic certifier in Europe and provided certification services internationally upon request. Lady Eve Balfour was on the founding board of the organization; therefore, it was a natural step for both organizations to work together. IFOAM sought to develop standards to which organic products should be grown; the Soil Association continued to lead the world in research on the benefits of organic agriculture to soil and human health.

As IFOAM and the Soil Association were providing guidance worldwide, several U.S.-based certification agencies were established, many of which continue to operate: California Certified Organic Farmers (1973), Oregon Tilth Certified Organic (1974), and Farm Verified Organic based in North Dakota (1979). Although similar, the organic standards developed by these organizations lacked uniformity. Thus, consumers could not be sure if "organic" on a label meant the same thing from state to state or from product to product.

These organizations, largely by happenstance, pioneered two critical steps in continuing to grow the industry. First, they introduced the need for standardization. In so doing, they provided the industry with a prescriptive road to access the emerging market. Second, they required third-party verification. This step provided consumers with assurance that the products that carried the "organic" label were, in fact, verified to comply with the standards. The circle was complete—industry had a clear path to follow and consumers had assurance that products were legitimately carrying the "organic" label. These features are what set organic apart from other "green" agricultural industries (natural, grass-fed, etc.) and are largely responsible for its success.

Because demand for organic food was continuing to grow and organic products were being traded throughout the country, consumers and farmers began to request the development of a national standard to which all products labeled "organic" must adhere. A national standard would provide the industry with consistency and protect it from fraudulent claims.

3.5 ADOPTION OF NATIONAL STANDARDS

The Organic Foods Production Act (OFPA) was Title 21 of the 1990 Federal Farm Bill (Gold, 2007) and was the country's response to the growing need for consistency in standardization. As described by the act, its primary purposes were:

- to establish national standards governing the marketing of certain agricultural products as organically produced,
- to assure consumers that organically produced products meet consistent standards, and
- to facilitate interstate commerce in fresh and processed organic foods.

Individual states had the capacity to establish more restrictive standards than those in the act. A state could not, however, prohibit the sale of products that had been grown according to the OFPA but not under its own stricter standard from entering the market in that state. Additionally, the OFPA made mandatory the creation of a national organic program by the Agricultural Marketing Services section of the U.S. Department of Agriculture (USDA).

In addition to providing the first nationwide standard for organic food production, the OFPA offered the United States the first legal definition of the term. *Organic* to some farmers and consumers might still imply a desire to return to a simpler life, to nourish the environment, to protest chemical-intensive agricultural production methods, but legally *organic* would be a labeling term for agricultural commodities produced in accordance with the act.

More than 10 years passed between the establishment of the Organic Food Production Act of 1990 and the development and implementation of the USDA's National Organic Program. The final rule went into effect on April 21, 2001, and allowed for an 18-month transition period for the industry to bring itself into compliance. On October 21, 2002, all products sold in the United States carrying an "organic" label had to be fully compliant with the national rule.

The National Organic Program (NOP) requires that all certification agencies obtain accreditation through the USDA. Accreditation verifies that the certifier has the capacity to inspect and enforce the national rule. All organic food products sold in the United States must be certified by an NOP-accredited certification agency, whether the food was produced on U.S. soil or abroad.

Under the USDA-NOP, producers of organic food may not use genetically modified organisms, must enhance or maintain the quality of the natural resources of their farming operations, must record all material applications, must maintain an organic production system plan that outlines the practices being employed on the farm, and may not use synthetic chemicals on the organic crops.

3.6   GROWTH AND CHANGE IN ORGANIC AGRICULTURE SINCE 1992

Driven by increasing demand, production of organic food has become one of the fastest-growing international industries. In the United States, the area of organic cropland quadrupled from 1992 through 2005, going from 935,450 to 4,054,429 acres (U.S. Department of Agriculture, 2006).

Growth in cropland was matched by increasing growth in consumer interest. Through the 1990s, consumer demand increased by about 20% a year. The connection between the increase in demand and the increase in production is obvious. As demand grew, farmers responded and production followed. Accordingly, organic farming has become a multibillion-dollar industry in the United States.

Historically, organic food products were sold primarily in health and natural food stores. In 1991, these stores accounted for approximately 68% of sales, while conventional retail outlets sold only 7% of organic food products. By 2000, figures began to tell a different story, with conventional grocery stores selling 49% of organic products, and health and natural food stores selling 48% (Dimitri and Greene, 2001).

Sales of organic food products in the United States exceeded $9 billion in 2001, and the USDA predicted continued growth as a result of the harmonization of organic standards. In 2000, for the first time, according to the U.S. Department of Agriculture–Economic Research Service (USDA-ERS), more organic food was purchased in supermarket chains than in health food stores and other venues. It was estimated that at the time, 73% of all conventional grocery food stores offered organic food products (Dimitri and Greene, 2001).

Growth in the international market for organic food products has also begun to see stupendous growth. The 27 members of the European Union, Japan, and Korea are among the nations with their own organic standards. More than 100 countries are now producing organic food on more than 59 million acres of land. The international market for organic food products in 2003 was estimated at $23 billion.

Multinational corporations such as Heinz, General Mills, and Pepsi have seen the investment value in expanding their portfolios to include organic options. As more countries and governing bodies continue to adopt organic standards, the easier it will be both for farmers to have access to this fast-growing market and for the organic food industry to feel the pressures of a globalized food system.

In the United States, therefore, organic agriculture has clearly established itself as an integral, growing part of the American food system, even though the amount of farmland certified as organic remains small—about 0.5% overall. In vegetable production, 4.7% of the land is organic, and 2.5% of land in fruit is organic. For most grains, the amount of certified organic land remains at less than 1%. At 5.8%, carrot production has the largest proportion of land certified organic (U.S. Department of Agriculture, 2006).

Even though organic remains a small component of the overall American food system, the growth and changes in organic production have ignited alarm bells among some proponents of organic agriculture. Buck et al. (1997) started to analyze the effects of increased commercial interest in organic agriculture before the National Organic Program set national standards. Based on interviews with approximately 70 organic vegetable growers in Northern California in 1995, they foresaw that organic agriculture would likely incur a process of penetration by capital, meaning that wage labor (rather than owner-operator labor) would eventually prevail at the site of production. Publication of the article in 1997 set off a debate among social scientists about whether organic agriculture would "conventionalize" or become more like conventional agriculture.

Halberg et al. (2006) concurred and saw that the conventionalization of the organic sector leads to a "lack of transparency and trust among producers and consumers, increasing food miles and dilution of the 'nearness' principle, specialization and concentration of production at the cost of smallholders and reduction in diversity in crops and farm types" (p. 8).

The conventionalization debate took on overtones of the sharply polarized arguments among natural scientists that occurred during what Heckman (2006) called the polarization phase (1940–1978) of the development of organic agriculture. Recall that during that earlier period, Sir Albert Howard characterized scientific disputes about soil science as a war in the soil. Just as Howard believed that conventional agriculture would lead to ruination of the soil and ultimately human health, those who lamented conventionalization believed that the penetration of capital into organic agriculture would betray and ruin the ability of organic agriculture to point the way toward an economically feasible, environmentally sustainable, and socially just food system (Guthman, 2000, 2004a, 2004b).

Other scholars who participated in the conventionalization debate shared the concerns about betrayal, but they did not necessarily agree that organic agriculture really would succumb to capital penetration, or that if it did, capital would subvert the ideology of progressive reform that underlay much of organic agriculture's history (Darnhofer, 2005; Lockie and Halpin, 2005; Obach, 2007; Best, 2008; Constance, 2008; Guptil and Welsh, 2008).

It remains too early to see the ultimate effects of national standards for organic produce, but Heckman (2006) and others have clearly indicated that already those committed to the ideology of progressive reform through organic agriculture are seeking new concepts to ensure that the social agenda remains strong. For example, Allen (2004) sees linking sustainability in agriculture with community food security as a productive agenda.

Lyson (2004) sees local production or "civic agriculture" as important to keeping reform vital. Morgan et al. (2006) amplify the importance of place for small-scale growers by analyzing how they can achieve quality, high-value production in specially defined areas.

Jaffee (2007) argues that "fair trade" has helped small, organic growers, particularly in coffee and bananas. Raynolds et al. (2007) see commercial success in fair trade commerce but also, as with organic agriculture's commercial successes, a growing tension between those motivated by social justice concerns and those seeking to enhance the market successes of fair trade. Just as in organic agriculture, these tensions arise from mainstreaming into major retail outlets, an increase in scale, and arguments about the governance of fair trade standards and which organization should set the standards.

3.7 CONCLUSIONS

What was once a market niche has firmly rooted itself in the mainstream, piquing the interest of farmers wanting to collect price premiums for their food, consumers seeking what they believe to be a healthier food product, and corporations wanting to capitalize on the increased demand for organic products. As demand continues to grow, all involved will be asked to evaluate their perceptions of the industry and its future.

Organic began as a socioenvironmental movement aimed at protecting human and environmental health. Today, organic is seen as a lucrative international industry that provides many farmers with hope for survival. This growth has resulted in a building tension between those who want organic to continue to exist primarily with the small farmer and those who are apparently concerned with capitalizing on the price premium organic products offer.

No reason exists to think that the geographic, biological, social, and economic forces that were so powerful in shaping agriculture in the past might not also have impacts on organic production. If organic production continues to grow and take an increasing share of the agricultural market, how will organic farmers distribute themselves across the landscape? Will "belts" emerge in which organic growers in particular areas specialize in a small range of crops, for which the area has low production costs?

More poignantly, will the agricultural "production treadmill" begin to operate among organic producers? Will the supply of organic produce begin to exceed the effective market demand, leading to intense competition among organic growers and a search for cost-cutting new technology? Will we see some organic growers driven from the field by competition from their more technically progressive peers, as the treadmill thesis argues? Will the conventionalization thesis lead the industrial organic sector to mimic the conventional food system? Although chemicals and farm practices are different, the issues of market choices and scale are not addressed when organic farms become industrial-scale food production systems with audiences worldwide.

Organic had previously provided small farms with an access point to the wholesale food market. As larger farming operations and corporations have moved into organic production, this access has been put in jeopardy. Farmers' markets and community-supported agriculture (CSA) are more and more becoming the small organic farmer's only hope for survival. Even these markets could be jeopardized as mainstream grocery outlets continue to expand their organic selections. This shift has set off alarm bells for those who believe that the social and cultural elements found in the early days of the organic food movement are as critical as the farming methods themselves.

> Why should food, of all things, be the linchpin of [the rebellion against a globalized food system]? Perhaps because food is a powerful metaphor for a great many of the values to which people feel globalization poses a threat, including the distinctiveness of local cultures and identities, the survival of local landscapes, and biodiversity. (Pollan, 2006, p. 162)

Organic food production, at its inception, was focused on improving soil and, later, on enhancing human health. Agriculturists, activists, and consumers now have to decide whether or not culture and justice are equally as important to organic agriculture as the environmental and health considerations. From the standpoint of soil vitality and the ability of large numbers of people to have access to organic food, one could argue that it is imperative that corporate farms engage, as they have direct control over more acreage than their smaller counterparts. From the standpoint of social and cultural justice, however, one could argue that by allowing industrial farms to participate in the market, we will only see the perpetuation of the already rapid loss of family farms and exploitation of agricultural labor and rural culture.

These are the issues that promoters, enablers, and participants in organic production are grappling with now. That more American agriculture will become organic seems a foregone conclusion. How that production will ultimately shape the organic industry we know much less clearly. There is no reason to think, however, that organic producers will be immune from the pressures that have shaped American agriculture in the past.

The resolution of these pressures will determine whether organic agriculture remains an acceptable proxy for sustainable agriculture or diverges in too many ways from the sustainable path. As detailed in the preceding chapters, conversion from conventional to organic production points in a sustainable direction only if the resulting systems satisfy an array of environmental, economic, and social justice criteria. It is doubtful that agroecosystems that meet the legal requirements for organic labeling but produce at a massive scale, put small farmers out of business, rely on low-wage labor, and depend on a fossil-fuel-based transportation infrastructure for getting their products to far-flung markets, can qualify as sustainable.

REFERENCES

Albrecht, William A. 1938. Loss of soil organic matter and its restoration. In *Soils and men, yearbook of agriculture 1938*, ed. U.S. Department of Agriculture, 347–60. Washington, DC: U.S. Government Printing Office.

Allen, Patricia. 2004. *Together at the table: Sustainability and sustenance in the American agrifood system*. University Park: Pennsylvania State University Press.

Balfour, Lady Eve. 1943. *The living soil*. London: Faber and Faber.

Balfour, Evelyn Barbara. 1976. *The living soil and the Haughley experiment*. New York: Universe Books.

Best, Henning. 2008. Organic agriculture and the conventionalization hypothesis: A case study from West Germany. *Agriculture and Human Values* 25:95–106.

Buck, Daniel, Christina Getz, and Julie Guthman. 1997. From farm to table: The organic vegetable commodity chain of northern California, *Sociologia Ruralis* 37:3–20.

Carson, Rachel. 1962. *Silent spring*. Boston: Houghton Mifflin Company.

Cochrane, Willard W. 1979. *The development of American agriculture, a historical analysis*. Minneapolis: University of Minnesota Press.

Colborn, Theo, Dianne Dumanoski, and John Peterson Myers. 1996. *Our stolen future*. New York: Plume, The Penguin Group.

Conford, Philip. 2001. *The origins of the organic movement*. Edinburgh: Floris Books.

Constance, Douglas H. 2008. The emancipatory question: The next step in the sociology of agrifood systems? *Agriculture and Human Values* 25:151–55.

Darnhofer, Ika. 2005. Organic farming and rural development: Some evidence from Austria. *Sociologia Ruralis* 45:308–23.

Dimitri, C., and C. Greene. 2001. *Recent growth patterns in the U.S. organic foods market*. Washington, DC: Economic Research Service, U.S. Department of Agriculture.

Electronic Code of Federal Regulations. 2008. Title 7, Agriculture, Part 205, National Organic Program. August 7. http://ecfr.gpoaccess.gov/cgi/t/text/text-idx?type=simple;c=ecf r;cc=ecfr;sid=4163ddc3518c1ffdc539675aed8efe33;region=DIV1;q1=national%20 organic%20program;rgn=div5;view=text;idno=7;node=7%3A3.1.1.9.31#7:3.1.1.9.31.1 .336.2 (accessed August 10, 2008).

Gieryn, Thomas F. 1999. *Cultural boundaries of science: Credibility on the line*. Chicago: University of Chicago Press.

Gold, Mary V. 2007. Organic production/organic food: Information access tools. http://www. nal.usda.gov/afsic/pubs/ofp/ofp.shtml (accessed August 8, 2008).

Guptil, Amy, and Rick Welsh. 2008. Is relationship marketing an alternative to the corporatization of organics? A case study of OFARM. In *Food and the mid-level farm: Renewing an agriculture of the middle*, ed. Thomas A. Lyson, G.W. Stevenson, and Rick Welsh, 55–78. Cambridge, MA: MIT Press.

Guthman, Julie. 2000. Raising organic: An agro-ecological assessment of grower practices in California. *Agriculture and Human Values* 17:257–66.

Guthman, Julie. 2004a. *Agrarian dreams: The paradox of organic farming in California*. Berkeley: University of California Press.

Guthman, Julie. 2004b. The trouble with 'organic lite' in California: A rejoinder to the 'conventionalisation' debate. *Sociologia Ruralis* 44:301–16.

Halberg, Niels, et al., eds. 2006. *Global development of organic agriculture: Challenges and prospects*. Wallingford, UK: CABI.

Haystead, Ladd, and Gilbert C. Fite. 1955. *The agricultural regions of the United States*. Norman: University of Oklahoma Press.

Heckman, J. 2006. A history of organic farming: Transitions from Sir Albert Howard's *War in the Soil* to USDA National Organic Program. *Renewable Agriculture and Food Science* 21:143–50.

Howard, A. 1943. *An agricultural testament*. London: Oxford University Press.

Howard, Sir Albert. 1946. *The war in the soil*. Emmaus, PA: Organic Gardening.

Jaffee, Daniel. 2007. *Brewing justice: Fair trade coffee, sustainability, and survival*. Berkeley: University of California Press.

Jamison, Rachael. 2003. Certified organic farming in Washington State: A feminist empiricist examination. Master's thesis, Evergreen State College, Olympia, WA.

Lobao, Linda, and Katherine Meyer. 2001. The great agricultural transition: Crisis, change, and social consequences of twentieth century U.S. farming. *Annual Review of Sociology* 27:103–24.

Lockie, Stewart, and Darren Halpin. 2005. The 'conventionalisation' thesis reconsidered: Structural and ideological transformation of Australian organic agriculture. *Sociologia Ruralis* 45:284–307.

Lyson, Thomas A. 2004. *A civic agriculture: Reconnecting farm, food, and community.* Medford, MA: Tufts University Press.

Morgan, Kevin, Terry Marsden, and Jonathan Murdoch. 2006. *Worlds of food: Place, power, and provenance in the food chain.* Oxford: Oxford University Press.

Obach, Brian K. 2007. Theoretical interpretations of the growth in organic agriculture: Agricultural modernization or an organic treadmill? *Society and Natural Resources* 20:229–244.

Perkins, John H. 1997. *Geopolitics and the green revolution: Wheat, genes, and the Cold War.* New York: Oxford University Press.

Pollan, Michael. 2006. *Omnivore's dilemma: A natural history of four meals.* New York: Penguin Books.

Raynolds, Laura T., Douglas L. Murray, and John Wilkinson, eds. 2007. *Fair trade: The challenges of transforming globalization.* London: Routledge.

Smith, Timothy J. 1994. Successful management of orchard replant disease in Washington State, United States. In *Proceedings of the International Society for Horticultural Science Symposium on Replant Problems*, pp. 161–67.

U.S. Department of Agriculture. 2006. Organic production, Tables 2 and 3. http://www.ers.usda.gov/Data/Organic/index.htm#tables (accessed August 8, 2008).

Section II

Global Perspectives

4 Northern Midwest (U.S.)
Farmers' Views of the Conversion Process

Paul Porter, Lori Scott, and Steve Simmons

CONTENTS

4.1   INTRODUCTION

Agriculture has always been an activity characterized by tension. It is a human enterprise, which means the practices used are the outgrowth of discussion and disagreement regarding the best approaches for growing crops and animals. Although biophysical constraints within agricultural systems are important, the reality is that humans make the decisions about what will be produced within agroecosystems—and when and how. Making the decision to move toward agricultural sustainability is no exception.

Agriculture is heavily influenced by tradition. Because of the intergenerational nature of agricultural systems, tradition serves as both a source of stability and a source of tension. Although the current discussion about sustainability in agriculture is more complex than just tradition versus nontradition, such oppositions certainly are present. In this chapter, after defining and describing some of the agricultural systems of the northern Midwest, we explore changes that farmers in the northern Midwest of the United States have made—and are making—in their efforts to move toward a sustainable agriculture. We also examine some of the inherent tensions that these farmers have encountered in the process of making these changes.

We believe that humans—and the countervailing forces they face when making decisions related to their agricultural systems—are critically important in the process of moving agriculture in the northern Midwest toward sustainability. Thus, in preparing this chapter we utilized qualitative approaches to focus on farmers and their assumptions, perspectives, motivations, and decisions as they endeavored to improve their farms. In writing this chapter we interviewed people from eight farms that represent some of the agricultural systems in the northern Midwest. Through these interviews, we learned of changes that have been made, as well as the processes by which those changes occurred. The farmers' stories also provided insights regarding the internal and external tensions that have been part of these changes.

We conclude our chapter by summarizing what we have learned about the processes of change from these interviews—and about the "transformational learning" that has occurred for the farmers. We trust that these farmers' stories, and the conclusions that we have drawn from them, can serve as useful guides for others within this region and beyond.

4.2   AGRICULTURAL SYSTEMS IN THE NORTHERN MIDWEST

Although it is difficult to delineate precise spatial boundaries for the northern Midwest, for the purposes of this chapter we define the northern Midwest as an area centered on the state of Minnesota and including parts of the adjoining states of Iowa, North Dakota, South Dakota, and Wisconsin.

The agricultural systems in these states vary greatly. We classified these systems according to the principal crops and animals that are produced. We recognize that some novel agricultural systems in the area, such as agroforestry and community-supported agriculture, are not amenable to such categorizations. But most of the land is utilized for systems that are described reasonably well by terms such as corn and soybean production or dairy and forages.

The variation in the agroecosystems corresponds to the major soil associations and topographical differences across the area. Such differences have been broadly classified by the USDA-NRCS (2005) into four land resource regions. The prevailing systems have evolved over a century and a half of agricultural and economic history. As a rule, agricultural systems within the southern and southwestern regions of the area are less complex in terms of the numbers of crops grown and the diversity of individual farms. Other areas have retained traditional practices such as companion cropping for the establishment of forages in dairy systems (Simmons et al., 1992) and swathing of wheat (*Triticum aestivum* L.) or barley (*Hordeum vulgare* L.) prior to harvest in more northerly small grain cropping systems.

The most dominant cropping systems in the northern Midwest, particularly in parts of Minnesota, Wisconsin, South Dakota, and Iowa, involve a two-crop rotation of corn (*Zea mays* L.) and soybean (*Glycine max* Merrill) (Figure 4.1a,b). The

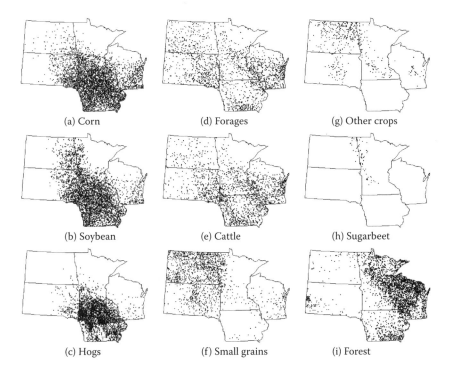

(a) Corn　　　(d) Forages　　　(g) Other crops

(b) Soybean　　　(e) Cattle　　　(h) Sugarbeet

(c) Hogs　　　(f) Small grains　　　(i) Forest

FIGURE 4.1 Crop and livestock distribution in the northern Midwest. Dots on maps are distributed by county; each dot represents 10 acres or 10 animals per square mile. (Data are from the 2003 Census of Agriculture, except for the data on hogs, which are from the 2002 census [USDA-NASS, 2004].)

principal livestock associated with these systems is hogs (*Sus scrofa* L.) (Figure 4.1c). Regions dominated by these agricultural systems tend to be in the subhumid climate zone with 450 to 750 mm precipitation, much of which falls during the growing season (Table 4.1). The deep soils that are characteristic of this system in the northern Midwest were mostly formed under prairie before the advent of extensive agriculture and are quite productive. Manure from the hogs is used to fertilize some of the fields. Large amounts of external resources such as synthetic fertilizers and pesticides are imported into the system, and the crop and livestock products are mostly exported from the system.

Another set of agricultural systems in the northern Midwest involves production of forages and beef and dairy cattle (*Bos taurus*) (Figure 4.1d,e). Forages include alfalfa (*Medicago sativa* L.) and other legumes, grasses, and corn harvested for silage. Again, manure produced by dairy or beef cattle is usually used as a source of nutrients and organic matter for some fields. Some agricultural systems involve grazing of livestock, and one noteworthy system, termed rotational grazing, is becoming an important pasture management strategy, especially for sustainable dairy and beef farmers. The northern Midwest is home to the largest organic dairy cooperative in the United States—Organic Valley, based in La Farge, Wisconsin.

TABLE 4.1

Climate Characteristics of Land Resource Regions in the Northern Midwest

Region	Precipitation Range	Mean Temperature Range	Freeze-Free Period (days)
Central feed grains and livestock	625–900 mm 25–35 in.	6–13°C 43–55°F	140–180
Northern Great Plains spring wheat	250–550 mm 10–22 in.	4–9°C 39–48°F	100–155
Northern lake state forest and forage	500–825 mm 20–32 in.	2–7°C 36–45°F	95–145
Western Great Plains range and irrigated	275–600 mm 11–24 in.	7–16°C 45–61°F	100–200

Source: Data from PSU ESSC (1998).

These agricultural systems are found within the Minnesota and Mississippi river valleys and in much of southern Wisconsin, southeastern South Dakota, and northeastern Iowa. These are often lower-valued, hilly lands that are subject to erosion and not suited to intensive corn or soybean production. Perennial crops such as forages are better suited to such landscapes than are row crops. However, contour strip cropping (alternating strips of corn with strips of perennial forage) is sometimes used to suppress erosion in this terrain. The regions where these agricultural systems predominate receive higher amounts of precipitation than the former prairie regions to the west (Table 4.1). Much of the land was originally wooded. Dairy tends to be concentrated in southern and southwestern Wisconsin, in central and southeastern Minnesota, and in northeastern Iowa. Lands in the western parts of the northern Midwest (west of the Missouri River) receive low amounts of precipitation (usually less than 450 mm annually), but still support rangeland and livestock grazing.

A third array of agricultural systems in the area is based on small grains such as wheat, oats (*Avena sativa* L.), barley, and rye (*Secale cereale* L.), as well as other crops (Figure 4.1f). These systems predominate in North Dakota, central South Dakota, and northwestern Minnesota, with limited areas also in southern Minnesota and central Wisconsin. These systems tend to be found where there are lower amounts of precipitation and cooler growing season temperatures (Table 4.1). Wheat, barley, and oats were once commonly grown in regions that are now dominated by corn and soybean, but their acreage has diminished greatly over the past 70 years (Cardwell, 1982). Some of the areas where small grain-based agricultural systems are now found were formerly tallgrass and shortgrass prairie and have rich, productive soils. Since the early 1990s, yield reductions caused by diseases such as scab (*Fusarium* sp.) have led to a dramatic decline in the acreage of small grains in eastern parts of the northern Midwest (Corselius et al., 2003). Although oats have traditionally been an important crop in the northern Midwest for companion cropping, as well as for bedding straw and feed for livestock, the oat acreage has also declined markedly in recent decades.

A number of other crops are grown within agricultural systems in the northern Midwest, including flax (*Linum usitatissimum* L.), canola (*Brassica napus* L.), sunflower (*Helianthus annuus* L.), dry edible beans (*Phaseolus vulgaris* L.), and sorghum (*Sorghum bicolor* L. Moench) (Figure 4.1g). Snap beans, peas (*Pisum arvense* L.), sweet corn, sugarbeets (*Beta vulgaris* L.), and potatoes (*Solanum tuberosum* L.) are also grown on limited acreages. Sugarbeets are a crop worthy of special mention because the northern Midwest ranks in the top echelon of beet sugar production areas in the United States. The acreages of this crop, although relatively small, are clustered within close proximity of the sugarbeet processing facilities in southern and northwestern Minnesota and in easternmost North Dakota (Figure 4.1h). All of these processing facilities are farmer-owned cooperatives and farmers must own shares in the cooperative to have a market for their sugarbeets.

Two extensive parts of central Wisconsin and Minnesota, characterized by sandy soils, are the locations for production of specialty crops such as potatoes and canning vegetables. Although these areas receive moderate amounts of precipitation, the low water-holding capacity of the soils dictates the use of irrigation for profitable and market-quality crops. Other pockets of irrigated agriculture are found in localized areas in the northern Midwest.

In rounding out the characterization of agricultural systems in the northern Midwest, it is important to highlight extensive regions of both deciduous and coniferous forests, particularly in parts of Minnesota, Wisconsin, and Iowa (Figure 4.1i). Timber and pulp are harvested from many of these woodlands, and maple sugar is also produced in some areas. Other uses for these landscapes include recreation and wildlife habitat. There also are extensive areas of forested land in the western part of South Dakota associated with the Black Hills.

Certified organic acreage of cropland and pasture increased in the northern Midwest between 1997 and 2003 (Table 4.2). North Dakota has the largest acreage, followed by Minnesota, Wisconsin, Iowa, and South Dakota. In 2003, these five states accounted for 22% of the total U.S. certified organic farms, 32% of the total U.S. certified acreage of cropland, and 24% of the total U.S. certified acreage of cropland plus pasture. That same year they accounted for 24% of the total U.S. certi-

TABLE 4.2

Certified Organic Agriculture in the Northern Midwest, 1997–2003

State	Farms 2003	Total Certified Cropland and Pasture Acreage			
		1997	2000	2001	2003 (Cropland Only)
Iowa	448	35,769	68,939	80,354	74,985 (67,717)
Minnesota	392	63,685	81,953	103,297	123,923 (115,470)
North Dakota	145	90,790	153,737	159,300	147,780 (128,963)
South Dakota	84	32,319	46,532	57,417	59,286 (53,772)
Wisconsin	659	47,622	80,285	91,619	120,643 (91,906)
Total	1,728	270,185	431,446	491,987	526,617 (457,828)
Percent of U.S. Total	22%	20%	24%	24%	24% (32%)

Source: Data from ERS-USDA (2005).

TABLE 4.3

Certified Organic Livestock in the Northern Midwest, 2003

State	Milk Cows	Total Cows, Pigs, and Sheep	Total Chickens and Poultry
Iowa	2,222	6,592	323,103
Minnesota	5,215	7,387	33,660
North Dakota	—	784	—
South Dakota	—	1,133	—
Wisconsin	24,884	28,103	569,429
Total	32,321	43,999	926,192
Percent of U.S. Total	43%	35%	11%

Source: Data from ERS-USDA (2005).

fied cows; 35% of the total U.S. certified cows, pigs, and sheep; and 11% of the total U.S. certified chickens and poultry (Table 4.3) (ERS-USDA, 2005).

Farmers in this region of the United States were relatively quick to adopt organic production practices for several reasons. One important factor was that the winter climatic conditions limit weed, pest, and disease pressures. Another reason may have been related to the fact that the region is on the northern edge of the traditional corn belt, where land prices tend to be lower and cropping systems tend to be more diverse. No doubt the presence of a large number of certification agencies, including chapters of the Organic Crop Improvement Association, Farm Verified Organic, and the now defunct Organic Growers and Buyers Association, played a key role in promoting organic production in the region. The northern Midwest is home to the largest organic farming conference in the United States. Held annually in February in La Crosse, Wisconsin, the Upper Midwest Organic Farming Conference, conducted by the Midwest Organic and Sustainable Education Service (MOSES) of Spring Valley, Wisconsin, will attract over 1,500 participants.

While land grant universities in the region have had a limited number of faculty involved with organic research over the years (OFRF, 2003), increased interest by these institutions in organic research and extension has occurred since national standards came into effect in 2002. The various state Departments of Agriculture, followed by the National Resource Conservation Service, are likewise beginning to have an increased presence with respect to activity involving organic production and marketing. Publications including *The Upper Midwest Organic Resource Directory* (MOSES, 2005), *The Status of Organic Agriculture in Minnesota* (MDA, 2001; MDA 2003), and *Organic Certification of Crop Production in Minnesota* (MISA, 2001) have aided farmers in transitioning to organic production.

4.3   AGROECOLOGICAL CONSTRAINTS

Agroecological constraints to conversion or transition to organic and other sustainable systems in the northern Midwest can be categorized three ways—environmental, social, and economic. These three categories are complex and interrelated and represent some of the reasons why tensions have arisen around defining the best approaches in moving toward sustainable agriculture.

4.3.1 ENVIRONMENTAL LIMITATIONS

There are a number of environmental factors that can constrain sustainability within agricultural systems. Situated at the place where the areas of former tallgrass and shortgrass prairies and the areas of northern coniferous and eastern deciduous forests came together, the northern Midwest is characterized as an ecological transition zone. In general, precipitation declines from east to west, and there are gradually fewer cumulative growing degree units (lower average growing season temperatures) from south to north (Table 4.1). Soil types and climatic conditions play a strong role in determining how various agricultural systems are distributed.

Soil types and their relative productivities can vary markedly from location to location within the northern Midwest. Much of this landscape is artificially drained, which has been a key development for increasing productivity of the land for crops (Fausey et al., 1995; Skaggs et al., 1994). In some cases, however, the intensity and types of artificial drainage have had negative consequences for water quality within various watersheds, such as the Minnesota River Basin (Alexander et al., 1995; Rabalais et al., 2001). Artificially drained agricultural lands are highly valued, which further fuels the viewpoint that only high-income-producing crops can be profitably grown on them.

The continental climate that is characteristic of the northern Midwest accounts for extremely variable weather patterns. Arctic cold fronts in winter lower soil temperatures to well below freezing for several months each year. Snowfall begins as early as October and may not end until April or May. While such climatic conditions can be beneficial for suppressing certain crop pests, they also limit the number of cropping options that farmers can use. The window for crop establishment in the spring is short, and lower temperatures hinder production of warm-season crop species. In portions of the northern Midwest, soil moisture recharge by fall and spring precipitation is crucial. Furthermore, snowmelt and spring rainfall can lead to large amounts of spring runoff and flooding in riparian areas. The establishment of fall cover crops to counter soil erosion after harvest can be difficult because of the short time period available for establishing them. And typically, the desire to optimize yields of cash crops outweighs environmental benefits of sowing cover crops since farmers usually receive little or no direct income from them.

Perhaps the principal ecological constraint over the longer-term in almost all areas of the northern Midwest is the reduction in species diversification within the agroecosystems. Many crop rotations in the region consist of two—or at most three—crops in rotation over time. And sometimes these crops are similar in terms of their pest susceptibilities, particularly for diseases (Corselius et al., 2003). Historically, crop rotations within the region were more diverse and often included a perennial legume when livestock were a more common component on most farms. Such perennial legumes are now much less common. Thus, the simplification of agroecosystems within the region over the past decades, which has often been touted as more specialized or efficient, has diminished their ecological integrity. Some recent ventures, such as the Green Lands, Blue Waters initiative described in Section 4.5, have attempted to address this underlying fallacy of oversimplified agroecosystems with inadequate crop species diversity.

4.3.2 SOCIAL LIMITATIONS AND THE PROCESS OF CHANGE

Social factors limiting adoption of alternative agricultural practices include family, community, and institutional pressures. Adoption of such practices requires farmers to recognize a *need* for change. Sometimes this might involve hearing about a neighbor's farm failing or perhaps experiencing an economic downturn of one's own farm. Such situations (termed disorienting dilemmas) can cause questions to be raised about prior assumptions concerning farms and practices (Percy, 2005). This questioning may lead to a change in attitudes and perspectives that results in altered approaches to crop and livestock management.

Once farmers have concluded that some kind of change is needed, it is important for them to communicate with others, such as a partner or spouse, and thereby help validate their observations and conclusions. If their observations and conclusions are not well received or supported, the momentum for change can be stifled. It is rare that farmers "go it alone" when making substantive changes to their farming operations.

A shift toward sustainability on one's farm usually prompts an initial acceptance or rejection by one's neighbors and peers. Whether or not one implements new practices before discussing them with neighbors is partly a function of the level of trust that exists between them. It is almost certain that discussions of evolving change will eventually occur within the community (at the cafe, the grocery, the co-op, or social events), and such conversations can be important for defining the kinds of responses a farmer receives to his or her on-farm innovations. Pressures to conform to peer or community norms in the northern Midwest can be large and can hinder adoption of new practices.

4.3.3 ECONOMIC LIMITATIONS

Economic pressures come in a variety of forms. A move toward sustainable practice may impact a farmer's relationship with implement, chemical, and seed dealers. In certain locations within the northern Midwest, the infrastructure no longer exists to handle and process crops other than the principal commodities. Transporting an alternative or uncommon crop to an appropriate market is complicated by this lack of infrastructure. Marketing such a crop is further complicated by volatile prices and consumer demand.

Lending institutions often have to be convinced of the soundness of a new practice, and those who are employed at such institutions don't always know how to evaluate practices that are new to a locale. It is sometimes easier for farmers to just continue practices that they have been using than to convince lenders that adopting little known or understood practices is in their best interests (Hamilton, 1990).

If farmers are under financial stress, that situation can, in itself, serve as a stimulus to consider making substantive changes. The farmer may conclude that changes will help alleviate adverse economic circumstances. However, lending institutions or other farmers may observe the same economic situations and conclude that the best future direction is to "stay the course" while assuming that the circumstances will get better. Thus, adverse financial circumstances are often a constraint in moving toward sustainable agriculture, but they also are an impetus for change (LSP, 2003).

In recent years, farm profits have been closely linked to participation in governmental farm programs (Nordquist et al., 2004). Multiperil crop insurance, loan deficiency payments, preventative planting payments, historical yields, and base acreages all play a part in decisions that influence farm operations. Government farm programs can be dynamic (e.g., the vagaries of the Conservation Security Program), and farmers seeking greater financial stability through such programs may favor the status quo so that they can remain positioned for participation.

In this region, the North Central Sustainable Agriculture Research and Education (SARE) Program, as well as the Minnesota Department of Agriculture, have had both producer and researcher on-farm demonstration grant programs that have aided in organic research and outreach efforts.

4.4 PERSPECTIVES AND PRACTICES OF FARMERS MOVING TOWARD SUSTAINABILITY

The environmental, social, and economic challenges farmers face constitute agroecological constraints that can affect their ability to move toward change. Nonetheless, in their efforts to move toward sustainability within their agricultural systems, the farmers we interviewed have made substantial changes in their practices. Their eight farms are distributed within various agricultural systems (Table 4.4) and regions of Minnesota (Figure 4.2). Their farms range in size from a few hundred acres to several thousand acres, and they produce a wide array of crops, livestock, and other products (Table 4.5).

The interviews of the farmers were organized around the following topical areas:

- The farmers' perceptions of the processes they have followed in moving toward sustainability.
- The kinds of factors that prompted them to make changes in how they farm.
- The farmers' perceptions of constraints they have faced during the change process.
- Surprises they have encountered during the process.
- Examples of specific changes that farmers have made on their farms.
- How they regard the change process they have made (e.g., is it a "redesign?").
- How the farmers perceive the influences of neighbors and community during the process.
- Their perceived needs for further knowledge as they continue the process of change.

The specific questions used to guide the interviews are provided in the Appendix at the end of this chapter.

TABLE 4.4
Luverne and Mary Jo Forbord's A to Z "Product" List

Alfalfa and art

Beef, bees, beets, birds, and butterflies

Cows, calves, corn, and carbon sequestration

Deer, ducks, dogs, and dried distiller grains

Eggs, education, and ethanol

Flax, forbs, fish, and fun

Geese and grass (native and not)

Home, history, hens, and hunting

Inspiration and insight

Jobs for everyone!

Kinship and kohlrabi

Ladyslipper, leadplant, and leaves

Musicians and mystery

Native prairie, nutrients, and next generation

Opportunities and oddities

Photosynthesis, ponds, and ponderings

Quality of life

Rye, recreation, and relationships

Straw, scenery, skinks, spirituality, and sensory delights

Trees, teff, tomatoes, and thankfulness

Unity and understanding

Vodka

Wheat, wildlife, and water filtration

X-asperation!

Yarrow and yawns

Zest for this way of life and all of its rewards for us and everyone

Farming Forever! Amen.

The farmers we interviewed are active within their communities and various agricultural organizations. For example, Mary Jo Forbord has served as executive director of the Sustainable Farming Association (SFA) of Minnesota. Jaime DeRosier has chaired a local SFA chapter and has authored a book about his experiences with organic and sustainable production methods. Another farmer, Carmen Fernholz, has used his background as a teacher to share his knowledge of organic systems at field days, workshops, and guest lectures at universities. He has been a leader in the SFA and has served on an organic marketing board (Organic Farmers' Agency for Relationship Marketing [OFARM]). He also contributes to his community beyond agriculture by directing local high school and community theater plays. Tony Thompson's Willow Lake Farm regularly hosts an agroecology summit designed to help participants gain a greater understanding of agroecological concepts. He has also been an advocate for innovative strategies to manage wildlife within agricultural

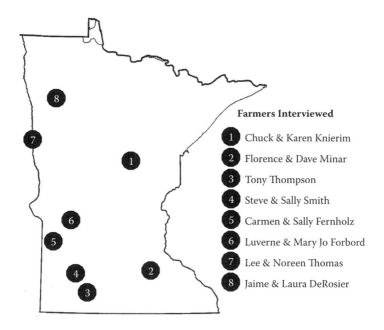

Farmers Interviewed

1. Chuck & Karen Knierim
2. Florence & Dave Minar
3. Tony Thompson
4. Steve & Sally Smith
5. Carmen & Sally Fernholz
6. Luverne & Mary Jo Forbord
7. Lee & Noreen Thomas
8. Jaime & Laura DeRosier

FIGURE 4.2   The location of the farmers chosen for interviews.

systems. Steve Smith has served as a community advisor for a research/outreach center of the state university.

The farmers we interviewed have strong convictions regarding the importance of serving the public good through their farm operations. For example, Chuck and Karen Knierim's Wildrose Farm Web site states, "At Wildrose Farm there is no million dollar designer at the top and there are no kids in sweatshops at the bottom.... The people that work with us are talented artists and sewers.... Everyone earns a decent wage." Similarly, Luverne and Mary Jo Forbord's "alphabet product list" at their Prairie Horizons Farm includes intrinsic values as well as products (Table 4.4). Dave and Florence Minar of Cedar Summit Farm are committed to "provide clean, healthy, locally grown and safe meat and dairy products." Similarly, Tony Thompson has stated, "I would like be recognized as much for the increased numbers of upland plover birds that are on my farm as for the bushels of corn I produce." Clearly a healthy environment and food supply are high priorities for all of the farmers we interviewed. They possess a strong work ethic and an absence of inflated or unrealistic financial expectations from their farms.

4.4.1 Characteristics of the Farms and Changes Made

Of the eight farmers interviewed, seven have some certified organic acreage, two have a diversified organic crop/livestock operation, and two others have organic crops and conventional livestock (Table 4.5). Two would be considered pasture-based systems, two mixed cropping, three crops only, and one a sustainably managed timber

TABLE 4.5
Characteristics of the Farms Operated by Interviewees

Farm Name	Acres[a]	Operators	Crops	Livestock	Other	Duration and Transitions	Use Estimates
Thomas Farm, Moorhead	1,200 (1,200)	Lee and Noreen Thomas and three children	Wheat, corn, soybean, rye, alfalfa	None	Certified organic	Began farming conventional sugarbeet-based system in 1986; began transition to organic in 1999	100% resale 60% for human consumption
Smith Farm, Southern Minnesota	390 (358)	Steve and Sally Smith[b]	Corn, soybean, wheat, red clover; fruit and vegetables for own use	Organic pasture-fed cows/calves; chickens for own use; conventional hogs	Certified organic	Began farming conventionally in 1973; transitioned to organic in 1998	90% for resale 60% for human consumption
Cedar Summit Farm, New Prague, McGrath, and Kerkhoven	871 (821)	Dave and Florence Minar and adult children	Pasture and hay for winter feed	Grass-fed cows, pasture-raised hogs and steers	Milk processing; direct marketed meat	Began conventional dairy in 1969; transitioned to rotational grazing–based dairy in 1993	Crops for animal consumption; milk and meat for human consumption
Prairie Horizons Farm, Benson	480 (250 in transition)	Luverne and Mary Jo Forbord and three children	Alfalfa, grass, corn, flax, wheat, rye, native prairie plants, vegetables, trees	Beef cows, chickens, eggs	Rotational grazing	Conventional dairy before 2002, when transitioned to rotational grazing–based beef	99% for resale Crops for human and animal consumption and industrial use Animals for human consumption

Farm	Acres[a]	Crops	Livestock	Production system	History	Market/consumption
A-Frame Farm, Madison Carmen and Sally Fernholz	400 (350)	Corn, soybean, wheat, oats, flax, interseeded legumes	Conventional feeder pigs	Certified organic	Started farming in 1972; began pursuing sustainability in 1973; entire farm certified organic in 1994	100% for resale. Wheat, flax, pigs for human consumption. Beans for seed. Corn, oats, alfalfa for animal consumption
DeRosier Farm, Red Lake Falls Jaime and Laura DeRosier and five children	1,000 (1,000)	Wheat, rye, barley, oats, soybeans, flax, millet, corn, peas, sunflowers, alfalfa, buckwheat, popcorn	Chickens, turkeys, pheasants, pigs, calves, horses (for own use)	Certified organic	Began farming in 1985; received organic certification in early 1990s	100% resale. Mostly for human consumption
Willow Lake Farm, Windom Tony Thompson	3,700 row crops; 400 wetlands; 2,000 prairie (12)	Corn, soybean, native prairie plants, buckwheat, popcorn	None	Ridge-till planting system, conventional with some certified organic crop acres	Ridge-till system since 1991	99% for resale, animal consumption
Wildrose Farm, Breezy Point Chuck and Karen Knierim	21 (0)	Sustained-yield timber, herbs and vegetables	Chickens, turkeys	Organic cotton clothing and rugs, recycled fiber, timber made into furniture	In business for 30 years; expanded to organic clothing about 10 years ago	Furniture and clothing for resale/human consumption. Herbs and vegetables for own use

[a] Certified organic acres in parentheses.
[b] Identity disguised to preserve anonymity.

operation that purchases off-farm organic cotton for the manufacturing of clothes. There were no vegetable producers surveyed.

The 1,200-acre farm of Lee and Noreen Thomas, which is in the Red River Valley of northwestern Minnesota, has been in Lee's family for multiple generations. From the mid-1980s (when Lee started farming) until the late 1990s, this farm depended on sugarbeets as the principal cash crop. However, a combination of sugarbeet diseases, human and food safety concerns, and changes in the global sugar economy caused the Thomases to reevaluate their operation. This resulted in their transition to certified organic production in the early 2000s. They now commonly grow food-grade soybean, wheat, and corn.

The 390-acre farm of Steve and Sally Smith, located in southwestern Minnesota, had been a typical corn and soybean operation since the Smiths began farming in the early 1970s. Both Steve and Sally had grown up on conventional farms in the area. By the late 1990s it had become apparent to them that something had to change if they were going to remain viable. Other farmers in their area were expanding their acreages in order to try to remain profitable. But similar to the Thomases, the Smiths chose in the late 1990s to transition their crop acreage toward a certified organic system involving food- and feed-grade corn, soybean, and wheat. They have retained their conventional confined hog production operation; however, in early 2004 the Smiths' son began raising cattle organically on the farm.

Dave and Florence Minar's Cedar Summit Farm in east-central Minnesota began as a conventional dairy in 1969. In 1974 they chose to discontinue pesticide use on their farm, and in 1993 they changed to milking entirely grass-fed cows. In the early 2000s they established an on-farm milk and cheese processing facility, and began to change their entire operation to certified organic production. Detail regarding their business development plan is available from the Minnesota Institute for Sustainable Agriculture (MISA, 2003). Their farm is approximately 440 acres with 180 milk cows.

Luverne and Mary Jo Forbord's 480-acre Prairie Horizons Farm in west-central Minnesota was also formerly a conventional dairy operation. In 2002, they changed their farm operation to rotationally grazed beef.

Carmen and Sally Fernholz, of A-Frame Farm in west-central Minnesota, began farming in 1972. They have tried to farm their 400-acre operation sustainably since 1973, and all of their crop acreage has been certified organic since 1994. As with the Smiths, the Fernholzes have continued to use a conventional confinement approach to raise feeder pigs, and they do so in partnership with Carmen's brother.

Jaime and Laura DeRosier's farm, located in northwestern Minnesota, began in 1985. Their 1,000 acres of cropland have been certified organic since the early 1990s. They grow a variety of crops and market what they can as food grade, with the remainder being sold as feed grade. They have no livestock except what they raise for their own use.

The land that forms the backbone of Tony Thompson's Willow Lake Farm, located in southwestern Minnesota, has been in his family for several generations. Thompson farms in partnership with a neighbor and together they manage approximately 4,000 acres. The main crops are corn and soybean, but there are also a number of wetlands on the farm, as well as riparian buffer strips along most of the waterways. A portion of Thompson's agricultural income is derived from harvest of seed from native prairie species, which is sold to public agencies and others who do prairie restoration and right-of-way work. He also has a limited number of certified organic acres on which he has grown soybean, wheat, rye, and alfalfa. Thompson utilizes ridge-till planting, which is a reduced-tillage system that conserves soil on his corn and soybean acreage. He began using ridge-till planting in the early 1990s.

Chuck and Karen Knierim operate the Wildrose Farm in a deciduous forested area of central Minnesota. They have been in the lumber business for 30 years and recently have diversified their business by manufacturing and marketing high-quality organic cotton (*Gossypium hirsutum* L.) clothing.

4.4.2 PERCEPTIONS OF CHANGE

In our interviews of the farmers, we were interested in understanding how each regarded the substantial changes they had made to their farm management practices. Farmers in the northern Midwest tend to view their farms as dynamic, e.g., in a state of constant transition. And probably all farmers, whether or not they consider their farms sustainable, have made substantial changes in the approaches they use. For example, the average farm in the northern Midwest has increased its number of acres or the numbers of animals produced per year (USDA-NASS, 2004). The size of machinery has also increased. In addition, many farmers in the northern Midwest no longer use a moldboard plow and have adopted various forms of conservation tillage that use no-till or field cultivators and chisels for primary tillage to reduce soil erosion. Most farmers would have considered such substantive changes to be unwise only a few decades ago.

While such statistical shifts and changes in tillage are of interest, we were most interested in documenting the fundamental changes in the assumptions, attitudes, and perspectives of farmers that led to their changes in practice. One premise that informed our approach to our interviews was that substantial transitions in farming practice, such as the move from conventional to organic, require transformation of one's assumptions, perspectives, and attitudes. The stories that we collected from our interviews appear to support this view.

So how did the interviewed farmers view the transitions they have made on their respective farms? To answer this question, we need to begin with semantics. Some of the farmers were quite comfortable with describing the changes they have made as a type of conversion. For example, in evaluating the possibility that her neighbors would also shift to organic production, organic farmer Noreen Thomas used the word *convert*: "I don't think any of my neighbors will be converting to organic." Although Steve and Sally Smith did not use the word *conversion* per se in their interview, they did describe their transition to organic crop production in terms of a dramatic change: "We just went right straight through with it—jumped in—didn't look back."

But other farmers were less certain about using the concept of conversion to describe the process of change for their farms. Aversion to this term may have had to do with negative connotations around the word itself. For example, Mary Jo Forbord stated, "*Conversion* sounds immediate and kind of religious. *Transition* is more descriptive of what we are doing." Similarly, Carmen Fernholz (also an organic farmer) preferred to use the term *transitioning* when describing the course of change for his farm over the past 30+ years. He felt that this term better described the "never ending" nature of the changes he had made and the reality that much of his innovation as a farmer involved "tweaking" from year to year rather than making dramatic, all-or-nothing changes. An example of a continuing transition in Fernholz's crop production practice is his ongoing modifications of his crop cultivator, an implement that he considers to be one of the most important on his farm. And while agreeing that the term *conversion* fit his shift to rotational grazing–based dairy, Dave Minar saw this change as anything but "immediate." He stated, "Going from conventional to sustainable is definitely a conversion. Our conversion started in 1974 and it culminated in 1994."

It's fair to say that, regardless of the terms farmers used to describe the process, all of them regarded the changes they had made on their farms as gradual or evolving. For example, Jaime DeRosier stated, "Rather than converting to sustainable agriculture, my farming could [best] be described as having evolved … in that direction." Similarly, Tony Thompson described his farm as "in development"—in other words, an unfinished work, which supports his long-range view of his farm as multigenerational. This is consistent with systems changes in farming, as opposed to input substitution.

All of the farmers we interviewed regarded the modifications that they had instituted on their farms, whether occurring over a short or a long period of time, as fundamental changes that arose from their core beliefs and values. Some, like Carmen Fernholz and Dave Minar, were comfortable with describing the changes that had been made as a fundamental redesign. Fernholz said, "I would call it a total redesigning." He further noted that it was a redesign that was "site specific and management specific." Although Jaime DeRosier did not use the term *redesign* to describe the overall changes that he made to his farm, he did state that he was in a process of redesigning the farm from year to year, "making the good changes when I can, and learning from the mistakes." Mary Jo Forbord further raised the semantics ante when she described the changes that she and Luverne had instituted: "We have reinvented our farm." Tony Thompson and his brother went so far as to imagine a "dream farm," and they drew diagrams on large pieces of butcher paper that were later converted to a computer spreadsheet plan and presented to prospective lenders as they executed their new vision.

4.4.3　Motivations for Change

A historical perspective helped motivate Carmen Fernholz to begin changing to an organic approach to farming in 1972. Fernholz stated, "I reasoned that if they were able to do it [farm without chemicals or major purchased inputs] prior to the 1960s, there was no reason we couldn't do it [in the 1970s]. And so that was sort of a motivation that got me going and it has been a continuous motivation." He also vividly recalled that his father had used him as a human "marker" in the field when he applied pesticides in order to keep track of where he had already applied chemicals. In hindsight, Fernholz realized that since he had made the shift to an organic crop production system relatively early in his farming career, he never really allowed the chemical-intensive agricultural model to shape his image of farming. Despite his father's adoption of chemical-based agriculture in the 1960s, he continued to subscribe to *Organic Gardening and Farming*, a magazine published by Rodale Press, which Fernholz described as an important "imprint" during his upbringing on the farm.

For some farmers, the motivation to change came from economics. As previously noted, Steve Smith initiated his organic crop production approach because of perceived economic advantages. Not willing or able to increase the size of their farm, and increasingly aware that "you can't make a living on 300 conventional acres," Steve and Sally switched to organic production out of economic necessity. His and Sally's social and ecological motivations grew in importance only after they made the change. Now they would say that they farm organically "for reasons we *should* be doing it—the right reasons—instead of just economics." Similarly, Jaime DeRosier noted that, although he had other motivations for limiting chemical use on his farm, the strongest factor was economics. He said, "I try to use the 'no-cost' or the 'low-cost' solutions; the cost of chemical and synthetic inputs can be staggering!"

Association with and support of other farmers was also a powerful motivator for some. For example, Tony Thompson and his brother mutually committed to become "the best farmers [they] could be" in 1990. This precipitated preparation of a visionary farm plan, which included unconventional practices such as the adoption of ridge tillage. Tony was motivated to adopt ridge tillage both to reduce tillage machinery and fossil fuel costs and to save time spent tilling his fields. As a prairie enthusiast, conservationist, and naturalist, Thompson also regarded moldboard plowing and other primary tillage operations as disruptive of natural ecosystems and detrimental to wildlife. Peer farmers were important for Thompson in his tillage transitions since some of his neighbor farmers had already adopted ridge-till planting and encouraged him to do likewise. He learned from these farmers that it was possible to make this change without sacrificing yields, which was an important consideration for him.

Lee and Noreen Thomas had attended an organic farmers' conference in Wisconsin in 1999. Their first impressions of the other farmers who attended this conference were not positive, but as the conference progressed they found that they had an affinity for much of what those in attendance were saying. They listened, asked questions, and learned. As a consequence, they were motivated to change to an organic cropping system on 200 of the 1,200 acres on their farm.

Another motivation for the Thomases was their concern for the health of their children. They had noted that several children among their neighbors had been born with birth defects. When it was suggested to Lee that part of Noreen's motivation for shifting to an organic system had involved health concerns, he replied, "Nope—*all* of it." By his own admission, Noreen's strong convictions about the necessity of approaching the management of their farm differently had been very important for him. He concluded that if it had not been for her insistence, he probably would never have made the transition to organic farming.

Finally, some farmers were motivated to change their approaches to farming because of their interest in creating more favorable opportunities for "the next generation" of farmers. Mary Jo Forbord noted, "We don't know how farming might look in our children's lifetime, but we do know that the era of our farm succeeding by marketing into the commodity system is drawing to a close."

4.4.4 CONSTRAINTS TO CHANGE AND SUSTAINABILITY

When the farmers we interviewed were asked to reflect on what they thought were the primary constraints or factors hindering their move toward sustainability on their farms, they responded with a wide array of considerations.

Workload and labor availability were common themes. For example, although reducing labor costs was a factor in the Minars' decision to transition from a conventional to a grass-based, organic dairy, they now find it an ongoing challenge to find enough employees "who have the same values that we have." They also have had difficulties acquiring adequate financial backing to support their transition. Since converting to an organic system, the Thomases have found that the workload can sometimes be overwhelming. They note that weather and their short growing season "cause everything to happen at once," especially with respect to timely weeding of their organic fields.

Chuck Knierim noted the difficulties posed by the limited knowledge of some of his customers and of consumers generally, who often don't understand the concept of sustainability or the many problems posed by conventional production. He observed, "[By] selling direct to consumers, we continue to be amazed at the amount of false information given to consumers. We find ourselves doing a large amount of education."

For Carmen Fernholz an ongoing problem is the decline of premiums in the organic market. He stated, "The margins are narrowing for organic producers just as for conventional ones.... I just can't sit on this one. Organic [production] is not low input, its alternative input." Carmen also noted that his weed species and pressures had shifted since converting to organic production, and this remains a continuing challenge.

The DeRosiers, who had sought to minimize chemical use as a means to find "low-cost solutions," still find that economic factors can be major constraints, and can be made worse by weather and other uncontrollable conditions that differ very much from one growing season to the next. "Those inconsistencies can be difficult," said Jaime, "one has to be able to ride the waves.... We [sustainable] farmers might not be storing up a lot of riches on earth, but we've got a little piece of heaven here, haven't we?"

Several farmers have found that there are a variety of constraints to a complete conversion to organic production. When asked why their hog operation is run conventionally, Steve Smith responded, "The hog barns own us, we don't own the hog barns. They have to be paid for. We cannot leave them sit idle, and they are a very good source of manure." The Smiths had already committed to the hog operation and invested in the barns prior to 1998, when they began converting to organic. Without the existing hog barns, they believe they most likely would be producing organic hogs at this time. They noted, however, that with organic hogs in the pasture, they would not have as many hogs and would not have the manure supply they now rely on. (Although others farm organically without manure, the Smiths use manure exclusively as their nutrient source.) Carmen Fernholz's reason for maintaining a conventional hog operation was that the organic market for pork has not been consistent enough over the years to justify remodeling or building new facilities to meet organic standards. And the economics of feeding the high-valued organic grains do not "pencil out" given the unreliable markets and the market value achievable for organic pork.

4.4.5   ROLE OF COMMUNITY IN THE CHANGE PROCESS

When farmers elect to deviate from the generally accepted practices used by the majority of farmers in their locale, it does not escape notice. Because the production decisions made by the farmers we interviewed were out of the ordinary, their relationships with their extended families and communities were sometimes affected. Having once been conventional farmers themselves, the Smiths still respected their conventional neighbors and regarded their own attitudes toward their neighbors as not having changed since they converted to organic production. But they were not as certain about what their neighbors thought of them. The Smiths had to deal with a stigma associated with having weedy fields. Steve also noted, "There is a joke that says you only convert to organic if you aren't good enough to farm conventionally. [But] according to conventional farmers, you aren't good enough unless you get big enough. Success is based on size."

"Most of our farmer neighbors are very supportive [of what we are doing], while some think we are crazy and wonder why we have not yet gone broke," remarked Dave and Florence Minar. After a sufficient time period of apparent success with the new approach, the reaction of neighbors can progress to a hesitant curiosity. For example, Mary Jo Forbord noted that her neighbors' attitudes are "slowly becoming more positive. Now just about everyone seems curious, but would rather ask someone else what we are doing instead of us."

Most of the interviewed farmers had seen improvements in the attitudes of peer farmers and others in their communities toward their altered approaches to farming. For example, Jaime DeRosier stated, "I have seen a change in my neighbors' attitudes toward organic or sustainable farming since I first started around 1988." At first, he said, he was considered "a bit of an oddball." Eventually respect and acceptance can emerge. Over the past 10 years, people have become more aware of Carmen Fernholz's organic production systems and neighbors have expressed interest. Tony Thompson noted that his neighbors have been responsive and creative, especially his neighbors who have cooperated in enhancing migratory waterfowl habitat within the watershed. Noreen Thomas's father-in-law was initially skeptical of her advocacy for organic crop production, but has since become much more accepting of her approach when he realized that she could sometimes sell organic soybeans for $18 to $20 per bushel—almost four times the price of conventional soybeans.

Most often, conventional neighbors remain as they have been, but on rare occasions a neighboring farmer may alter his or her own farming practices based on the example provided by the alternative farmer. "By the time I organized my organic notes together into a booklet and was giving a few presentations at meetings here and there, the concepts and practices of sustainable farming were coming closer to home, and [now] I have neighbors with similar operations," observed Jaime DeRosier.

The interviewed farmers sometimes found that they received good support from nonfarm neighbors or those who were newer to their communities. "Those that have been [most] encouraging since the beginning tend to be people who have not always lived in our community. Right now, urban consumers are our greatest champions," stated Mary Jo Forbord. Dave Minar further observed, "Our urban neighbors are excited to see our animals on the land again and are thrilled with the fact that they can come to our farm store and buy our products." Echoing this theme, Carmen Fernholz noted, "Consumers are more discriminating. Some farmers' 'city cousins' are saying that these [organic] guys are 'okay.'"

How farmers communicate with each other can influence the future of farming. For example, Mary Jo Forbord remarked, "In rural areas, we talk mostly about hard work, low prices, misery, loss, and decline. It becomes a self-fulfilling prophecy and doesn't do much to attract young people. We need more success stories, more support for diversity in farming, more communication, more confidence, more innovation, and more fun. Maybe then we will attract a next generation to rural areas." Steve and Sally Smith have noticed a difference between their conventional farmer neighbors and themselves. Their neighbors are often upset about low commodity prices and many of them are advising their children not to pursue a farming career. But Steve and Sally have encouraged their son to return to their locale and farm organically. He likely would not have chosen to continue to farm if he had been required to take a conventional approach.

4.5 RESEARCH AND OUTREACH EFFORTS

There are numerous research and outreach efforts under way in the northern Midwest to assist farmers in their transition toward a more sustainable agricultural production system. Some examples of these efforts include the Minnesota Institute for Sustainable Agriculture (MISA), the Land Stewardship Project (LSP), the Sustainable Farming Association of Minnesota, the Leopold Center for Sustainable Agriculture, the Practical Farmers of Iowa, the Michael Fields Agricultural Institute, the Center for Integrated Agricultural Systems at the University of Wisconsin–Madison, the Wisconsin Women's Sustainable Farming Network (WWSFN), and the Northern Plains Sustainable Agriculture Society (NPSAS).

There is increased awareness that organic agriculture is growing in importance. More researchers are conducting research focused on organic agriculture practices, and plant breeders are also taking a closer look at the specific needs of this sector of agriculture. Groups such as the Midwest Organic and Sustainable Education Service (MOSES), the Minnesota Organic Farmers Information Exchange (MOFIE), and increasingly, the Land Grant Extension Service help extend information gained from both farmer-to-farmer networks and land grant agricultural research institutions.

There is awareness that a move toward sustainability is linked in part to establishing more perennial vegetation within the agricultural landscape. A multi-institutional project known as Green Lands, Blue Waters (GLBW) is a long-term comprehensive effort whose objective is to support development of and transition to a new generation of agricultural systems in the Mississippi River Basin that integrate more perennial plants and other continuous living cover into the agricultural landscape (GLBW, 2005). The project's goal is to keep lands working while developing new (and expanding existing) cropping options, such as using alfalfa or perennial native legumes and promoting the use of annual plants to provide ground cover in corn and soybean fields. The hope is for alternatives that are economically viable and improve the environment.

Nongovernment organizations such as the Institute for Agriculture and Trade Policy (IATP) provide avenues for future research. IATP's Environment and Agriculture program seeks to enhance the quality of life in rural agricultural communities by promoting conservation-based economic opportunities (IATP, 2005). This group provides an Internet-based periodical known as "The Third Crop," which promotes alternatives to corn-soybean production.

In Minnesota, the Regional Sustainable Development Partnerships (RSDP), funded through the state legislation, offers rural citizens opportunities to engage in sustainable development. The mission of the RSDP is to support sustainable development in greater Minnesota through community and university partnerships in outreach, education, and research. The three bedrock principles of this initiative are: (1) develop and sustain a richer and more vibrant partnership with the citizens of each region and their land grant university; (2) address agriculture, natural resources, and tourism issues consistent with sustainable development principles identified as central to RSDP's work; and (3) promote the concept of active citizenship, which calls on us to think first and foremost as citizens with a commitment to working through issues and exploring opportunities in an integrated and democratic manner (RSDP, 2005).

4.6   TRANSFORMATION AND SUSTAINABLE AGRICULTURE

One can characterize the northern Midwest of the United States as possessing unique natural, climatic, and social tensions. As previously noted, the northern Midwest is a place where three major biomes come together (Tester, 1995). It is also where three major air masses from the west, north, and south meet. And it is where several American Indian and postsettlement European cultures came together—and sometimes clashed.

We began this chapter focusing on tensions that can often infuse the decisions that contribute to evolution of agricultural systems. The farmers we interviewed expressed a strong sense of urgency as they considered future directions for their farms. Ecological, social, and economic stresses heightened their concerns. A financially strapped farmer in Minnesota graphically expressed this: "Economics is coming to determine [crop] rotational plans more than agronomics. Under financial stress you see only to the end of the year, not to the end of the decade. I [have] cheated a little bit. I planted canola and sunflowers on all my acres, kind of breaking away from a sensible rotation because it seemed to be the quickest payback" (Corselius et al., 2003).

But where tensions exist, there is also the possibility for transformation. The concept of transformational learning was proposed by Mezirow in the late 1970s (Merriam and Caffarella, 1999; Percy, 2005) and concerns changes in how one works and lives. According to this learning theory, the process of transformational learning begins with tension—a "disorienting dilemma"—that arises from life events or personal experiences that cannot be resolved using previously held perspectives, assumptions, or problem-solving strategies. Such dilemmas can provoke self-examination and deeper questioning about one's prior assumptions and practices, which in turn may lead to recognition that such dilemmas and questions are shared within a broader community. The final stage of transformational learning results in formulation of new practices that address the original dilemma and are consistent with new assumptions, perspectives, and problem-solving approaches that were constructed during the transformation process (Merriam and Caffarella, 1999; Percy, 2005).

For each of the farmers we interviewed during the preparation of this chapter, it is possible to recognize elements of transformational learning. For example, the Minars called attention to the questioning of assumptions that began after they encountered the dilemmas of chemical-based agriculture: "We began questioning the use of chemicals after the [physical] reaction Dave had and adverse effects we could see happening on our farm, such as a lack of bees and many dead birds after planting in the spring." Similarly, Mary Jo Forbord acknowledged the importance of changes in Luverne's and her thinking prior to making major changes on their farm when she noted, "Our transition [in approach to farming] first had to occur in our minds." For Lee and Noreen Thomas, a disorienting dilemma seems to have been the decline in economic return from sugarbeets they experienced in the late 1990s. As they expressed it, sugarbeets were "by far the most profitable crop in our rotation, [but] were beginning to look less certain." Their conclusion that cane sugar could someday easily replace beet sugar in the United States caused them to have doubts about the future of the sugarbeet industry and called into question their prior assumptions about the dominant cropping system in their area.

Even when not faced with disorienting dilemmas, it appears that Carmen Fernholz has cultivated the habit of routinely asking himself the types of questions that are characteristic of transformational learning. For example, he often asks, "Do I really believe this?" This is a question that can lead to examining one's prior assumptions and perspectives. Similarly, Fernholz asks, "Why am I doing this?" This question prompts self-examination and deeper thinking about practices.

Carmen Fernholz also has another habit that leads to self-examination and transformational learning—record keeping. He recalled that his father urged him to "keep a narrative. If you don't write it down, it's gone." Although Fernholz's "narrative" is often practical and procedural, it contains elements of a personal journal and helps him to "remember the 'tuition' I've paid [through my mistakes]."

In conclusion, it is crucial to note that all of the interviewed farmers regarded the process of change—of transformational learning—on their farms as an ongoing one. A quote by Mary Jo Forbord best describes the overarching conclusions we can draw from the farmers we interviewed regarding future paths toward sustainable agriculture in the northern Midwest:

> It's not nearly complete, and we expect that it never will be. It's a biological system, alive and in need of observation, tending and rebalancing daily.... It takes time and a fair amount of courage and faith. We plan to make it up as we go. Sustainable agriculture is very site specific and depends so much on our farm ecosystems, resources, and goals ... but we are so much more optimistic about the future than we were three years ago.

And we are optimistic as well.

APPENDIX: QUESTIONS USED TO GUIDE
THE INTERVIEWS WITH FARMERS

- Describe the key elements of your farm (e.g., crops and acreages, livestock, other significant enterprises—income producing or otherwise).
- The title of the chapter that we have been asked to write is "Conversion to Sustainable Agriculture." How does this title resonate with you? Is it descriptive of your own experience in farming over the (period of time that is appropriate for the individual farm situation)? If so, how?
- What kinds of factors prompt reconsideration of how you farm? Ecological? Economic? Agricultural? Social and personal? Give a few examples.
- As you have looked at the management of your farm, what are the primary limits that have constrained you and placed the greatest stress on your farm?
- Give a couple specific examples of significant changes that you have made to your farm's practices and explain the circumstances that led to those (e.g., input or enterprise changes).
- Did you see what you expected in relation to these specific examples? What were the surprises—both positive and negative—that accompanied the changes that you made? Are there any data that you collected in the process of making—or tracking—these changes?
- To what extent would you consider the changes you have made in managing your farm a fundamental redesign? Explain.
- How have your attitudes and perspectives toward neighbors and your local communities changed as a result of your following the path of change and conversion that you have taken? How have your neighbor's (or community's) attitudes and perspectives changed toward you and your farm since you made the shift in your approach to farming?
- As you look ahead, what are the most pressing needs for further information, understanding, and knowledge as you look to continuing to change or redesign your farm? From what source(s) do you expect to receive this information, understanding, and knowledge?

REFERENCES

Alexander, R.B., R.A. Smith, and G.E. Schwarz. 1995. The regional transport of point and non-point source nitrogen to the Gulf of Mexico. In *Proceedings of the Hypoxia Management Conference*, New Orleans, December 5–6, 1995, pp. 127–132.

Cardwell, V.B. 1982. Fifty years of Minnesota corn production: Sources of yield increase. *Agronomy Journal* 74:984–90.

Corselius, K.L., S.R. Simmons, and C.B. Flora. 2003. Farmer perspectives on cropping systems diversification in northwestern Minnesota. *Agriculture and Human Values* 20:371–83.

ERS-USDA (Economic Research Service). 2005. The economics of food, farming, natural resources, and rural America. http://www.ers.usda.gov/Data/Organic/.

Fausey, N.R., L.C. Brown, H.W. Blecher, and R.S. Kanwar. 1995. Drainage and water quality in Great Lakes and corn belt states. *Journal of Irrigation and Drainage Engineering* 121:283–88.

GLBW (Green Lands, Blue Waters). 2005. http://greenlandsbluewaters.org/.

Hamilton, N.D. 1990. *What farmers need to know about environmental law*. Des Moines, IO: Drake University Agricultural Law Center.

IATP (Institute for Agriculture and Trade Policy). 2005. http://www.iatp.org/.

LSP (Land Stewardship Project). 2003. Getting a handle on the barriers to financing sustainable agriculture: The gaps between farmers and lenders in Minnesota and Wisconsin. http://www.landstewardshipproject.org/pdf/edsurvey.pdf.

MDA (Minnesota Department of Agriculture). 2001, 2003. *The status of organic agriculture in Minnesota*, ed. M. Moynihan. St. Paul: MDA. www.mda.state.mn.us.

Merriam, S.B., and R.S. Caffarella. 1999. *Learning in adulthood: A comprehensive guide*. San Francisco: Jossey-Bass.

MISA (Minnesota Institute for Sustainable Agriculture). 2001. *Organic certification of crop production in Minnesota*, ed. Lisa Gulbranson. St. Paul: University of Minnesota, MISA.

MISA (Minnesota Institute for Sustainable Agriculture). 2003. *Building a sustainable business*, ed. Beth Nelson. St. Paul: University of Minnesota, MISA. www.misa.umn.edu/publications/bizplan.html.

MOSES (Midwest Organic and Sustainable Education Service). 2005. *The Upper Midwest organic resource directory*. 5th ed. Spring Valley, WI: MOSES. www.mosesorganic.org.

Nordquist, D.W., L.L. Westman, and K.D. Olson. 2004. *Southeastern Minnesota Farm Business Management Association 2003 annual report*. Staff paper P04-5, Department of Applied Economics, University of Minnesota, St. Paul.

OFRF (Organic Farming Research Foundation). 2003. *State of the states*, ed. J. Sooby. 2nd ed. Organic Farming Systems Research at Land Grant Institutions 2001–2003. Santa Cruz, CA: OFRF. www.ofrf.org.

Percy, R. 2005. The contribution of transformative learning theory to the practice of participatory research and extension: Theoretical reflections. *Agriculture and Human Values* 22:127–136.

PSU ESSC (Pennsylvania State University Earth System Science Center). 1998. Soil information for environmental modeling and ecosystem management. Land resource regions. http://www.essc.psu.edu/soil_info/soil_lrr/.

Rabalais, N.N., R.E. Turner, and W.J. Wiseman, Jr. 2001. Hypoxia in the Gulf of Mexico. *Journal of Environmental Quality* 30:320–29.

RSDP (Regional Sustainable Development Partnerships). 2005. http://www.regionalpartnerships.umn.edu/.

Simmons, S.R., N.P. Martin, C.C. Sheaffer, D.D. Stuthman, E.L. Schiefelbein, and T. Haugen. 1992. Companion crop forage establishment: Producer practices and perceptions. *Journal of Production Agriculture* 5:67–72.

Skaggs, R.W., M.A. Breve, and J.W. Gilliam. 1994. Hydrologic and water quality impact of agricultural drainage. *Critical Reviews in Science and Technology* 24:1–32.

Tester, J.R. 1995. *Minnesota's natural heritage*. Minneapolis: University of Minnesota Press.

USDA-NASS (National Agricultural Statistics Service). 2004. Historical data queries by U.S., state, or county for crops, livestock and census. http://www.usda.gov/nass/pubs/histdata.htm.

USDA-NRCS (Natural Resources Conservation Service). 2005. State soil geographic (STATSGO) database. http://www.ncgc.nrcs.usda.gov/products/datasets/statsgo/.

5 Pacific Northwest (U.S.)
Diverse Movements toward Sustainability Amid a Variety of Challenges

Carol Miles, David Granatstein, David Huggins, Steve Jones, and James Myers

CONTENTS

5.1   OVERVIEW OF PACIFIC NORTHWEST AGROECOLOGICAL ZONES

The Pacific Northwest region of the United States extends across three states—Washington, Oregon, and Idaho (Figure 5.1). Four agroecological zones can be defined within the region based on major climatic and agricultural characteristics: the coastal maritime zone, the irrigated crop zone, the dryland grain zone, and the livestock rangeland zone. Both Oregon and Washington contain all four zones, while Idaho contains all but the maritime zone. Average annual precipitation across the region varies considerably, ranging from 760 to 1,300 mm (30 to 50 inches) in the maritime zone to 150 to 815 mm (7 to 32 inches) in the dryland zone. More than 200 crops are commercially produced in the region; the best known agricultural products include Washington apples, dairy products, wheat, Idaho potatoes (also grown in Washington and Oregon), red raspberries, and vegetable, grass, and flower seeds. In this section we present a brief overview of each agroecological zone, including its major cropping or production systems and key sustainability issues.

On the west coast of Oregon and Washington, the temperate, moist maritime climate is well suited to berries (raspberries, blueberries, strawberries, cranberries), vegetables, dairy, seed crops, and nursery crops. The agricultural potential of the Willamette Valley in Oregon was grasped early on, as settlers traveling over the Oregon Trail put its rich soil to the plow in the 1850s. Winter temperatures

FIGURE 5.1   The Pacific Northwest region of the United States encompasses Washington, Oregon, and Idaho. (University of Texas Libraries, 2006.)

are generally above freezing, so plant growth is possible year-round. Soils mostly formed under coniferous forests and are typically acidic (pH 5.2 to 5.5) and weathered. Key sustainability challenges for agriculture in this zone include crop diseases, water quality, and urban development pressures. Proximity of farmland to populated urban centers provides value-added and direct marketing opportunities, which have enabled small farmers to prosper through production of diverse crops that are often direct marketed through farmers' markets, community-supported agriculture (CSA) arrangements, or other outlets. Global competition has induced a decline in many berry crops, while urbanization has encouraged dairy operations to move to the eastern part of the state.

The irrigated zone lies east of the Cascade Mountains in central Washington, in parts of central Oregon, and in the Snake River Valley of eastern Oregon and Idaho. The hot, dry, long-day-length growing season is suitable for a wide variety of crops under irrigation, including potatoes, carrots, sweet corn, onions, asparagus, forages (e.g., alfalfa hay), tree fruit, hops, mint, and rotation crops such as field corn and wheat. A food processing and storage infrastructure has developed to support many of these crops, for which the Pacific Northwest is a major national or global supplier. Much of the production from this zone is sold through wholesale channels and exported out of state and overseas. In recent years, many dairies have relocated to this zone to escape urbanizing areas in western Washington and California, to expand operations for economy of scale (dairies of several thousand cows are not uncommon), and to be closer to key feed production sites. Soils in this zone tend to be younger alluvial soils of sandy or silt loam texture with neutral pH (pH 7.0). Production of most crops currently relies on substantial inputs of fertilizers and pesticides. However, this zone is also the most important for production of commercial organic crops, as disease and pest pressures are often lower than in other zones. Water for irrigation is primarily snowmelt runoff from neighboring mountains, and although it is relatively abundant, its use is being impacted by endangered species issues with native salmon. Many improvements in water use have led to substantial increases in water conservation and a reduction of irrigation-induced soil erosion. Sustainability issues include lack of profitability due to global competition, reduced labor supply for labor-intensive crops, nitrate contamination of groundwater, consumer concerns regarding pesticide use, constraints in water supplies, and reliance on fossil fuel inputs (e.g., natural gas for fertilizer, diesel fuel). This zone produces a substantial portion of agricultural revenue in the region (about 55%) and accounts for approximately 52% of the region's crop acres.

Dryland crop production comprises the largest area of agriculture in the Pacific Northwest—approximately 2.4 million acres in 2002. Annual precipitation, growing degree-days, and soil depth define major production zones. Where annual precipitation averages less than 16 inches and soils are greater than 3 feet deep, summer fallow is used once in three years or every other year to conserve water and stabilize crop yields (Douglas et al., 1990). During the fallow year, precipitation is stored within the soil profile with an efficiency approaching 30%, thereby supplying water for crop use in the following year. Winter wheat has been the predominant cash crop in this zone for more than a century, and some of the highest dryland wheat yields in the world have been achieved near Pullman, Washington. Other rotation crops include spring wheat, barley, dry pea, lentil, chickpea, and rapeseed/canola. Rolling to hilly topography and deep soils that formed in silt-sized materials deposited by wind (loess) are typical of more productive areas. Preagricultural soil fertility derived from native steppe vegetation (grasses and forbs) was high, but intensive tillage, fallow, and synthetic fertilizers have led to major soil degradation from erosion by wind and water, accelerated biological oxidation of soil organic matter, and soil acidification (McCool et al., 2001). Soil conserving practices have been a research and extension priority since the 1930s, and the current focus is on conservation tillage systems that can virtually eliminate soil erosion, conserve water, increase yield potential, rebuild soil organic matter, and reduce tractor use and fuel consumption. Other sustainability challenges include lack of profitability and reliance on government subsidies, lack of agricultural diversity including economically viable rotation crops, invasive weeds and herbicide resistance, water quality degradation from sediment and agrochemicals, and reliance on synthetic fertilizer as a major source of nitrogen, phosphorus, and sulfur.

The rangeland zone, located east of the Cascade Mountains, encompasses land with topography or soils unsuited to crop production, but which is able to produce quality forage for livestock, albeit at low stocking rates (e.g., one animal per 40 to 70 acres per year). Extensive livestock grazing, primarily for beef cattle, occurs in this zone. Cattle ranchers rely on their own private land to produce hay or pasture, and they lease extensive federal rangelands for summer grazing. Fragile rangeland is susceptible to soil degradation and loss of native plant biodiversity, particularly in riparian areas. Poor herd management can lead to sediment and manure deposits in water, invasive weed problems, and reduced forage productivity. Ranchers and public land agencies have worked to develop innovative range management practices such as management-intensive grazing to restore ecosystem function and productivity. Projects such as Holistic Resource Management, weed control with insect biocontrol agents, watershed collaborations among ranchers and environmental groups, and multispecies grazing for weed control are providing pathways for increased sustainability. Natural beef production (e.g., Oregon Country Beef®) and accompanying labeling schemes offer a value-added option that provides market support for sustainable production. While some people object to cattle grazing on rangelands, this practice represents a sustainable system for converting sunlight into high-quality protein when proper management is used.

Across the Pacific Northwest region, major expansion and progress is evident in direct seeding, biocontrol, organic systems, and water conservation. In addition, all the agroecological zones share the same difficult-to-surmount barriers to sustainability, including reliance on fossil fuel energy, urbanization, climate change, and global competition. With the diversity of agriculture present in the region, however, sustainability is understandably moving forward in different ways and at different speeds, depending on the crop and region. In the remainder of this chapter, therefore, we examine the progress toward conversion to sustainable practices in the region by focusing separately on three major production systems: cropping systems in the coastal maritime zone, tree fruit production systems in the irrigated crop zone, and wheat-based cropping systems in the dryland zone (specifically in the Palouse and Nez Perce Prairies Land Resource Area). Each section follows the same structure—after introducing the system, we discuss the primary factors limiting sustainability, some experiences with conversion, the lessons learned in these efforts, and the indicators of sustainability for the system.

5.2   MARITIME CROPPING SYSTEMS

The Pacific Northwest maritime agroecological zone stretches from the Pacific Ocean to the Cascade Mountains and is characterized by a Mediterranean climate that includes mild winters, relatively cool summers, and high winter rainfall. Historically, the area has been known for dairy, vegetable, fruit, and flower and vegetable seed production. Dairy cows in western Washington have had one of the highest rates of productivity (gallons of milk production per cow) in the country due in part to the low-stress environment that results from the moderate temperatures. River valleys with their loamy clay soils, ample irrigation water, and moderate temperatures have been the production and processing home for an assortment of vegetable crops, including green peas, sweet corn, carrots, green beans, cauliflower, and broccoli (Oregon Department of Agriculture, 2005; Washington Agricultural Statistics Service, 2002). Valleys and hill slopes were planted with orchards of filberts, prunes, plums, cherries, and pears, as well as strawberries, raspberries, and blueberries. Apples do well in this climate, although most commercial production has shifted to the east of the Cascades. The old Fort Vancouver, located in present-day Vancouver, Washington, is home to the oldest apple tree in the Pacific Northwest—the tree was planted in 1826 by the Hudson Bay Company (Luce, 1975). Wine grapes with associated wineries, nursery production, and Christmas trees have also become important in the region. In addition, the climate is ideally suited to seed production of overwintering perennial and biennial crops as well as many annuals. Seed crops were first produced in the region in 1885 in the Skagit Valley (Rackham, 2002), and today the entire maritime region is known for its high seed quality, including high vigor and purity. Weather-related crop failures are rare. While winter rains cause periodic flooding and may be accompanied by occasional high winds, the region almost never receives violent summer thunderstorms with the hail, high winds, torrential rains, and tornados that can devastate crops in the interior United States.

5.2.1   Primary Factors Limiting Sustainability

Many issues have directly impacted the crops that have been historically produced in the region. Filbert blight has been a major disease problem; consumers have turned away from dried prunes; California has taken over the strawberry market; some processed vegetable production has moved to the Columbia Basin; and tree fruit production has moved to the central Pacific Northwest due to lower disease pressure. Although the maritime region receives copious precipitation in the winter, summers are dry, and irrigation is required to grow most crops throughout the summer months. Away from the major rivers, water for irrigation can be difficult to extract; wells are deep and water-bearing sediments are fine-grained, which limits the pumping rates. Another limiting factor has been distance to markets that are predominantly in the central and eastern United States. Although the maritime Pacific Northwest is known to produce high-quality vegetables, businesses that ship their products east cannot easily compete on price because Midwestern vegetable processors have lower transportation costs. Other agricultural commodities have similar constraints due to distance to markets, most notably nursery crops, Christmas trees, and fruit crops. A problem that may be shared with other regions is that of foreign competition. U.S.-based vegetable processors increasingly contract with overseas producers who have access to cheap labor (Oregon and Washington have some of the highest minimum wages in the United States). Overseas producers can sell into U.S. markets at lower prices than U.S. growers, despite the distance foreign produce must be shipped.

Currently, perhaps the single greatest factor limiting sustainability in the maritime region is urbanization pressure. Just as agriculture was attracted to the maritime Pacific Northwest in the early 1900s, there was a large influx of people in the late twentieth century. As the primary cities and towns in the region expanded, they engulfed much of the prime agricultural land surrounding them. This urban expansion is continuing.

In his book *The End of Agriculture in the American Portfolio*, Blank (1998) espouses the idea that in a free market, land use will change to whatever activity creates the highest value for the land. As populations increase, increased demand for land for residences and businesses drives up land prices. Individuals who bought the land when it was relatively cheap rural farmland have tremendous incentive to sell to developers, or to develop their land directly. In Blank's view, there will be a "natural" progression of rural land converted to urban residences and businesses so that eventually, the only agriculture that is left in a region is that involved in directly servicing urban communities. The agricultural operations that remain are primarily turf farms, golf courses, and nurseries. Food production moves to where there is less pressure to urbanize land and land prices and labor costs are lower. This ultimately means that food production for the United States will move to the developing countries that have the lowest labor costs. Even within the developing world, there has been an evolution in that as infrastructure and wages improve, food production shifts to other developing countries that are less developed and pay lower wages. This has been the case with fresh-market winter vegetable production, which has moved from the United States to Mexico and more recently to Central America.

It is instructive to compare the very different approaches to land use that have been employed by Washington and Oregon. Washington has put minimal restrictions on land use, whereas the Oregon approach has been to implement zoning laws that preserve some areas as public space, as well as preventing the conversion of forest and agricultural land to residential uses. As an example of preserving land as a public good, the Pacific Ocean shores of Oregon are public land with development restricted within 100 yards of the shoreline. Cities have developed zoning plans that set urban growth boundaries in an attempt to control land development. Within rural areas, some land has been set aside for residential development, mainly land that is not considered of good quality for agriculture or forestry. Prime agricultural and forest land cannot be developed for housing, with the exception of building a residence on the land if one can show that it is necessary for the management of that land.

While public land use policy has protected farmland from development in Oregon, there is a downside to Oregon's approach. Because of unmet demand, land prices are high. Some individuals chafe at being unable to sell their land to the highest bidder because doing so would require an impossible-to-get zoning variance. As a consequence, Oregon voters in 2004 passed Measure 37, which requires that individuals be compensated if a zoning law enacted after the purchase of their land impacts its value. If the landowner cannot be compensated monetarily, the zoning restriction must be waived. The intent of Measure 37 was to allow the landholder who had invested as a retirement security strategy to develop his or her land for retirement income, or to build additional dwellings for heirs. However, in some cases, land developers used the law to develop large housing subdivisions. In 2007, Measure 49 was passed, which closed the loophole allowing large-scale development. This legislative seesaw reveals the strong polarization within Oregon society regarding private ownership rights versus attempts to manage the landscape in a way that benefits the general public.

In Washington, as even the smaller towns have grown, little has been done to preserve the farms in the area. This scenario is typified in the Kent Valley outside of Seattle, where once highly productive vegetable fields are now covered with warehouses and box stores. It is a great challenge for farmers to remain in agriculture when they are being offered a price for their land for development purposes that is greater than they could hope to earn in a lifetime of backbreaking work. Faced with this situation, it is hard for many farmers to justify staying in the farming business. Many farmers are selling their land in western Washington and relocating to the eastern region. This has been the case for many dairies and the vegetable processing industries that were once based here. Lower costs of water and land in the eastern region enable farmers to be more profitable, and they escape the nuisance complaints that arise from new developments that surround farmlands in the maritime region. Due to expanding population settlements and the resulting relocation of farmers, farm numbers and acreage have decreased dramatically throughout western Washington. For example, the number of milk cows declined 30% in 2005 compared to 1993, while the number of acres in sweet corn in 2003 declined 15% compared to 1989 and the number of acres in green peas declined 70% compared to 1970 (USDA National Agricultural Statistics Service, 2005; Washington Agricultural Statistics Service, 1994, 2002).

5.2.2 Conversion Experiences

Although urbanization has tended to push farmers out of the maritime zone, it has also been a force promoting conversion to more sustainable practices. This somewhat paradoxical situation arises out of the proximity between agricultural land and large human populations that results from urbanization of formerly rural landscapes.

One of the trends toward sustainability that has been promoted by the proximity of agricultural land and urban centers in the maritime zone is the shortening and localization of the food production–food consumption chain. As the pressure to urbanize increases, the costs of land and services increase substantially and farmers find they need to earn more per acre in order to remain economically viable. Due to the low economic return of traditional commodity crops and markets, farmers look for ways to capture more profit from their crops. Some farmers have been turning to direct marketing, niche market crops, and value added as a means to accomplish this. Direct marketing enables farmers to capture retail prices for their products, while communities are supportive because direct marketing tends to be family oriented and provides direct access to fresh, high-quality products at reasonable prices. Direct marketing contributes to sustainability in that goods are usually sold locally, or at least regionally, which reduces the use of fossil fuels needed for transportation, promotes agricultural bioregionalism, and forges a closer, more democratic connection between producer and consumer. One type of direct marketing that has gained in popularity in the region is farmers' markets.

Farmers' markets have been very successful in the Pacific Northwest and have increased in number and annual sales, especially in the maritime region. Farmers' markets appear to be most popular in urban communities where consumers have limited access to farm-fresh foods. In Oregon there were 77 state-registered farmers' markets in 2005, while in Washington there were 89 (Washington State Farmers' Market Association, 2006). The Oregon Farmers' Markets Association is relatively new and was established in 1987 by a small group of market managers from around the state. Today it is estimated that more than 1,000 farmers participate in Oregon farmers' markets each year, and that more than 90,000 people visit the markets each week during the peak summer months (Oregon Farmers' Market Association, 2002). The Washington State Farmers Market Association was formed in 1979 with five member markets. In 1997 (the year of the earliest recorded sales figures) 56 farmers' markets earned $5 million; by 2005, sales from 89 farmers' markets had increased to more than $25 million (Lyons, personal communication, 2006).

The maritime region has a long history with farmers' markets. The first public market was established in Portland in 1870, and the Seattle public market was established in 1907. Although the Portland Public Market (2006) has not existed for the last 50 years, an effort is being made today to reestablish the market in downtown Portland. In contrast, the Seattle public market, the Pike Place Market, is currently recognized worldwide as America's premier farmers' market. The Pike Place Market was formed in response to complaints of citizens/consumers who were outraged that the cost of onions had increased tenfold in a two-year period. The market was established so that consumers could connect directly with farmers and avoid "price-gouging middlemen" (Pike Place Market PDA, 2006). Today, 10 million visitors visit the Pike Place Market each year, where in addition to year-round businesses, 120 farmers rent table space by the day. The Pike Place Market is one of the most frequently visited destinations in Washington State, and this is a testament to the desire of urban consumers to connect with agricultural producers and marketers, perhaps especially when this connection is made in the heart of an urban area.

Urbanization has encouraged conversion to sustainable practices in a rather different way by forcing many farmers to find alternatives to practices that generate conflict with urban residents and municipalities. These conflicts generally arise over three issues—noise from agricultural equipment (especially at odd hours), odor from poorly managed manure, and the application of agricultural pesticides. While complaints about noise, odor, and toxics have caused some farmers to sell their land and relocate their farms to more rural areas in other parts of the region, other farmers have responded by changing the practices that cause the complaints. Most commonly they have reduced pesticide use—taking advantage of new spray technologies that improve the distribution of active ingredients—and many have moved to eliminate synthetic chemical pesticide use altogether, choosing to become organic growers. Although there can be many long-term environmental advantages attributed to organic production, there are also two important short-term advantages. First, farmers are able to avoid complaints by neighbors who otherwise object to synthetic chemical pesticide applications. And second, farmers are able to capture a premium price for organic products. For those farmers who are able to solve their primary pest issues without the use of synthetic chemical pesticides, organic farming can be a profitable choice that enables them to be successful in an urban environment.

In general, urbanization and economic pressures have had mixed effects with regard to sustainability. On the negative side, economic necessity has driven farmers to intensify production and shorten rotations. This has led to increases in some insect pests and diseases that are normally kept in check by long rotations. For example, root rot pathogens are increasingly limiting yields of green beans and sweet corn. Farmers who grow processed vegetables in the Willamette Valley, however, are finding that rotating to grass seed for two or more years can reduce the pathogen load while increasing soil organic matter and improving soil structure for subsequent vegetable crops. Canola for biodiesel may become a significant crop in the Willamette Valley in the near future, and its introduction could help further diversify rotations.

Although economic factors drive farmers toward greater efficiency sometimes at the expense of sustainability, sustainable practices that also lead to economic benefits can be incorporated. Nearly all farmland in western Oregon is now covered with perennial or winter cover crops. This has had two benefits: reduction in soil erosion during the winter and fewer chemical fertilizer inputs because of the increased fertility and organic matter from incorporating the cover crop before the main crop is planted. A second sustainable practice has been the recent introduction of strip till, which maintains more soil cover during the cropping season. Farmers have had to adjust to these practices. For example, soils under cover crops do not dry out and warm up as quickly as bare soils. If a farmer plans to plant an early crop, he or she may need to resort to traditional soil preparation practices. Symphylan populations may increase in systems with continuous cover or higher organic matter and may cause root damage in a subsequent vegetable crop. In addition, it has been difficult with strip till systems to achieve well-worked and uniform seedbeds that allow uniform germination and emergence of green beans and small-seeded vegetable crops.

5.2.3 LESSONS LEARNED

Blank's (1998) vision of an America without traditional agriculture represents one extreme view of what will happen to agriculture in the United States in the future. We believe that while many changes are afoot and that the gradual erosion of farmland is inevitable, there will be bright spots for agriculture in our region. When contrasting changes to the landscape in western Washington and Oregon, it is apparent that land use policy plays a major role in the sustainability of agriculture throughout the region. To prevent the wholesale conversion of agricultural lands to housing and businesses, local governments must become involved in land use policy that creates and maintains urban growth boundaries. In addition, public officials and community leaders should uphold right-to-farm laws when nuisance complaints are made against farmers.

Small-scale farms that produce high-value products are proliferating in the region, whether these are U-pick blueberry operations or organic vegetable farms that sell in farmers' markets. The number of wineries and vineyards in the region also continues to grow. Many of the farm operations that are successful are able to provide products that attract and appeal to urban customers. In many cases, a large amount of the appeal is due to the fact that the product was produced locally. Farmers' markets provide consumers access to fresh, high-quality produce at affordable prices, while farmers are able to gain retail prices for their products. City and community leaders view a vibrant downtown farmers' market as a means to revitalize neighboring downtown businesses.

Much can be learned from the approach of various European countries to the preservation of their agricultural landscapes. Farmers receive subsidies to stay in business and maintain their farms because the majority of citizens want to keep farms as managed open spaces that provide some level of food security, wildlife habitat, water management areas (i.e., in periods of flooding), and pleasing views. However, there are enough differences between Europe and the United States to wonder whether such a system could work in the United States. Europe has a centuries-old tradition of people living closely together in cities surrounded by much less densely populated farmland. European culture has evolved in such a way that essentially urban growth boundaries are self-imposed. In addition, government policies throughout Europe tend to consider the public good above the rights of the individual. In contrast, U.S. culture encourages individual settlement in rural areas and individual property rights dominate in public policy. In the United States we are uncertain that with contemporary capitalistic attitudes, citizens would be receptive to the idea of paying farmers to farm in order to provide environmental and aesthetic benefits.

5.2.4 INDICATORS OF SUSTAINABILITY

As urbanization pressure continues to grow, local and state governments must decide whether or not to take an active role in safeguarding farmland. Development policies and incentives will to a great extent determine the future of agriculture in the region. An indicator of sustainability within each county will be that county's public policy regarding the development of agricultural lands.

At the community level, economic viability will be a primary indicator of the sustainability of agriculture in the region. As land values continue to increase, farmers must continue to expand the production of high-value crops or find ways to capture a greater share of the market price. Farmers will continue to diversify away from bulk commodity crops that are traditionally low value and will seek niche market crops, unique marketing outlets, and value-added products. High-quality products will continue to play a large role in the appeal and profitability of agricultural production in the region.

Direct marketing will continue to expand in the region, and the number of farmers' markets and their annual revenues will be a measure of their success. Urban consumers will likely continue to support local farmers as long as they feel that local farmers are implementing environmentally sound production practices. At the farm level, practices that protect soil and water quality will be of high importance. The number of acres planted to winter cover crops will be an indicator of sustainability. In addition, consumers will likely continue to demand reduced pesticide applications from local farmers. These demands stem from two concerns: first to protect water resources, and second to protect the health and well-being of urban communities that might otherwise be affected by pesticide drift.

5.2.5 CONCLUSIONS

The defining features of the maritime zone include a mild year-round climate that enables farmers to produce some the of the highest-quality agricultural products in the country. But since the mainstream food crop market in the United States continues to favor the lowest cost of production and does not reward crop quality, our farmers are no longer able to compete in the expanding global marketplace. In order to survive, they must find new markets and marketing opportunities. These may include niche market crops, local markets, direct marketing, and ecolabeling. Seed crops should remain a viable alternative because of their high value per unit area, and need for small but isolated acreages. Research is needed to promote growing seed crops sustainably. This research may be largely driven by the need for organically produced seed. Farmers' markets located in the heart of urban centers will continue to play a role in supporting small farmers throughout the region. Medium- and large-sized farmers must rely on mainstream market outlets to gain access to customers.

Urban pressure will continue to intensify in the maritime region in the next generations, and land use policy decisions made today by communities and their leaders will decide the future and fate of agriculture in the area. In western Washington, many of the best agricultural lands have already been paved over. Several counties have not adopted state growth management laws, and many citizens throughout the state (including farmers) do not support such laws. Through strict public policy, Oregon has maintained much of its prime farmland, though this could change in the near future due to citizen discontent with restrictions on development. As the value of land throughout the region continues to rise, communities may have to create monetary incentives for landowners to keep land in agriculture. Incentive programs may include purchasing of development rights and greater farmland tax incentives.

5.3 TEMPERATE TREE FRUIT PRODUCTION SYSTEMS

Washington State has long been known for its large, red apples. Tree fruit production is the largest agricultural crop in the state, yielding over $1 billion in sales of packed fruit from some 200,000 acres of orchard land (WASS, 2002). Apple is the largest tree fruit crop, followed by pear and then cherry, with other tree fruits produced in relatively small amounts. Tree fruit production is concentrated in central Washington, just east of the Cascade Mountains, largely due to the favorable semiarid climate with a xeric (winter) rainfall pattern and dry, sunny summers. Irrigation is necessary to grow crops, but this means that moisture can be controlled and many diseases of tree fruits can be avoided. In addition, the insect pest complex is relatively modest compared to other tree fruit production areas. During the winter season, snow is stored in the mountains and provides adequate summer runoff in most years to supply high-quality surface water for the region's irrigation needs.

5.3.1 PRIMARY FACTORS LIMITING SUSTAINABILITY

The three most significant limits to sustainability of fruit production are lack of profitability, shortages of labor, and limited water for irrigation. Tree fruit production has endured several cycles of prosperity and decline over the past century. The recent globalization of the fruit industry is perhaps the greatest challenge orchardists face in the near-term (O'Rourke, 2002). With labor generally accounting for about 40% of orchard production costs in Washington, and with Washington having the highest minimum wage law in the country (indexed annually to the rate of inflation), both cost of labor and its availability pose a threat to sustainability. Fruit producers in some countries have access to much cheaper labor and are developing the skilled workforce and infrastructure needed to deliver high-quality fruit to any market at a lower price than Washington can.

Aligning fruit production with consumer demand remains a challenge, as does returning profits to the grower. For example, Red Delicious had been the dominant apple variety for decades, accounting for some 70% of production. By the 1990s, however, the variety was falling out of favor with consumers, and prices began to erode. At the same time, new, more flavorful varieties appeared on the market. As a result, Red Delicious acreage has dropped substantially, while varieties unheard of 20 years ago, such as Fuji and Gala, are now major players. Growers now find themselves in a guessing game as to which new variety will catch on and prove profitable. This decision must be made each time an orchard is replanted at a cost of over $10,000 per acre. Whereas a planting might have lasted 20 to 50 years previously, growers now have no more than a 15-year period in which to recoup their investment.

Since rainfall is inadequate, reliable water supplies are needed for a perennial crop such as tree fruit. Certain irrigation districts have less reliable water supplies, particularly those that rely on runoff from mountain rivers with no reservoir storage. Farms watered from the Columbia River generally do not have problems, but increasing competition for water in the Columbia River and regulations regarding endangered salmon may lead to restrictions in the future. Global warming is expected to negatively impact the timing of water supply in the region, and this could impact fruit production. Also, due in part to new orchard systems that have reduced tree canopy density to allow more light penetration for fruit quality, water is now being used for evaporative cooling of orchards on extremely hot days to avoid sunburn damage. Climate change may also be exacerbating the sunburn problem. There is ample land with water rights on which to expand tree fruit production in the region if market demand increases.

Insect pests, particularly the codling moth (*Cydia pomonella*), have historically provided the greatest production challenge to apple growers (Beers et al., 1993). Codling moth is a pest in apple production in most regions of the world. Being an introduced pest, there are no effective natural enemies for its control in central Washington and continual pesticide intervention has been used for more than a century to prevent crop losses that can approach 100%. A succession of pesticides have been used, with many succumbing to insect resistance over time. These include lead arsenate, DDT, parathion, and azinphos-methyl. Newer pesticides that are more narrowly targeted and have lower human health concerns (e.g., insect growth regulators) are now available; however, insect resistance remains a challenge. The advent of pheromone mating disruption in the mid-1990s provided the first major nonpesticide control tool for codling moth, especially when adopted on an areawide basis over hundreds of contiguous acres (Calkins, 1998). Mating disruption seldom provides stand-alone control, but when augmented with other strategies such as codling moth granulosis virus, spinosad, and horticultural oil, it provides the basis for a highly effective and affordable control program that also meets the National Organic Standards. Mating disruption has been adopted as a pest management strategy on more than 60% of the apple acres in Washington (Brunner et al. 2001), and under areawide management, codling moth damage dropped to near zero while pesticide use declined.

5.3.2 Conversion Experiences

Orchards have undergone some dramatic design changes during the past 20 to 30 years, with several sustainability implications. A shift from furrow irrigation to impact sprinklers to microsprinklers today (Williams and Ley, 1994) has led to significant water conservation and improved soil quality. The filtration required for microsprinklers also prevents weed seed incursion in the water. Traditional orchards had tall trees grafted onto seedling rootstocks, which formed a dense canopy. This canopy reduced light penetration and thus fruit coloring, and also made complete spray coverage for pest control difficult. In addition, workers had to use tall ladders to pick the crop, and injuries were common and costs were high. Trees also took five to eight years to come into full production, with the commensurate loss of income during that time.

During the 1980s, extensive research was conducted on high-density orchard plantings using dwarfing rootstocks (Barritt, 1992). This type of system originated during the 1950s in Europe in response to economic pressures and later a move toward integrated fruit production (El Titi et al., 1993). The dwarfing rootstock controls tree size to create more of a pedestrian orchard, where much of the work can be done from the ground or low ladders, saving labor time and reducing accidents. The trees generally need support in the form of trellises or posts, and branches are trained to form a variety of canopy configurations that improve spray coverage, light penetration, and renewal wood. These systems will bear a commercial crop by the third year, with full production a year or two later, and a much higher-quality fruit is produced overall compared to the old system.

While the economic and social aspects of sustainability are addressed by this shift in orchard design, environmental gains are not a given. With more light reaching the ground, and with the reduced tolerance of the trees for competition with weeds, more weed control inputs are generally needed. Weed control is usually obtained through the use of herbicides or tillage. However, recent research on mulching systems (Nielsen et al., 2003) has shown that this weed control strategy can reduce water use and increase tree growth and yield up to 50% over the bare ground control. Although pesticide options evolved independently of changes in orchard design, the low, open canopy did prove particularly well suited to the use of mating disruption dispensers when that technology appeared.

Sustainability in apple orchards has been impacted by changes in pesticide choices and strategies. The widespread adoption of synthetic insecticides after World War II led to outbreaks of pests that had not occurred before, especially mites such as McDaniel spider mite (*Tetranychus mcdanieli*) and European red mite (*Panonychus ulmi*) (Beers et al., 1993). An integrated mite management program for apples was initiated in Washington during the late 1960s to deal with the situation (Hoyt, 1969). It focused on careful pesticide choice and timing, exploiting the fact that a key predatory mite (*Typhlodromus occidentalis*) had become resistant to organophosphate insecticides and thus was able to exert acceptable biological control if specific pesticides were used. This practice reduced pesticide costs from $85 per acre to $25 per acre and was widely adopted (Brunner, 1994). Subsequent research has developed insect phenology models that drive sampling and control decisions (Beers et al., 1993), track potential new biocontrol agents and enhance their habitat (Unruh and Brunner, 2005), and ultimately result in the implementation of "soft" pesticide programs (Dunley and Madsen, 2005).

More recently, production of organic tree fruit in Washington has grown dramatically. Organic apple acreage grew fourfold from 1989 to 1990 due to the Alar incident, but dropped off rapidly when growers were not able to adequately control codling moth (Granatstein, 2000) and prices plummeted due to a supply spike that the market could not absorb. As mating disruption provided control for codling moth, organic apple production grew from 1,300 acres in 1995 to 7,049 acres in 2004 (Granatstein et al., 2005). The semiarid climate makes the region particularly well suited for production of organic apples, pears, cherries, and other stone fruits. In a multiyear systems study comparing conventional, organic, and integrated apple production in the Yakima Valley of Washington State, Reganold et al. (2001) found that fruit yields were similar across all systems, with the organic and integrated systems exhibiting higher soil quality and less negative environmental impact based on a pesticide impact rating system. The organic system also produced improvements in fruit quality, higher profitability (with price premiums), and greater energy efficiency.

5.3.3 LESSONS LEARNED

Dramatic changes in orchard sustainability have occurred since the 1950s due to the development and adoption of integrated pest management, the change in planting design to high-density dwarf orchards, the adoption of newer, more desirable varieties, and more recently, the expansion of organic fruit production. Orchardists have proven their willingness to change and innovate, primarily for economic reasons, but with increasing appreciation for the environmental and social benefits that can accrue. While many orchardists have adopted organic production for economic reasons, they often experience other benefits from the sustainability of the organic system that they then extend to their conventional production acres. As conventional production evolves, in part due to societal pressures for sustainability, the distinction among systems is beginning to blur as growers mix and match the best practices to fit their situation. Overall, this is leading to reduced use of the most disruptive and toxic pesticides, better water and nutrient management, improved conditions for workers, and higher-quality fruit. However, consistent profitability remains elusive, impacted more by retail consolidation and global market forces than by choices a grower may make.

5.3.4 INDICATORS OF SUSTAINABILITY

Among several key indicators of sustainability in tree fruit production, there are both positive and negative signals:

- *Pest control*—Positive progress is being made in reducing pesticide use, but documented success in using beneficial insects within orchard systems to exert biocontrol of pests has proven more elusive.
- *Sources of nitrogen*—Growers still rely on external sources of nitrogen, be they synthetic or organic. Although it is biologically feasible to grow the crop requirement of N in the orchard with legume cover crops, the challenge is to integrate these cover crops into the orchard system without causing other problems, such as excess N during fruit maturation and outbreaks of potentially damaging rodents and insect pests.
- *Energy use*—All orchards rely on fossil fuels to run tractors, wind machines, and trucks, and the need to transport the fruit to distant markets leaves the tree fruit sector vulnerable to petroleum supply and cost impacts in the future.
- *Economic viability*—The current push to mechanize more orchard operations aims to reduce production costs and boost competitiveness, but will have community impacts such as reducing certain types of jobs permanently.
- *Social factors*—Workforce training programs such as the Latino Agricultural Education Program at Wenatchee Valley College represent an important investment in social sustainability. A key sustainability indicator is the number of remaining orchardists, and this number has steadily fallen, with both age and lack of profitability driving this decline.

5.3.5 CONCLUSIONS

Sustainability is being more widely discussed in the tree fruit sector, with integrated pest management, profitability, mechanization, and water use its more obvious manifestations. Growers will continue to make positive stewardship changes to the degree that economics will allow them. New products (varieties), new strategies ("club" varieties with restricted production), new health linkages (antioxidant content), and new ways for growers to participate in the value chain are all essential ingredients for economic sustainability. Simply cutting costs is unlikely to prove viable in the long run. Climate impacts on water supply and affordable energy may end up being the key determinants for the future of tree fruit production in the Northwest.

5.4 WHEAT-BASED PRODUCTION SYSTEMS IN THE DRYLAND ZONE

The Northwestern Wheat and Range Region includes portions of eastern Washington, northern Idaho, north–central Oregon, and southern Idaho (Figure 5.2). The seasonal precipitation pattern is Mediterranean, with 60 to 70% of the total annual precipitation occurring from November through April and only 5% in July and August (Kaiser, 1967; Papendick et al., 1995). Annual precipitation follows a steep east–west gradient, from 150 mm in the rain shadow east of the Cascade Mountains of Washington and Oregon to 815 mm along the eastern edge of the Palouse and Nez Perce prairies in Idaho (Daubenmire, 1988). The soil under agricultural production developed primarily from windblown deposits of silt-sized material (loess) and has a relatively high water-holding capacity (Busacca and Montgomery, 1992). The loess deposits are as deep as 75 m in some locations and overlie massive flows of basalt (Ringe, 1970). Much of the productivity of dryland crops in this water-limited region is dependent on the capacity of the soil to store winter precipitation (Busacca and Montgomery, 1992).

Agriculture in the region has historically been based on wheat; however, cultural practices associated with wheat production have seriously degraded soil resources and caused adverse air and water pollution (Saxton et al., 2000). Losses of topsoil from water and wind erosion threaten the long-term productivity of the region, as annual rates of erosion average 10 to 14 tons per acre (USDA, 1978; Scheinost et al., 2001). This is equivalent to annual losses of 12 bushels of topsoil for each bushel of wheat produced (Michalson et al., 1999).

Three major land use conversions have taken place in the region from 1870 to the present (Black et al., 1997). First, European-American settlement occurred from 1870 to 1900 and was accompanied by the rapid conversion of native prairie dominated by bunchgrasses to cropland, hayland, or pasture (Daubenmire, 1988). Agriculture was primarily horse-powered until the 1930s, and wheat and other annual cereals emerged as the major cash crops. Production systems were labor intensive, with the steepest hillsides and hilltops left as pasture for horses and cattle. The second conversion began in the 1930s and was driven by petroleum-based technology that replaced horse and human power. By 1970, most farms were mechanized and the remaining tillable acres were converted to annual crops of wheat, barley, pea, and lentil. Short rotations, intensive inversion-based tillage, and large inputs of synthetic fertilizers typified most farming operations in these systems. From the 1970s to the present, the third land use conversion has been under way; it includes population growth encroaching on agricultural lands (Black et al., 1997) as well as the evolution and gradual adoption of conservation cropping systems (Michalson et al., 1999). The latter shift has occurred to curb the degradation of air, soil, and water resources as well as to gain farm efficiencies in fuel use and labor.

Over the course of these land use conversions, wheat emerged as the dominant crop of the dryland region. In the western United States, wheat was first grown during the 1770s in California and in 1825 at Fort Vancouver, which lies across the Columbia River from present-day Portland, Oregon. To put this in perspective, wheat was first grown in the Pacific Northwest just 20 years after the Lewis and Clark expedition reached the area (Jones, 2002).

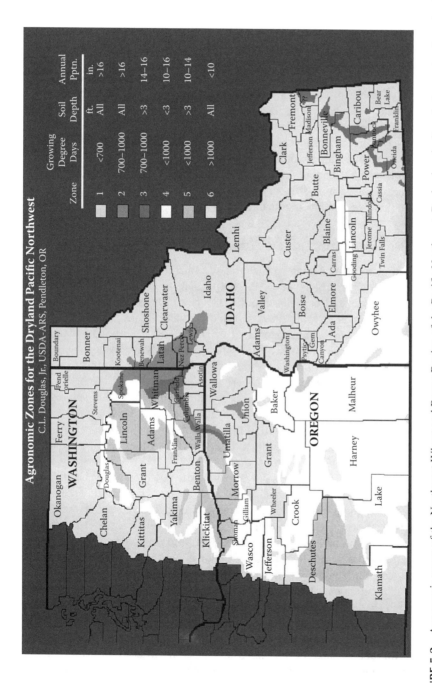

FIGURE 5.2 Agronomic zones of the Northwest Wheat and Range Region of the Pacific Northwest. Dryland agriculture is the predominant land use in Zones 2, 3, 5, and 6.

Early farmers in the Pacific Northwest planted wheat varieties that originated in Northern and Eastern Europe, and did what their ancestors had practiced for 10,000 years: saved their best seeds and replanted them the next season. This strategy works well when seed is replanted in the same environment for many years; however, it is not effective when saved seeds are moved every few years to new locations with differences in climate and soil, as was the case in the dryland Pacific Northwest. By the late 1800s, other areas of the United States had grown and selected wheat for nearly 100 years and produced varieties that were well adapted to specific locales. In contrast, the dryland Pacific Northwest lacked well-adapted varieties for the region's diverse environments until the early 1900s. In 1905, the regional land grant universities began breeding and releasing wheat varieties specifically for the Pacific Northwest. It soon became profitable to grow wheat in the region, and the dryland Pacific Northwest became one of the most productive wheat growing regions of the world. Wheat production is still very profitable in the region; however, conventional production practices are not sustainable.

5.4.1 PRIMARY FACTORS LIMITING SUSTAINABILITY

Biophysical factors threatening agricultural sustainability in the dryland region are wind and water erosion (Saxton et al., 2000), tillage-induced translocation of soil from upper to lower landscape positions (McCool et al., 1998), declining soil organic matter levels (Rasmussen et al., 1989), soil acidification (Mahler et al., 1985), limited biological diversity including economically viable crops (Elliott and Lynch, 1994), pest pressures, and lack of crop varieties bred specifically for sustainable systems. Reliance on mechanical tillage and short crop rotations that often include summer fallow (a season-long fallow period where the soil is routinely tilled, leaving the surface bare and exposed) have resulted in substantial degradation of soil resources and air and water pollution (McCool et al., 2001). Particularly evident are the adverse effects of long-term soil erosion. All of the original topsoil has been removed from 10% of the cropland, and one-fourth to three-fourths of the topsoil is gone from another 60% (USDA, 1978).

Low-input and organic systems have traditionally relied on tillage as an integral component of weed and nutrient management strategies. The hazard of soil erosion under these systems is similar to or greater than that found under current tillage-based farming systems. Conservation tillage systems, including direct seeding and no-till, conserve soil and water, but generally require herbicides for weed control, which increases the hazard for adverse pesticide effects on environmental and food system factors, while promoting the development of herbicide-resistant weeds.

Nitrogen requirements for soft white wheat are 2.7 lb N per bushel and are largely met through the use of synthetic N fertilizers. Consolidation of livestock production systems and the lack of animals on most farms have limited the availability of local animal sources of fertilizer (manure). In addition, lack of sufficient precipitation throughout most of the region precludes the widespread use of legumes as N-supplying green manure crops. Currently, the high cost and low N concentration of organic fertilizers make them cost prohibitive for relatively low-value crops like cereal grains. The eroded and nutrient-depleted soils of the region require high amounts of synthetic fertilizers to meet cereal crop nutritional requirements, and wheat is often overfertilized, particularly when the price of nutrients is low relative to wheat prices. This is particularly true for the hard wheats used for bread, pizza, and other products that require very strong dough. Strong dough is directly related to high-protein (gluten) grain content, which in turn requires more nitrogen (N) per bushel than soft wheat, in which a lower protein content is desirable. Greater N requirements of cereal crops are often associated with decreased N use efficiency of applied N (Huggins and Pan, 2003). Although the Pacific Northwest is primarily a soft wheat growing area, farmers are interested in hard wheats because of the historical price advantage of up to $1 per bushel as compared to soft wheat.

The vast majority of wheat, like all major crops throughout the United States, has been grown with increasing dependence on chemical inputs since the late 1940s. Plant breeding programs, both public and private, have not only bred varieties *for* high-input, tillage-based systems, but they have also bred them *in* high-input systems that rely on intensive soil tillage. This is a self-reinforcing system that works to the detriment of developing low-input, organic, no-till, and other sustainable agroecosystems. Breeding programs have not developed modern varieties that are well suited to these systems.

To address this issue, varieties of wheat are currently being developed at Washington State University on certified organic ground based on selections for yield, weed competitiveness, growth under nitrogen stress, disease resistance, and in some cases, the ability to regrow after harvest, thereby providing protection against soil erosion. This is the first time in over half a century that varieties are being developed and screened in a system that is free of synthetic chemical inputs.

Socioeconomic factors influencing sustainability include historically low commodity prices for wheat, barley, and grain legumes (peas, lentils, and chickpeas); rising costs for external farm inputs, including fuel, fertilizer, and pesticides; issues related to absentee landowners, land tenure, and government policy (Carlson and Dillman, 1999; Walker and Young, 1999); and lack of local and high-value markets. Currently, more than 90% of the wheat grown in the Pacific Northwest is exported out of the United States. At the same time, the majority of wheat consumed in the region is imported from other states and Canada. Currently, there is no mainstream market mechanism in place for farmers to sell their wheat within the region. There is a need to develop value-added markets for crops produced in the region and to take into account production practices and their socioeconomic impacts when calculating farm gate value.

Together, biophysical and socioeconomic factors create agroecosystem complexities that are difficult to assess, predict, and improve using traditional discipline-oriented approaches to research. Systems approaches to research may offer alternatives that lead toward more holistic advances in sustainable agroecosystems.

5.4.2 Conversion Experiences

Agricultural change during the last 30 years in the dryland Pacific Northwest has been a slow process that reflects the risk-averse strategies typical of the agricultural sector. Traditionally, agroecosystem performance in the dryland region has been assessed primarily by production and economic factors related to yield. Technologies and practices that increased yield were strongly supported by agribusiness, research and educational institutions, and government commodity-based support programs. Mechanization, pesticide and fertilizer use, biotechnologies, and crop variety improvements are examples of technologies that were often adopted by conventional farmers as their impacts on increasing yield and production efficiencies became evident. Since the 1960s, however, public and farmer perception of agriculture's role has broadened to include not only production of food and fiber but also the provision of multiple ecosystem services related to water and air quality, soil quality, wildlife habitat, and open space. These greater expectations of agriculture require the assessment of factors beyond yield to evaluate if a farming system is more sustainable. Unfortunately, these factors may not be readily assessed in the short-term, or such assessment may provide contradictory results. Thus, trade-offs among diverse measures of performance are common. For example, conservation tillage can reduce soil erosion to acceptable levels, but reducing tillage can have fundamental impacts, both negative and positive, on many production factors, including equipment requirements, pest management, crop rotation, nutrition, and virtually every aspect of production. Consequently, if major shifts in management as well as new technologies are required to meet sustainability goals such as erosion control, there is a greater likelihood that initial efforts will be economically risky, unpredictable, and incompatible with conventional practices. As a result, longer time periods are required for the more sustainable production system to develop, evolve, and be adopted.

Another factor is that public benefits resulting from improvements in ecosystem services may not be reflected in economic rewards to the farmer. This is the case with conservation tillage, for example. Consequently, farm competitiveness and economic viability may suffer, sending a negative signal to farmers making decisions about adopting more sustainable practices. A recent effort to address this issue was led by an innovative farmer group that developed a sustainable label for wheat produced in a no-till system for sale to restaurants and bakers who are willing to purchase a high-quality product that is locally grown (Kupers, 2003).

In 1997, scientists at Washington State University and with the U.S. Department of Agriculture–Agricultural Research Service (USDA-ARS) in Pullman, Washington, recognized that despite nearly 30 years of research and grower efforts, there had been little adoption of sustainable agricultural practices (such as direct seeding) in the dryland Pacific Northwest (CTIC, 2000). In addition, despite the heterogeneity of the regions' soils and topography and the availability of precision technologies such as yield monitors and variable rate applicators, there were no science-based, site-specific recommendations and little grower practice of precision agriculture. Given this situation, scientists initiated a systems research approach (e.g., Conway, 1985) in 1999 to design and establish a continuous direct-seed and precision agricultural cropping systems study in the dryland Pacific Northwest. The study is at the Washington State University Cook Agronomy Farm (CAF), located within 10 km of Washington State University, Pullman. The 57 ha research farm has soils and topography representative of the dryland annual cropping area of eastern Washington and northern Idaho.

The following describes our efforts in implementing this systems research. First, we established a diverse working group comprised of researchers, farmers, and people in agribusiness, government agencies, and nonprofit organizations to define the system of interest. The working group identified continuous direct-seed (no-tillage) and precision agriculture as the system of primary interest. Furthermore, the working group established that field-scale farming equipment similar to that used by surrounding growers would be used for all operations, and that other cultural practices would be used in accordance with the best university, grower, agency, and industry standards.

Next, pattern analysis was initiated to characterize the temporal and spatial variability of the biophysical and socioeconomic factors of the study area and the primary factors that influence resource management decisions, including those related to the flow of water, nutrients, energy, materials, and net profits. The intent of pattern analysis is to characterize the agroecosystem and its components, identify strengths and weaknesses, and provide insights into how the system can be improved. The focus remains on the agroecosystem as a whole system and disciplinary expertise is drawn in as needed to further understand the system. To enable pattern analysis in this study, a nonaligned, randomized grid sampling design with 369 geo-referenced points was established over a 37 ha field at the CAF. At each geo-referenced point, samples were collected and characterized to determine crop performance (yield, quality, profitability), soil and terrain attributes, water and nutrient use efficiencies, presence of pathogens and weeds, and other biophysical and economic variables of interest (samples continue to be collected as of 2008). Weather stations and other *in situ* instrumentation have been installed to facilitate process-oriented modeling efforts. These data are integrated into a relational database for system analyses, including characterization of field-scale hydrologic and soil erosion processes; cycling and flow of soil carbon, nitrogen, phosphorus, and sulfur; soil acidity; and spatial variability and temporal persistence of agroecosystem performance factors, including biomass production, grain yield, efficiency of use of nitrogen and water, soil-borne diseases, weed species, and economic returns. Complementing these analyses are efforts to develop process-oriented models to identify unknowns, extrapolate findings to appropriate areas, and develop decision support systems.

The third step is system reflection, where system elements are studied and understood sufficiently to recognize leverage points and to devise and prioritize alternative strategies for agroecosystem improvement. These data have provided the framework for a more holistic characterization of agroecosystem performance and enabled several additional studies to be launched that focus on crop suitability and rotational design and variable timing and application of nitrogen fertilizers. These studies are aimed at improving agroecosystem performance by increasing water and nitrogen use efficiency, soil organic carbon, and economic returns. Traditional hypothesis testing is often used at this stage of research.

Finally, strategies to improve the agroecosystem are selected and implemented as part of a new system and the process begins again. The agroecosystems approach, as described above, has been supported by growers, agribusiness, commodity groups, and researchers and administrators within Washington State University (WSU) and USDA-ARS. Although the research emphasis at the CAF constitutes first steps toward more sustainable agroecosystems, it is expected to evolve and change as signals from biophysical and socioeconomic determinants and indicators warrant.

5.4.3   LESSONS LEARNED

Sustainability should be viewed as a process rather than an endpoint. As such, sustainability for the dryland Pacific Northwest should be considered the ability to meet production goals and provide broader ecosystem services both now and in the future (Huggins, 2001; Jones, 2004). Using this criterion, current conventional dryland production systems are not sustainable. Considering sustainability at a landscape scale as opposed to single farms, the Pacific Northwest faces tremendous challenges to achieve sustainability of its dryland agroecosystems. On its current path, crop production is directly responsible for the decrease in the long-term health of the ecological system in the region.

Many conventional wheat farmers know their current production systems are not sustainable, but the great majority of them worry more about the next few years than the next few decades. One positive indicator is that there has been tremendous interest in recent years in transitioning away from high inputs due to the ever-rising cost of fossil fuel and the impact this has on the cost of inputs.

Public sector research into sustainable agroecosystems requires a systems approach in order to contend with the inherent complexities of agriculture. Whereas many technology-based advancements are comparatively simple to assess and easily incorporated into conventional agroecosystems, the development, evaluation, and adoption of sustainable agroecosytems is less certain and less predictable.

The success of sustainable agroecosystems research is dependent on developing an effective, diverse team. Compared to what is usually the case with more disciplinary-focused research, more time is required to nurture team participation and mutual respect. The team is also vulnerable to turnover in research faculty and changing research interests. Although systems approaches are intuitive to many agricultural researchers, few have been trained outside of disciplinary constraints. Furthermore, systems research may not be ideal for meeting short-term professional and institutional goals. Consequently, a team pursuing agroecosystem studies must achieve a balance between short- and long-term objectives.

The benefits of agroecosystem research become more apparent with time. Because the agroecosystem studies are located at a permanent site, each individual study builds on data, information, and knowledge that were previously collected and interpreted. This is a powerful research model. As the agroecosystem database accumulates and matures through time, it becomes attractive to various disciplinary efforts independent of the focus on agroecosystems. Consequently, the agroecosystem explicitly serves as the conceptual framework for guiding continued disciplinary research efforts. In the end, we believe that the systems approach will lead to more rapid conversion of current agricultural practices to practices that are more sustainable.

5.4.4   INDICATORS OF SUSTAINABILITY

In the dryland cropping systems of the Pacific Northwest, soil erosion by water, wind, or tillage translocation has to be eliminated to maintain soil resources and improve water and air quality. One important indicator of sustainable soil management is annual maintenance of surface cover (greater than 30%), which is achieved with the use of crops and crop residue management systems that result in minimal soil erosion. Current federal programs, including the Conservation Reserve Program (CRP), the Environmental Quality Incentives Program (EQIP), and the Conservation Security Program (CSP), have encouraged the adoption of conservation practices, including crop residue management systems, that help reduce soil erosion.

Levels of soil organic carbon are an important biophysical indicator of sustainability. Organic matter levels of tillage-based agroecosystems in the region have declined by 40 to 75% during the past 100 years of cultivation (Rasmussen et al., 1989). Recent concerns over rising levels of atmospheric carbon dioxide and global climate change have stimulated interest in soil carbon sequestration (Lal et al., 1995). The use of conservation tillage and cropping systems could annually increase soil organic carbon by 0.1 to 0.7 Mg C ha^{-1} until a new steady state is reached (Sperow et al., 2003).

Soil pH is a third indicator of sustainability. Soils formed under native Palouse and Nez Perce prairies had near neutral pH (7.0) when first cultivated. Mahler et al. (1985) reported that soil pH had declined to values less than 6.0 in over 65% of agricultural soils in the region by the early 1980s. More recently, Bezdicek et al. (1998) and Brown et al. (2008) have reported soil pH values as low as 4.0, indicating soil acidification has continued to the present. Soil acidification has been primarily a consequence of base depletion from crop removal, increased organic matter decomposition, and the application of ammonium-based nitrogen fertilizers. At soil pH levels below 5.0, grain yields of all major crops grown in the region can be adversely affected (Mahler and McDole, 1987). An additional concern about soil acidification is that there are currently no regionally available sources of lime.

Water and nutrient use efficiency are also critical indicators of cropping system sustainability. Summer fallow practices are at best 30% efficient in storage of over-winter precipitation, and this practice should be reduced or eliminated (Schillinger et al., 2003). Nitrogen uptake efficiency across the region's diverse soils and topography is highly variable, ranging from 15 to 60% (Huggins and Pan, 2003). Crop rotations need to be diversified and intensified so that water and nutrients are more efficiently utilized. The use of precision conservation practices and technologies could lead to greater efficiencies of water and nutrient use that will enable farmers to achieve specific resource conservation and environmental goals.

5.4.5 Conclusions

The dryland cropping region of the inland Pacific Northwest is facing severe challenges to its sustainability. At the same time, biophysical and socioeconomic factors create agroecosystem complexities that are difficult to assess, predict, and improve using traditional discipline-oriented approaches. The successful development of sustainable agroecosystems is dependent on developing and testing holistic systems. Agroecosystem studies that are long-term and relatively permanent better reflect the interactions and likely outcomes of sustainable practices. The end goal is that a systems approach will lead to more rapid conversion of current agricultural systems to more sustainable agroecosystems.

5.5 OVERALL CONCLUSIONS

Agriculture represents a critical land use throughout the Pacific Northwest. It makes important contributions to the region's economy and to the nation's food supply. As in many other regions of the United States, adverse environmental impacts, pressure from urbanization, and chronic lack of profitability are serious sustainability challenges for Pacific Northwest agriculture. Soil degradation and soil loss due to agricultural practices are especially prevalent throughout the dryland wheat production zone and may be the greatest factors limiting future production in these areas. There are no simple strategies for reversing these trends as each change in production management has far-reaching impacts on multiple production factors. As farmers throughout the region develop and test new sustainable practices, wider adoption may be achieved by increasing the visibility of sustainably produced agricultural products in the marketplace through ecolabeling. Ecolabels are provided by programs such as Food Alliance and Salmon Safe and use third-party verification of sustainable farming practices.

Throughout the Pacific Northwest region, important gains are being made in soil and water conservation, biocontrol and pesticide reduction, organic farming, and direct marketing. Genetically engineered crops and livestock have yet to have a major presence in the region, unlike in the Midwest where herbicide-resistant soybeans and *Bacillus thuringiensis* corn raise other issues in regards to sustainability. Agriculture in the region, as elsewhere, is heavily reliant on fossil fuels for field operations, fertilizer, transport, and processing. Opportunities for bioenergy production are being explored to both reduce this dependence and create new value-added opportunities for farms and rural areas. Factors such as climate change are likely to impose new sustainability challenges in the future. Given the natural attributes of the Pacific Northwest, agriculture will most likely find ways to adapt and sustain itself for future generations.

ACKNOWLEDGMENTS

The authors thank anonymous and named contributors who provided background information and knowledge that may not otherwise appear in printed sources. In addition, we thank the many colleagues and lab members who took the time to critically read the chapter content as it was being developed and offered suggestions for its improvement.

REFERENCES

Barritt, B.H. 1992. *Intensive orchard management.* Yakima, WA: Good Fruit Grower.
Beers, E., J. Brunner, M. Willett, and G. Warner. 1993. *Orchard pest management: A resource book for the Pacific Northwest.* Yakima, WA: Good Fruit Grower.
Blank, S. C. 1998. *The end of agriculture in the American portfolio.* Westport, CT: Quorum Books.
Brunner, J. 1994. Integrated pest management in tree fruit crops. *Food Reviews International* 10:135–57.
Brunner, J., W. Jones, E. Beers, J. Dunley, and J. Tangren. 2001. Pest management practices in Washington: A journey through time. In *Proceedings of the 97th Annual Meeting of the Washington State Horticultural Association,* Wenatchee, WA, pp. 177–84.

Busacca, A.J. and E.V. McDonald. 1994. Regional sedimentation of the late Quaternary loess on the Columbia Plateau: Sediment source areas and loess distribution patterns. Regional Geology of Washington State, *Washington Division Geol. Earth Resource Bull.* 80:181–190.

Busacca, A.J., and J.A. Montgomery. 1992. Field-landscape variation in soil physical properties of the Northwest dryland crop production region. In *10th Inland Northwest Conservation Farming Conference Proceedings: Precision Farming for Profit and Conservation*, Washington State University, Pullman.

Calkins, C.O. 1998. Review of the codling moth areawide suppression program in the western United States. *Journal of Agricultural Entomology* 15:327–33.

Dunley, J., and T. Madsen. 2005. The areawide organic project: Three years in Peshastin Creek. In *Proceedings of the 3rd National Organic Tree Fruit Research Symposium*, Chelan, WA, pp. 39–40.

El Titi, A., E.F. Boller, and J.P. Gendrier. 1993. Integrated production: Principles and technical guidelines. *IOBC/WPRS Bulletin* 16:13–40.

Granatstein, D. 2000. *Trends in organic tree fruit production in Washington State*. EB 1898 E, Washington State University Extension, Pullman.

Granatstein, D., E. Kirby, and C. Feise. 2005. Trends of organic tree fruit production in Washington State. In *Proceedings of the 3rd National Organic Tree Fruit Research Symposium*, Chelan, WA, pp. 3–5.

Hoyt, S.C. 1969. Integrated chemical control of insects and biological control of mites on apple. *Washington Journal of Economic Entomology* 62:74–86.

Jones, S.S. 2002. Wheat in the west. *Journal of the West* 41:44–46.

Jones, S.S. 2004. Sustainable agriculture: Ecological indicators. In *Encyclopedia of plant and crop science*, ed. F. Kirschenmann and R. Goodman, pp. 1191–94, Pullman, WA: Washington State University.

Kupers, K. 2003. Shepard's grain: A farmer's strategy for survival in the global marketplace. *Sustaining the Pacific Northwest* 2:5–7. http://csanr.wsu.edu/whatsnew/PNW-v2-n1.pdf.

Luce, W. A. c. 1975. *Washington State fruit industry: A brief history*.

Lyons, Z. 2006. Washington State Farmers Market Association director. Personal communication.

McCool, D.K., D.R. Huggins, K.E. Saxton, and A.C. Kennedy. 2001. Factors affecting agricultural sustainability in the Pacific Northwest, USA: An overview. Paper presented at Sustaining the Global Farm. Selected Papers from the 10th International Soil Conservation Organization Meeting, Purdue University, IN, May 24–29, 1999.

McCool, D.K., J.A. Montgomery, A.J. Busacca, and B.E. Frazier. 1998. Soil degradation by tillage movement. In *International Soil Conservation Organization Conference Proceedings: Advances in Geoecology*, Vol. 31, pp. 327–32, Purdue University, West Lafayette, IN.

Nielsen, G.H., E.J. Hogue, T. Forge, and D. Nielsen. 2003. Mulches and biosolids affect vigor, yield and leaf nutrition of fertigated high density apple. *HortScience* 38:41–45.

O'Rourke, D. 2002. Does Washington State have the weapons to win future world apple wars? In *Proceedings of the 98th Annual Meeting of the Washington State Horticultural Association*, Wenatchee, WA, pp. 18–27.

Oregon Department of Agriculture. 2005. Oregon county and state agricultural estimates. http://eesc.oregonstate.edu/agcomwebfile/EdMat/SR790-04.pdf.

Oregon Farmers' Market Association. 2002. http://www.oregonfarmersmarkets.org/.

Pike Place Market Preservation and Development Authority (PDA). 2006. http://www.pike-placemarket.org/.

Portland Public Market. 2006. http://portlandpublicmarket.com/.

Rackham, R.L. 2002. *History of specialty seed crop production in the Pacific Northwest*. Willamette Valley Specialty Seed Crops Association, Oregon State University Department of Horticulture, Corvallis, OR.

Scheinost, P., D. Lammer, X. Cai, T. Murray, and S. Jones. 2001. Perennial wheat: A sustainable cropping system for the Pacific Northwest. *American Journal of Alternative Agriculture* 16:146–50.

Reganold, J.P., J.D. Glover, P.K. Andrews, and H.R. Hinman. 2001. Sustainability of three apple production systems. *Nature* 410:926–30.

USDA National Agricultural Statistics Service. 2005. Washington State agricultural statistics. http://www.nass.usda.gov/Statistics_by_State/Washington/Publications/County_Estimates/index.asp.

University of Texas Libraries. 2006. Perry-Castañeda Library, United States Map Collection, University of Texas at Austin. http://www.lib.utexas.edu/maps/united_states.html#usa. February 22, 2006.

Unruh, T. and J. Brunner. 2005. Rose and strawberry plantings adjacent to orchard to enhance leafroller biological control. In *Proceedings of the 3rd National Organic Tree Fruit Research Symposium*, Chelan, WA, pp. 41–42.

Washington Agricultural Statistics Service. 1994. *Washington agricultural county data 1994*. Olympia, WA: WSDA.

Washington Agricultural Statistics Service. 2002. *Washington agricultural statistics 2002*. Olympia, WA: WSDA.

Washington State Farmers Market Association. 2006. http://www.wafarmersmarkets.com/.

Williams, K., and T. Ley. 1994. *Tree fruit irrigation*. Yakima, WA: Good Fruit Grower.

6 California (U.S.)
The Conversion of Strawberry Production

Stephen R. Gliessman and Joji Muramoto

CONTENTS

6.1 INTRODUCTION

The central coast of California, with its Mediterranean climate, is one of the most important strawberry growing regions in the world. On approximately 5,400 ha, Monterey and Santa Cruz counties together produced more than US$800 million worth of strawberries in 2007 (Monterey County Agricultural Commissioner, 2008; Santa Cruz County Agricultural Commissioner, 2008), about 60% of the total California crop. Conventional strawberry production here, as in many other locales, is highly dependent on expensive, energy-intensive, and often environmentally harmful off-farm inputs.

Strawberry production, therefore, is an excellent target for conversion. Researchers at the University of California, Santa Cruz recognized this in the mid-1980s. Located not far from the most intensive strawberry growing areas on the central coast, they initiated a research program aimed at helping local strawberry growers move their production systems toward sustainability. This ongoing effort, now more than 20

FIGURE 6.1   Conventional strawberry field fumigated with methyl bromide near Watsonville, California. Vaporized MeBr is held under the plastic for several days. Conversion to organic management involves replacing this very toxic and expensive chemical with a variety of alternative inputs and practices. (Photo by S. Gliessman.)

years old, demonstrates that is possible to make fundamental changes in agricultural systems that were once firmly wedded to unsustainable conventional practices.

6.2   CONVENTIONAL STRAWBERRY PRODUCTION IN CALIFORNIA

Chemical fumigation with methyl bromide (MeBr) has been the core technology for the development of large-scale strawberry production in California (Figure 6.1). In the 1950s and 1960s, Wilhelm and Paulus demonstrated that preplanting soil fumigation with a mixture of MeBr and chloropicrin was effective in controlling Verticillium wilt, the most lethal soil-borne disease of California strawberries (Wilhelm and Paulus, 1980). Fumigation with MeBr allowed growers to plant disease-sensitive strawberries continuously in the same fields without crop rotation or diversification. Until that time, growers treated strawberries as a perennial crop, with each field requiring rotation out of strawberries for several years. Use of MeBr allowed growers to manage strawberries as an annual crop, planted year after year on the same piece of land, or on any given field regardless of its soil-borne disease pressure. Fumigation became the key feature of a system that employed high-yielding cultivars, improved irrigation systems, plastic mulch, and the application of chemical pest management schemes. This intensive system allowed growers to dramatically increase the yield of strawberries beginning in the 1960s.

In the most common conventional system, in use for the last four decades, strawberry plants are removed each year following the end of the season in late summer or early fall, then the soil is cultivated and fumigated before being replanted with new plants for the next season. Another common conventional system among large-scale specialized growers is crop rotation of strawberries and vegetables in a fumigated field. In this case, strawberries are removed in late fall by a strawberry grower; then, taking advantage of the fumigated field, a vegetable grower comes in and grows vegetables in the field for a season or two. After harvesting vegetables in late summer or early fall, the strawberry grower comes back, fumigates the soil, and plants strawberries for the next season. In either system, intensive systems of drip irrigation and plastic mulch are required along with fumigation.

MeBr fumigation technology has been so central to strawberry production systems in California that everything else has evolved around it. For example, with fumigation fully addressing disease problems, the University of California strawberry breeding program has—until the past few years—focused for four decades on improving traits such as yield, firmness, and flavor, and not on soil-borne disease resistance (Shaw and Larson, 2001).

In recent years, attention has been devoted to finding other fumigants that could be used in place of MeBr. As of October 2008, three alternative chemical fumigants were registered for California strawberries: chloropicrin, metam sodium, and 1,3-dichloropropene (Telone). Although use of these alternative chemicals has increased, 33% of California strawberry growers were still using MeBr in 2006 (USDA National Agricultural Statisitics Services, 2007).

6.3 THE NEED FOR CONVERSION

Production problems inherent in growing strawberries conventionally, as well as positive incentives associated with alternative systems, are pushing growers to consider converting conventional, high-input strawberry production systems to reduced-input or organic systems. Soil erosion, nutrient leaching, pest and disease resistance to conventional agrichemicals, escalating production costs (especially for any input based on fossil fuel, from plastic drip systems to tractor fuel), and more stringent environmental regulations are just a few of the problems strawberry growers face. At the same time, growing consumer demand for organic products has created good price premiums for organic strawberries, giving farmers a strong incentive to shift their production practices.*

6.3.1 Restrictions on the Use of Fumigants

Perhaps the most compelling force for change in strawberry production in California is the increasing likelihood that growers will be denied the use of MeBr in the future and face increasingly stringent regulations for all other fumigants (Trout, 2005). The serious environmental impacts of both MeBr and its chemical alternatives have made them the targets of increasingly tight restrictions. The major environmental impacts

---

* No solid statistics on the price premium of organic strawberries over their conventional counterparts is available. Limited data on the wholesale price at Boston (April to November 2007) and San Francisco (June to December 2007) markets indicated that the premium for organic strawberries averaged 48% and 43%, respectively (USDA Agricultural Marketing Service, 2008).

TABLE 6.1

Methyl Bromide Use under the Critical Use Exemption

Country	MeBr Baseline (tons)	Critical Use Exemption (tons)				
		2005	2006	2007	2008	2009
Australia	704	147	75	49	48	38 (1%)
Canada	200	62	54	53	42	34 (1%)
European Union	20,873	4,393	3,537	689	245	0
Israel	3,580	1,089	880	966	861	0
Japan	6,107	748	741	636	444	306 (7%)
United States	25,529	9,553	8,082	6,749	5,355	4,262 (92%)
Total	56,993	16,050	13,418	9,160	6,995	4,640 (100%)

Source: Porter et al. (2007).

of MeBr and alternative fumigants are very different, but interestingly enough, both have something to do with ozone (O_3).

The problem with MeBr is its deleterious effect on the atmosphere's ozone layer. This layer of ozone, located high in the stratosphere, protects organisms on the earth by absorbing harmful ultraviolet waves from solar radiation. In the 1980s, it was discovered that reactive chemicals such as MeBr had created an "ozone hole" above the Antarctic (Farman et al., 1985). In 1997, to protect the ozone layer from further deterioration, what is known as the Montreal Protocol banned the use and production of MeBr in developed countries in 2005 and in developing countries in 2015 (United Nations Environment Programme, 1997). This was the major impetus for developing alternatives to MeBr. None of the MeBr substitutes, however, are as effective as MeBr in strawberry production (Thompson, 2004). To continue using MeBr after the deadline of 2005, the United States and some other countries applied for—and were granted—critical use exemptions (CUEs) from the United Nations (see Table 6.1). In 2009, the United States accounted for 92% of the total amount of CUE MeBr use worldwide.

To obtain the CUE, a government must submit CUE nominations to the United Nations every year. Generally, as alternatives are introduced into markets, the number of nominations are reduced from earlier years. However, this is not always the case. For example, in 2009, the California Strawberry Commission requested 953 tons of CUE nomination for 2011, the same amount as for the previous year, due mainly to two new soil-borne diseases: charcoal rot by *Macrophominia phaseolina* and Fusarium wilt by *F. oxysporum*. These diseases are increasingly damaging strawberry plants grown in fields fumigated with alternative fumigants for more than two years. Because of this, some growers who have used alternative fumigants are going back to MeBr to "clean up" their fields (Finman, 2009). Efforts to completely ban MeBr and do away with the CUE process continue to face many challenges.

Use of the alternative fumigants in strawberry production has an uncertain future as well. The three chemical fumigants mentioned above—chloropicrin, metam sodium, and 1,3-dichloropropene—are categorized as volatile organic compounds (VOCs), which are under strict regulation by the U.S. Clean Air Act due to their effects on air quality. VOCs are known to react with oxides of nitrogen to form ground-level ozone, which has harmful oxidizing effects on living tissue (and never reaches the upper protective ozone layer). In California, five geographic regions (the Sacramento Valley, the south coast, the San Joaquin Valley, the southeast desert, and Ventura County) were designated as ozone nonattainment areas failing to meet federal air quality standards for ozone by the U.S. Environmental Protection Agency (USEPA). In particular, the Ventura area—the center of southern strawberry production in the state, and where fumigants for strawberry production constitute a major source of VOC emissions—became subject to a state regulation that a 20% VOC reduction must be attained by 2012 (Warmerdam, 2008). In the San Joaquin Valley and the southeast desert, low-emission fumigation methods, such as reduced application rates and applications through drip irrigation, are required during the period of May through October (California Department of Pesticide Regulation, 2008). To develop strawberry production systems that can avoid the complexities of highly regulated fumigants, the California Strawberry Commission started to fund Farming without Fumigants initiatives in 2008 (Finman, 2009).

6.3.2 ALTERNATIVES TO CHEMICAL FUMIGATION

Seeing the clouded future of fumigation, some growers have attempted to totally eliminate soil fumigation from their production practices while maintaining other pest management practices that rely in part on the use of synthetic pesticides. The most common way of doing this is through a crop rotation–based Integrated Pest Management (IPM) approach. Approximately 20 to 30% of strawberries in the world are produced this way, without using chemical fumigation (Porter and Mattner, 2002; Svensson, 1997). In Europe (Bevan et al., 2001), the Northeast and Midwest United States, and eastern Canada (Pritts and Handley, 1998), a minimum of a three-year rotation is recommended for conventional strawberries that do not use chemical fumigants. However, the rotation-IPM approach has yet to gain the attention of most California strawberry growers. Because of the high cost of leasing land (approximately US$5,000 per ha annually) and the highly specialized nature of the production system, most California strawberry growers cannot justify developing a crop rotation system that does not yield a price premium for the fruit.

6.3.3 ORGANIC STRAWBERRY PRODUCTION

Organic strawberry production is the only system currently used by strawberry growers in California that does not involve chemical fumigation. Weeds, soil-borne pests, and diseases in organic strawberry systems are managed using a combination of organically acceptable methods, including crop rotation, cover cropping, use of plastic mulch, compost applications, cultural controls, and careful management of naturally occurring beneficial predators with supplementary releases of beneficial arthropods when needed.

Because of the price premium for organic berries and other positive factors, organic strawberry production has become a viable alternative system for some California strawberry growers. The first commercial-scale organic strawberry farm in California was certified in 1987. For the first 10 years of organic production, growth in the sector was very slow. By 1997, the total acreage of organic strawberry in California was little more than 50 ha. However, the organic market began to expand rapidly at this time in response to consumer demand (Dimitri and Greene, 2002). In the next 10 years, the land area devoted to organic strawberry production grew over 14-fold to 710 ha. In 2007, organic strawberry acreage represented 4.9% of total California strawberry acreage (California Department of Food and Agriculture, 2008). Organic strawberry sales in California increased from US$2.0 million in 1997 to $46.5 million in 2007.

6.3.4 Problems with Organic Strawberry Production

Despite these positive trends, several sustainability issues are connected with the dramatic growth in organic strawberry production. For example, significant nitrogen loss has been observed in some organic strawberry fields during the winter rainy season (Muramoto et al., 2004). What might be called level 4 thinking should include consideration of such issues, as part of a concern for the health of the entire system. In addition, since organic strawberries usually require more labor, issues of worker health, safety, and pay equity must be also considered.

There is also the issue of consolidation and viability for small growers. As can be seen in Table 6.2, the number of organic strawberry producers has recently declined, even as the acreage planted has increased. The number of California organic strawberry growers peaked in 2000 with 122, and has remained below this level since, despite increasing sales. This pairing of trends indicates that large portions of sales are beginning to be dominated by a relatively smaller number of larger-scale growers. The same general trend has been noted in the organic industry as a whole both in California (Guthman, 2000; Tourte and Klonsky, 1998) and worldwide (Scialabba, 2005).

A more fundamental issue, perhaps, is that organic strawberry production may not be a feasible alternative for all growers at this time (Martin and Bull, 2002). Yields in organic research plots have been shown to be 65 to 89% of the yields achieved by conventional practices that include fumigation (Anonymous, 1996; Gliessman et al., 1996a; Sances and Ingham, 1997). Recent cost studies of organic (Bolda et al., 2003) and conventional (Bolda et al., 2004) strawberries on the central coast of California indicated that the yield of organic strawberries ranged from 22 to 36 tons per ha (20,000 to 32,000 pounds per acre), which corresponds to 40 to 65% of the average conventional yield for this area. The typical total cost of organic strawberry production is US$71,997 per ha (US$29,136 per acre), which is 90% of total cost for conventional strawberry production, but the labor cost for hand weeding in organic strawberry systems is about US$5,120 per ha, more than twice that of conventional systems (Bolda et al., 2003, 2004). Furthermore, when the typical yields of each system (33.6 tons per ha for an organic system and 58.6 tons per ha for a conventional system) are compared, the cultural cost of organic strawberries per acre is 28% higher than for conventional strawberries. Consequently, the break-even price per tray for organic strawberries is currently 30% higher than the break-even price per tray for conventional strawberries. This is consistent with the

TABLE 6.2

Growth of Organic Strawberry Production in California, 1997–2007

Year	Area in Organic Production (ha)[a]	Gross Declared Value (US$ in millions)	Number of Organic Producers
1997	54	2.0	n/a
1998	95	2.3	83
1999	312	8.3	101
2000	206	9.7	122
2001	278	9.5	120
2002	506	12.8	111
2003	522	17.5	103
2004	804	31.2	111
2005	569	25.1	112
2006	772	27.4	n/a
2007	710	46.5	n/a

Source: 1997, 2006, and 2007: California Department of Food and Agriculture, California Organic Program, http://www.cdfa.ca.gov/is/i_&_c/organic.html. 1998–1999: Klonsky and Richter (2005); 2000–2005: Klonsky and Richter (2007).

[a] Area may tend to be an overestimate since it may also include fallow or unplanted land set aside for future plantings.

finding, in a cost study of organic strawberry growers in the northeastern United States (Pritts and Handley, 1998, p. 129), that organic strawberries must receive a price premium of 35 to 40% in order to be profitable. After 10 years of research, Sances (2005) listed four criteria for successful organic production in California: (1) the soil-borne disease inoculum is low at planting, (2) the cultivar or plant type chosen is tolerant to diseases, (3) plastic mulch is used to suppress weed germination, and (4) the market is such that a premium price is paid for fruit.

Because of the greater labor costs and lower yields associated with organic production, and because a considerable price premium is needed in order to turn a profit, a large shift to organic strawberry production within a short-term in California is very likely impossible under current conditions. If the supply of organic strawberries grows faster than demand, price premiums and profitability will decline, pinching growers with no ability to lower their higher production and labor costs (Oberholtzer et al., 2005).

The cropping system design for organic strawberries is essentially the same as its conventional counterpart—it is still a monoculture of strawberries planted with plastic mulch and drip irrigation, with the conventional fertilizers and pesticides replaced with methods that meet national organic standards. Because of this basic similarity to conventional production, organic growers still face the same problems that confront conventional growers. Speaking in terms of the levels of conversion identified in Chapter 1, organic strawberry production in California in general represents conversion at level 2. To address the roots of the problems that still shadow organic strawberry production, it is necessary to make the more profound and difficult transition to conversion level 3, and then beyond that to level 4.

6.4 SUSTAINABLE STRAWBERRY PRODUCTION RESEARCH PROJECTS

The Agroecology Research Group at the University of California, Santa Cruz has been leading the effort to convert conventional strawberry production systems in this more thorough way—to go beyond input substitution to create redesigned, sustainable agroecosystems. For more than 20 years, the research group has been carrying out a variety of farmer-centered research projects designed to gradually build the knowledge and expertise needed for conventional strawberry growers to progress through all four levels of conversion.

After focusing for many years on conversion to organic production and demonstrating that even systems strongly invested in conventional practices can be changed, the research group has in more recent years turned to confronting the challenges involved in complete system redesign (level 3) and transformation of the food system in which strawberry production is embedded (level 4). Although the research group's projects have made it clear that conversion faces enormous obstacles, they have achieved many successful outcomes and generated optimism for the future.

The year-by-year evolution of the strawberry conversion research project is a story that reveals many of the most important aspects of the challenges inherent in making fundamental changes in an $800 million industry. It shows the broad scope of the conversion effort and demonstrates how conversion efforts at the different levels build on each other. Below, we provide brief descriptions of the conversion project's activities, organized by conversion level. These are followed by a chronological summary of those activities (Table 6.3).

6.4.1 LEVEL 1 CONVERSION

The first efforts related to conversion, carried out before the involvement of the Agroecology Research Group, were focused as much on increasing yields and profitability as on changing the nature of the production system. Extensive research was carried out to discover more effective ways of controlling pests and diseases so that inputs could be reduced and their environmental impacts lessened. For example, different miticides for control of the common pest two-spotted spider mite (*Tetranychus urticae*) were tested with the goal of overcoming the problems of evolving mite resistance to the pesticides, negative impacts on nontarget organisms, pollution of ground water, persistent residues on harvested berries, and health impacts for farmworkers (Sances, 1982). In addition, some work was done with mixed cover crops for erosion control and weed management on lands that were temporarily removed from strawberry production (Gliessman, 1989).

TABLE 6.3
Chronology of Strawberry Conversion Research Activities[a]

Date	Activity or Milestone	Conversion Level
1986	Contact with first farmer in transition	Level 1 to level 2
1987–1990	On-farm comparative conversion study	Level 2
1990	First conversion publication (*California Agriculture* 44:4–7)	Level 2
1990–1995	Refinement of organic management	Level 2
1995–1999	Rotations and crop diversification	Initial level 3
1996	Second conversion publication (*California Agriculture* 50:24–31)	Level 2
1997–1999	Alternatives to MeBr research projects	Level 2
1998	BASIS (Biological Agriculture Systems in Strawberries) work group established	Levels 2 and 3
1999	Soil health/crop rotation study initiated	Levels 2 and 3
2000–2006	Strawberry agroecosystem health study	Levels 2 and 3
2002–2003	Pathogen study, funded by NASGA (North American Strawberry Growers Association)	Levels 2 and 3
2001–2005	Poster/oral presentations at American Society of Agronomy meetings	Level 3
2003–2006	Alfalfa trap crop project	Level 3
2004	Organic strawberry production short course	Levels 2 and 3
2004–2009	USDA–Organic Research Initiative project: Integrated network for organic vegetable and strawberry production (building organic strawberry research networks among researchers, farm advisors, growers, NGOs, and industries in California)	Levels 2 and 3
2005–2006	Local organic strawberries in UC Santa Cruz dining halls and local farmers' markets	Level 4
2006	California Strawberry Commission and NASGA fund organic rotation system research	Level 3

[a] Carried out by the Agroecology Research Group at the University of California, Santa Cruz.

6.4.2 LEVEL 2 CONVERSION

In the early 1980s, as interest in organic food became a potential market force in agriculture and issues of pesticide safety and environmental quality came to the fore, farmers began to respond. It was in this environment that researchers at the University of California–Santa Cruz (UCSC) and a local farmer formed a partnership for conversion. In 1987, this partnership became a comparative strawberry conversion research project. For three years, strawberries were grown in plots using conventional inputs and management side by side with strawberries grown under organic management. In the organic plots, each conventional input or practice was substituted with an organic equivalent. For example, rather than control the two-spotted spider mite with a miticide, beneficial predator mites (*Phytoseiulus persimilis*) were released into the organic plots. Over the three-year conversion period population levels of the two-spot were monitored, releases of the predator carried out, and responses quantified. By the end of the third year of the study, ideal rates and release amounts for the predator—now the norm for the industry—had been worked out (Gliessman et al., 1996b).

After the three-year comparison study, researchers continued to observe changes and the farmer continued to make adjustments in his input use and practices. This was especially true in regard to soil-borne diseases. After a few years of organic management, disease organisms such as *Verticillium dahliae*, a source of Verticillium wilt, began to affect plants with greater frequency. The response was to intensify research on input substitution. Initial experiments with mustard biofumigation took place, adjustments in organic fertility management occurred, and mycorrhizal soil inoculants were tested. But the agroecosystem was still basically a monoculture of strawberries, and problems with disease increased.

6.4.3 Level 3 Conversion

It was at this point that a whole-system approach began to come into play. Based on the concept that ecosystem stability comes about through the dynamic interaction of all the component parts of the system, the researchers and farmer conceived of ways to design into the system resistance to the problems created by the simplified monoculture. The farmer realized he needed to partially return to the traditional practice of crop rotations that had been used before the appearance of MeBr. The researchers used their knowledge of ecological interactions to redesign the strawberry agroecosystem so that diversity and complexity could help make the rotations more effective and, in some cases, shorter. Testing of these ideas is ongoing. For example, mustard cover crops were tested for their ability to allelopathically reduce weeds and diseases through the release of toxic natural compounds. Broccoli is being tested as a rotation crop (Muramoto et al., 2006) since it is not a host for the *Verticillium* disease organism, and broccoli residues incorporated into the soil reduce the presence of disease organisms (Subbarao et al., 2007; Xiao et al., 1998).

Rather than rely on predator organisms, which still have to be purchased outside the system and released, the researchers and farmer have undertaken redesign approaches intended to incorporate natural control agents into the system, keeping them present and active on a continuous basis. For example, they tested the idea that refugia for the *P. persimilis* predator mite could be provided, on either remnant strawberry plants or trap crop rows around the fields, allowing the predator to remain in the system without continual reintroduction.

Perhaps the most novel redesign idea is the introduction of rows of alfalfa into the strawberry fields as a trap crop (Figure 6.2) for the western tarnished plant bug (*Lygus hesperus*). The pest can cause serious deformation of the strawberry fruit, and because it is a generalist pest, it is very difficult to control through input substitution. By replacing every 25th row in a strawberry field with a row of alfalfa (approximately 3% of the field), and then concentrating control strategies on that row (vacuuming, biopesticide application), it was possible to reduce Lygus damage to acceptable levels without sacrificing net returns (Swezey et al., 2007). The ability of these alfalfa rows to also function as reservoirs of beneficial insects for better natural pest control is now being tested as well (Pickett et al., 2009).

FIGURE 6.2   Sampling arthropods in alfalfa rows used as a trap crop for pests and refugia for beneficials in a strawberry agroecosystem. Such field-scale diversification is an example of level 3 conversion. (Photo by Diego Nieto. With permission.)

Outcomes from the studies have been shared and disseminated through a USDA-funded research network focused on organic strawberry and vegetable agroecosystems. Researchers are a key component of the network, but, more importantly, there are also farmers, extension agents, and nonprofits involved as well. Through the network, the environmental impacts and economics of alternative practices are evaluated, and results are made accessible to a broader audience that includes smaller, family farm operations. The network has also facilitated the sharing of knowledge between farmers and the development of training materials for culturally diverse and resource-limited farmers.

6.4.4   LEVEL 4 CONVERSION

Consumers are a very important force in the conversion of agroecosystems to more sustainable design and management. Not only are changes in consumer habits and preferences a necessary part of building a sustainable food system, but they can also be a potent force in actually driving the transformation of agriculture.

Consumers are increasing the demand for organic produce, allowing organic farming to become increasingly important. Although there are problems with, and limits to, organic production (noted above for strawberries in Section 6.3.3), the increase in consumer demand for organic food is an indicator that a culture of sustainability is beginning to take shape. In the two central coast counties where so many strawberries are grown, there were a total of 7,589 organic-certified hectares in 2005, more than four times the organic acreage recorded in 1998. The total farm gate revenue from organic farming in these counties was $100 million in 2005, representing a dramatic increase of more than 600% from 1998 (Klonsky and Richter, 2005, 2007). A parallel increase in organic strawberry production occurred over the same time period, as seen in Table 6.2.

The fourth level of conversion—the emergence of a culture of sustainability—made its debut in the strawberry conversion research program in 2005. Students at the UC Santa Cruz campus, with support from the Agroecology Research Group, convinced campus dining service managers to begin integrating local, organic, and fair-trade items—including organic strawberries—into the meal service. During the last several years, we have also seen a dramatic increase in both the number and use of local, certified farmers' markets in the Santa Cruz County region (http://www.santacruz.com/Farmers_Markets). Local farmers, including those who sell straw-berries that are grown organically, are able to form direct relationships with the customers who come to the markets in search of organic produce. Consumers are also able to develop relationships with the growers and in the process learn how their strawberries are grown, where, and by whom. This direct relationship between the grower and the consumer is an integral part of level 4 thinking in sustainability, and an important driver of the conversion process.

6.5 CONCLUSIONS

Strawberries are a remarkably intensive crop, regardless of whether they are grown conventionally or organically. When the ultimate goal of the transition process is sustainability, there is a need to go beyond organic. Work must be done to integrate aspects of alternative agroecosystem designs that promote stability and maintain yields, yet at the same time give the agroecosystem more internally derived resis-tance to the problems that are encountered at conversion levels 1 and 2. Ultimately, however, a different way of thinking that values other aspects of sustainability—such as worker safety, living wages for workers, and tighter links between farmers and consumers—must be developed and become the basis of a culture of sustainabil-ity. When a culture of sustainability guides everything from the choice of farming practices to the choice that a consumer makes when purchasing strawberries, the transition process is making the move to level 4. At this level, sustainability is para-mount and becomes the context within which all other concerns—such as capturing a growing market—are considered.

REFERENCES

Anonymous. 1996. *Organic strawberry production as an alternative to methyl bromide.* EPA-430-R-96-021, EPA Case Study, Methyl Bromide Alternatives.

Bevan, J., Knight, S., Tolhurst, I. 2001. *Organic strawberry production—Grower's guide.* Coventry, UK: HDRA Publishing.

Bolda, M., Tourte, L., Klonsky, K., Bervejillo, J.E. 2003. Sample costs to produce organic strawberries: Central coast, Santa Cruz and Monterey Counties. University of California Cooperative Extension. http://www.agecon.ucdavis.edu/uploads/cost_return_articles/strawborgcc03.pdf.

Bolda, M., Tourte, L., Klonsky, K., De Moura, R.L. 2004. Sample costs to produce strawberries: Central coast region. University of California Cooperative Extension. http://www.agecon.ucdavis.edu/uploads/cost_return_articles/strawcc2004.pdf.

California Department of Food and Agriculture. 2008. 2004 organic sales reports. Gross yearly sales, acreage by commodity for 2007. http://www.cdfa.ca.gov/is/docs/COMMYearlysalesACREAGE2007Producers.pdf.

California Department of Pesticide Regulation. 2008. Reducing VOC emissions from field fumigation. November 2008 update. http://www.cdpr.ca.gov/docs/dept/factshts/voc_rules_11_08.pdf.

Dimitri, C., Greene, C. 2002. *Recent growth pattern in the U.S. organic foods market.* Washington, DC: U.S. Department of Agriculture Economic Research Service.

Farman, J.C., Gardiner, B.G., Shanklin, J.D. 1985. Large losses of total ozone in Antarctica reveal seasonal Clox/Nox interaction. *Nature* 315: 207–210.

Finman, H. 2009. Methyl bromide critical use nomination for preplant soil use for strawberries grown for fruit in open fields. http://www.epa.gov/ozone/mbr/CUN2011/CUN2011Strawberry.pdf. [Submitted in 2009 for 2011 use season]

Gliessman, S.R. 1989. Allelopathy and agricultural sustainability. In *Phytochemical ecology: Allelochemicals, mycotoxins, and insect pheromones and allomones*, ed. C.H. Chou and G.R. Waller, 69–80. Taipai, Taiwan: Institute of Botany.

Gliessman, S.R., Werner, M.R., Allison J., Cochran J. 1996a. A comparison of strawberry plant development and yield under organic and conventional management on the central California coast. *Biological Agriculture and Horticulture* 12:327–38.

Gliessman, S.R., Werner, M.R., Swezey, S.L., Casswell, E., Cochran, J., Rosado-May, F. 1996b. Conversion to organic strawberry management: Changes in ecological processes. *California Agriculture* 50:24–31.

Guthman, J. 2000. Raising organic: An agro-ecological assessment of grower practices in California. *Agriculture and Human Values* 17:257–66.

Klonsky, K., Richter, K. 2005. *Statistical review of California's organic agriculture 1998–2003*. Agricultural Issues Center, University of California, Davis. http://aic.ucdavis.edu/research/StatisticalReview98-03f8.pdf.

Klonsky, K., Richter, K. 2007. *Statistical review of California's organic agriculture 2000–2005*. Agricultural Issues Center, University of California, Davis. http://aic.ucdavis.edu/publications/Statistical_Review_00-05.pdf.

Martin, F.N., Bull, C.T. 2002. Biological approaches for control of root pathogens of strawberry. *Phytopathology* 92:1356–62.

Monterey County Agricultural Commissioner. 2008. Monterey County crop report 2007. http://www.co.monterey.ca.us/ag/pdfs/CropReport2007.pdf.

Muramoto, J., Gliessman, S.R., Koike, S.T., Schmida, D., Stephens, R., Swezey, S.L. 2006. *Maintaining agroecosystem health in an organic strawberry/vegetable rotation system (Part 5): The final result.* Annual Meeting Abstracts. Indianapolis, IN: Agronomy Society of America, Crop Science Society of America, and Soil Science Society of America.

Muramoto, J., Gliessman S.R., Schmida, D., Stephens, R., Shennan, C., Swezey, S.L. 2004. Nitrogen dynamics in an organic strawberry production system, California organic production and farming in the new millennium: A research symposium. University of California Sustainable Agriculture Research and Education Program (UC-SAREP). http://www.sarep.ucdavis.edu/organic/CCBCposters.pdf.

Oberholtzer, L., Dimitri, C., Greene, C. 2005. *Price premiums hold on as U.S. organic produce market expands.* Electric outlook report from the Economic Research Service, USDA-ERS. http://www.ers.usda.gov/publications/vgs/may05/VGS30801/VGS30801.pdf.

Pickett, C.H., Swezey, S.L., Nieto, D.J., Bryer, J.A., Erlandson, M., Goulet, H., Schwartz, M.D. 2009. Colonization and establishment of *Peristenus relictus* (Hymenoptera: Braconidae) for control of *Lygus* spp. (Hemiptera: Miridae) in strawberries on the California central coast. *Biological Control* 49:27–37.

Porter, I.J., Mattner, S.W. 2002. Non-chemical alternatives to methylbromide for soil treatment in strawberry production. In *Proceedings of International Conference on Alternatives to Methyl Bromide*, Sevilla, Spain, March 5–8 2002, pp. 39–48. http://europa.eu.int/comm/environment/ozone/conference/.

Porter, I.J., Pizano, M., Besri, M. 2007. Impact of the Montreal Protocol on preplant soil uses of MB and trends in adoption of alternatives. Paper presented at 2007 Annual International Research Conference on Methyl Bromide Alternatives and Emissions Reductions. http://mbao.org/2007/PDF/Preplant/PP12/Porter(65).pdf.

Pritts, M.P., Handley, D., eds. 1998. *Strawberry production guide for the Northeast, Midwest, and eastern Canada*. Ithaca, NY: Natural Resource, Agriculture, and Engineering Service (NRAES).

Sances, F.V. 1982. Spider mites can reduce strawberry yields. *California Agriculture* 36:6–9.

Sances, F.V. 2005. Ten years of methyl bromide alternatives research and development: Lessons learned. *Proceedings of the 2005 Annual Research Conference on Methyl Bromide Alternatives and Emissions Reduction* 18:1–2. http://mbao.org/2005/05Proceedings/018SancesF%20mbao%202005%20abstract%208-31-05.pdf.

Sances, F.V., Ingham, E.R. 1997. Conventional and organic alternatives to methyl bromide on California strawberries. *Compost Science and Utilization* 5:23–37.

Santa Cruz County Agricultural Commissioner. 2008. Santa Cruz County crop report 2007. http://www.agdept.com/content/cropreport_07.pdf.

Scialabba, N.E. 2005. Global trends in organic agriculture markets and countries' demand for FAO assistance. Paper presented at Global Learning Opportunity—International Farming Systems Association, Roundtable: Organic Agriculture, Rome, November 1, 2005.

Shaw, D.V., Larson, K.D. 2001. Relative performance of strawberry genotypes over four cycles of cultivation on fumigated and nonfumigated soils. *Journal of American Society for Horticultural Science* 126:78–82.

Subbarao, K.V., Kabir, Z., Martin, F.N., Koike, S.T. 2007. Management of soilborne diseases in strawberry using vegetable rotations. *Plant Disease* 91:964–972.

Svensson, B. 1997. Integrated production, IP, of strawberries in Sweden. *Vaxtskyddsnotiser* 61:56–59.

Swezey, S.L., Nieto, D., Bryer, J. 2007. Control of western tarnished plant bug, *Lygus hesperus*, Knight (Hemiptera: Miridae) in California organic strawberries using alfalfa trap crops and tractor-mounted vacuums. *Environmental Entomology* 36:1457–65.

Thompson, J.E. 2004. Methyl bromide critical use nomination for preplant soil use for strawberries grown for fruit in open fields on plastic tarps. http://www.epa.gov/ozone/mbr/2004_USStrawberryFruit.pdf.

Tourte, L., Klonsky, K.M. 1998. *Statistical review of California agriculture 1992–1995*. Davis: University of California Agricultural Issue Center.

Trout, T. 2005. Fumigant use in California. *Proceedings of the 2005 Annual Research Conference on Methyl Bromide Alternatives and Emissions Reduction* 12: 1–5. http://mbao.org/2005/05Proceedings/012TroutT%20mb-fumuse-05.pdf.

United Nations Environment Programme. 1997. Report of the ninth meeting of the parties to the Montreal Protocol on substances that deplete the ozone layer. http://ozone.unep.org/Meeting_Documents/mop/09mop/9mop-12.e.pdf.

USDA Agricultural Marketing Service. 2008. Wholesale fruit prices, Boston and San Francisco, 2007. Organic farmgate and wholesale prices. Dataset. http://www.ers.usda.gov/data/OrganicPrices/.

USDA National Agricultural Statisitics Services. 2007. Agricultural chemical usage 2006 vegetables summary. http://usda.mannlib.cornell.edu/usda/current/AgriChemUsVeg/AgriChemUsVeg-07-25-2007_revision.pdf.

Warmerdam, M. 2008. DPR sees opportunity to clear the air. http://www.cdpr.ca.gov/docs/pressrls/2008/dir_comment_8-08.pdf.

Wilhelm, S., Paulus, A.O. 1980. How soil fumigation benefits the California strawberry industry. *Plant Disease* 64:264–70.

Xiao, C.L., Subbarao, K.V., Schulbach, K.F., Koike, S.T. 1998. Effects of crop rotation and irrigation on *Verticillium dahliae* microsclerotia in soil and wilt in cauliflower. *Phytopathology* 88:1046–55.

7 Ontario, Canada
Lessons in Sustainability from Organic Farmers

E. Ann Clark and Jennifer Sumner

CONTENTS

7.1   INTRODUCTION

To discuss sustainable agriculture, we must first engage with the concept of sustainability itself. A clear definition of this contested term will help us to understand what is involved when making the conversion to sustainable agriculture.

While some understand sustainability as maintaining economic growth, others take a broader view. Given that sustainability would not even be an issue without perceived threats to ourselves and to the environment, any understanding of sustainability must move us beyond short-term profitability in ways that are environmentally sensitive, socially inclusive, and economically constructive. Philosopher John McMurtry (2003) approaches the concept of sustainability through what he terms "life capital," which he bases on the generic sense of capital as "any goods or wealth which produce more goods or wealth." Thus, life capital is "life-wealth that produces more wealth not just by sustaining it, but by 'value-adding' to it through providing more and better life goods." Life goods include such means of life as breathable air, nutritious food, clean water, adequate shelter, health care, and healthy ecosystems. He links life capital to what he calls "the real economy," which he describes as the production and equitable distribution of life goods otherwise in short supply, as opposed to its truncated financial subsystem, the money economy, whose growth imperative attacks the real economy at various levels. According to McMurtry, life capital subsumes all other forms of capital, even money capital, as long as it is steered or regulated to produce and distribute goods that enable rather than disable life systems.

If we understand sustainability as maintaining and enhancing life capital, then sustainable agriculture becomes a form of agriculture that maintains and enhances life capital, including ecological, social, and economic capital. Making the conversion to sustainable agriculture involves changing to a life values perspective and working to maintain or enhance ecological, social, and economic capital in ways that do not exploit nature, people, or the real economy.

Opinions differ as to whether organic farming should be considered a form of sustainable agriculture. Some would agree with Lampkin (1994), who argues that sustainability "lies at the heart of organic agriculture," while others might prefer the skepticism of Guthman (2004), who warns that organic farming in California (as an example) is farther away from sustainability than many people might think. Defining sustainability as maintaining and enhancing life capital, we ask in this chapter whether organic farmers in the province of Ontario are on the road to sustainability or whether they are being diverted onto other paths.

We start with an overview of the type, magnitude, and distribution of agriculture in Ontario to establish context, and then contrast the factors limiting progress toward sustainability by both institutions and organic farmers, acknowledging ecological, social, and economic dimensions. We then review farmer experiences with the process of conversion from conventional practices and analyze some of the lessons learned from the process. We finish with consideration of possible indicators for ecological, social, and economic sustainability, and add some synthetic concluding comments.

7.2 PRIMARY AGROECOSYSTEMS IN ONTARIO

Canada is divided into 15 terrestrial ecozones, which are subdivided into 53 ecoprovinces, and again into 194 ecoregions (Marshall and Schut, 1999). Most of Ontario agriculture occurs in four ecoregions within the Mixed Wood Plains ecozone (Figure 7.1.), although agricultural activity extends into the Boreal Shield ecozone.

Ontario agriculture is concentrated largely in an ecozone that supports (or supported) trees as the climax vegetation, with mean annual temperatures ranging from 5 to 8°C and annual precipitation in the 800 to 1,000 mm range. Cropping or mixed farming accounts for half to two-thirds of land use in the primary areas, with soil parent material being either carbonaceous (in the southern and western regions) or granitic (in much of the central, eastern, and northern regions) (Table 7.1).

In a recent Census of Agriculture (2001), Ontario had almost 60,000 census farms (grossing in excess of $2,500 per farm per year), with an average farm size of 91 ha (OMAF, 2002). Total land under annual crops was 2.6 million ha, with another 1.0

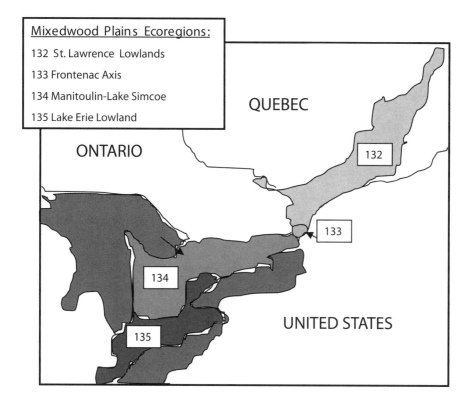

Mixedwood Plains Ecoregions:

132 St. Lawrence Lowlands

133 Frontenac Axis

134 Manitoulin-Lake Simcoe

135 Lake Erie Lowland

QUEBEC

ONTARIO

132

133

134

135

UNITED STATES

FIGURE 7.1   Ecoregions within the Mixed Wood Plains ecozone of Ontario.

and 0.85 million ha under hay and pasture, respectively. Total greenhouse area was 8.85 million m^2.

Farm cash receipts from Ontario crops totaled $3.7 billion in 2003, with a fairly even split between field crops, floriculture/nursery, and vegetable crops (Figure 7.2) (OMAF, 2004a). In the same year, the livestock industry earned $4.2 billion, which was divided among dairy, beef, hogs, and poultry (Figure 7.3).

Ontario is divided into five agricultural regions, with most agriculture in the intensively farmed southern and western regions, which are encompassed by Ecoregions 134 and 135 (Table 7.1). Among regions, on-farm income in 2001 ranged from $7,833 to $14,544 per operator, and was highest in the eastern dairy region (Figure 7.4), corresponding to Ecoregion 132. With the exception of the eastern region, off-farm income was at least twice as high as on-farm income for Ontario farmers in 2001, consistent with Canada as a whole (Martz, 2004).

Ontario agriculture is stratified, with the southern and western regions supporting most of the dairy and hog operations as well as much of the annual cropping of both field and horticultural crops (OMAF, 2004b, 2004c, 2004d, 2004e). The intensive grape, peach, and tender fruit operations on the Niagara Peninsula as well as the densely developed greenhouse industry around Leamington are all within the southern and western regions. A second dairy cluster is found in several counties in the eastern region. Beef is found primarily on grass-based and confinement operations

TABLE 7.1

Climatic and Edaphic Descriptors for Agricultural Land in Ontario, According to the National Ecological Framework for Canada

Ecoprovince	Ecoregion	Climate	Mean Annual T, °C	Mean Summer, °C	Mean Winter, °C	Mean Annual ppt, mm	Dominant Land Uses and Major Urban Centers	Climax Vegetation	Dominant Soil Types
Mixed Wood Ecozone (No. 6)									
Huron Erie Plains	Lake Erie Lowland (135)	Humid, warm to hot summers; Mild, snowy winters	8	18	−2.5	750–900, evenly distributed	65% cropped to corn, soy, tobacco, and tender fruit; Toronto, Hamilton, St. Catharines, Niagara Falls, and Windsor	Sugar maple, beech, white and red oak, shagbark hickory, black walnut, butternut	Carbonate-rich, Paleozoic bedrock with deep glacial deposits; or fine-textured lacustrine deposits
Great Lakes-St. Lawrence Lowlands	Manitoulin-Lake Simcoe (134)	Warm summers; Mild winters	6	16.5	−4.5	750–1000, evenly distributed	56% in mixed farms, dairy, and cash crops as small grains, corn, soy, hay, and fruit; Kingston, Peterborough, Kitchener-Waterloo, Barrie, and Stratford	Sugar maple, beech, eastern hemlock, red oak, and basswood	Carbonate-rich, Paleozoic bedrock with deep glacial deposits
	St. Lawrence Lowlands (132)	Warm summers; Cold, snowy winters	5	16.5	−7	800–1000	60% intensively farmed with corn, dairy, and mixed farming; Brockville and Ottawa, and several in Quebec	Sugar maple, yellow birch, eastern hemlock, and eastern white pine	Low-lying flat, Paleozoic strata on Canadian Shield, gleysolic soils on poorly drained, clayey deposits
Boreal Shield Ecozone (No. 8)									
Southern Boreal Shield	Algonquin Lake Nipissing (98)	Warm summers; Cold winters (humid, cool temperate)	3.5	15.5	−8.5	900–1000	Forestry, mining, and tourism; low intensity farming, primarily grazing accounts for about 5% of the area; Elliot Lake, Huntsville	Sugar maple, yellow birch, eastern hemlock, and eastern white pine	Massive, crystalline, acidic, Archean bedrock; broad-sloping uplands and lowlands; strongly glaciated; ridged to hummocky rock outcrops

Source: Adapted from Marshall and Schut (1999).

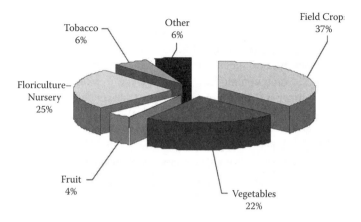

FIGURE 7.2   Distribution of Ontario farm cash receipts from crops ($3.7 billion), 2003.

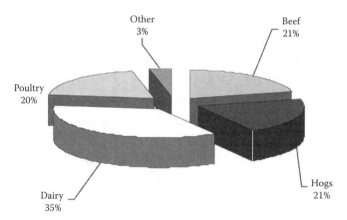

FIGURE 7.3   Distribution of Ontario farm cash receipts from livestock ($4.2 billion), 2003.

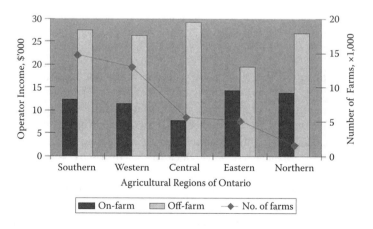

FIGURE 7.4   Regional distribution of Ontario farms, on- versus off-farm income, 2001.

in the western, central, and eastern regions. The sparser agriculture of the northern region emphasizes beef and some field crops concentrated in the Lesser Clay Belt north of New Liskeard, corresponding to Ecoregion 98.

Most Ontario farms are specialized, although degree of concentration varies among enterprises. Some commodities are dispersed over many farms while others are grown by comparatively few. Enterprises such as beef cattle, small grains, and dairy account for at least 51% of total potential receipts on 25, 23, and 12% of Ontario farms, respectively (OMAF, 2002). Conversely, vegetables, fruit, poultry, and pigs account for at least 51% of total potential receipts on just 2.2, 3.1, 2.9, and 4.5% of farms, respectively. Livestock combinations or other combinations account for just 2.9 and 2.3% of farms. Most dairy, beef, and pig farmers also grow some or all of their feed requirements.

In sum, most of the sown land and income from Ontario agriculture is concentrated in the southernmost regions bounded by Lakes Huron, Erie, and Ontario. The relatively benign growing conditions and favorable soil support a diverse range of field, horticultural, and livestock enterprises, including apples, tender fruit, grapes, and greenhouse crops. Nonetheless, for myriad reasons common to the rest of Canada and many other countries, this modern, highly specialized, $8.5 billion industry no longer yields sufficient net income to support full-time employment for most Ontario farmers.

7.3 PRIMARY FACTORS LIMITING SUSTAINABILITY IN ONTARIO

In addition to being unprofitable, Ontario agriculture is also based on practices that have proven to be ecologically unsound and socially unjust. In an assessment of institutional efforts to promote sustainability in the Great Lakes Basin, the federal commissioner of the environment and sustainable development concluded:

> Current farming practices are not sustainable. In spite of conservation efforts, close to half of Ontario's agricultural soil is at risk of washing away faster than new soil can form. Livestock operations in Ontario and Quebec—often "factory farms"—generate manure equal to the sewage of 100 million people…. Ottawa is not working effectively with the provinces to manage the problem, nor has it any formal plan in place. (Gelinas, 2001, p. 7)

Constraints to sustainability or, as we would term it, building life capital, occur in the ecological, social, and economic spheres. Some constraints are universal, because production systems that are not designed to be fully accountable for costs of production necessarily evolve practices that are unmindful of sustainability. In Ontario, as elsewhere, agriculture continues to be vulnerable to consolidation and to the tensions of the export economy, which further threatens both the social and economic sustainability of rural communities.

In the present section, we contrast the factors constraining both institutional and organic approaches to sustainability, and focus the remainder of the section on constraints to ecological, social, and economic sustainability within the organic farming sector.

7.3.1 Institutional versus Organic Approaches to Sustainability

To speak of constraints to sustainability presumes a shared understanding not simply of what sustainability is, but also that the goal of enhancing sustainability is sufficiently important to justify the effort to change. The primary constraint to institutional efforts to promote sustainability is that neither of these preconditions is fulfilled.

A case in point is the voluntary Ontario Environmental Farm Plan (EFP), which is widely cited as a successful, farmer-based, and participatory initiative to enhance on-farm sustainability. It was started in 1993 by the Ontario Farm Environmental Coalition, composed of four mainstream agriculture groups,* which intentionally excluded the Ecological Farmers Association of Ontario (EFAO) (Grudens-Schuck, 2004). The roughly $75 million,† 11-year program was federally funded, but was delivered by the Ontario Soil and Crop Improvement Association together with the Ontario Ministry of Agriculture and Food (OMAF). According to Grudens-Schuck (2004), the unambiguous if unstated agenda was to forestall societal demands for genuine change. Direction and power were retained by the leadership of the coalition, which was "composed of the same men and women who took defensive action against environmental initiatives in years prior."

The criterion of success for the EFP program was the head count at meetings and the number of grants awarded. A recent update for this program announced that 27,000 people had attended workshops, with 11,500 producers receiving $15 million over the past 11 years to implement EFP Action Plan initiatives on their farms.‡

Absent from the assessment was evidence of actual environmental improvement. Consistent with the lack of shared understanding of sustainability issues, environmental monitoring to gauge improvement was intentionally excluded from the EFP program. Yet, the Ontario EFP is nonetheless considered such a success that completing an EFP is a precondition for eligibility for some other Ontario programs, and was the model for a similar Alberta EFP.§

Actually improving sustainability through voluntary institutional initiatives such as the EFP was compromised by both the unstated intent of program developers and its failure to attract participation by more than a minority of farmers—many of whom were already careful stewards of the land. The voluntary nature of such programs also allowed major polluters to continue business as usual, leading to the mandatory Nutrient Management Act of 2002, discussed below.

A report commissioned by one of the four coalition partners sought to understand the motivations for farmers' declining participation in voluntary agri-environmental programs. McCallum (2003) found that more than half had never heard of agri-environmental programs such as the EFP, while 28% cited too many attached conditions,

---

* OFA, Ontario Federation of Agriculture; CFFO, Christian Farmers Federation of Ontario; AGCARE, Agricultural Groups Concerned about Resources and the Environment; OFAC, Ontario Farm Animal Council.

† According to Higgins (1998), the total cost of the program for workshops, training manuals, and grants of up to $1,500 per producer was $8 million per year for 1993–1997, and $5.8 million per year from 1998 onwards, or roughly $75 million to date.

‡ http://www.ontariosoilcrop.org/EFP.htm.

§ http://www.agfoodcouncil.com/serve/aesi02.html.

26% said too much time or paperwork, 13% felt it was not economically worthwhile, and the remainder felt either that they did not want others involved in managing their land or that they were already providing the best possible stewardship. Unless and until there is a shared consensus on both the meaning and importance of sustainable farming practices, institutional initiatives will be ineffectual.

In contrast, organic farmers embody both of the above preconditions—a shared understanding of sustainability and a willingness to bear the associated costs. Organic farming is, by default, a farmer-based and participatory initiative because until very recently, "both levels of government ignored organic farming by effectively excluding it from their definition and funding of sustainable agriculture" (Hall and Mogyorody, 2001). Farmers choosing to adopt organic practices are not responding to the free dinners or "rhetoric of participation" chronicled for the EFP program (Grudens-Scheck, 2004) but rather to heartfelt concerns about personal and environmental health (see Section 7.4). Thus, constraints to sustainability for organic farmers are of an entirely different nature than those affecting institutional initiatives.

Canadian Organic Agriculture

Hill and MacRae (1992) reported that organic agriculture in Canada emerged in the 1950s, gathered strength in the 1960s, and began to consolidate in the 1970s. During the 1980s, the first comprehensive surveys of organic farmers in Canada were carried out, several certification programs were established, some fledgling government support was initiated, and organic farming courses were introduced in some schools. While consumer interest in organics blossomed in the 1990s, institutional support continued to lag. It was not until 2005 that an institution of higher education—the University of Guelph—offered the first academic major in organic agriculture in Canada.

Consumer expenditures on organic products in Canada were $1.3 billion in 2003. Although 85% of organic goods consumed in Canada are imported, most of the organic grains grown in Canada are exported (OMAF, 2003b), with roughly half to the European Union (EU), 40% to the United States, and 5% to Japan (Macey, 2004).

Macey (2004) reported that Canada had 3,134 certified producers in 2003, accounting for 1.3% of all farms. Excluding transitional land, crown land used for range cattle, natural areas on organic farms, and wild lands for maple syrup, Canada's certified organic land base was just over 390,000 ha in 2003. The above excluded land totaled almost 120,000 ha, making a total of just over half a million hectares under organic management (Macey, 2004). In the same year, the certified organic Canadian livestock industry consisted of 15,600 beef cattle on 613 farms, 7,100 dairy cows on 102 dairies, as well as 32 sheep and 23 pig operations. A total of 305,660 meat chickens were produced by 62 farms, with 2.6 million dozen eggs produced on 89 layer operations.

Ontario Organic Agriculture

In 2003, Ontario had 487 certified organic farms or 0.8% of the 60,000 census farms in the province (Macey, 2004). Certified organic land in Ontario totaled 36,861 ha, with an additional 3,000 ha from 41 farms in transition. Unlike conventional operations, organic farms were typically diverse (Table 7.2). Vegetable and fruit farms tended to be small, servicing local farmers' markets, community-supported agriculture (CSA), and natural food stores, with a few large enough to supply the mainstream retail sector. Grain, oilseed, and forage crops accounted for the largest share of both farm number and hectarage sown to organic crops. Both spelt and soy were important cash grains, primarily for the export market. Other grain and forage crops were grown to support the Ontario organic livestock industry, which consisted of at least 47 beef and 46 dairy cattle enterprises, and a few sheep, goat, and pig enterprises (Macey, 2004). A total of 66,340 meat chickens were produced on 23 farms, with 288,000 dozen eggs produced on 25 layer operations. Outlets for sales of organic crops include the 127 farmers' markets in the province, which register gross sales of $500 million per year (Cummings et al., 1999) and, to a lesser extent, CSA, which are roughly estimated to number between 30 and 60 (Lewis, 2004).

TABLE 7.2

Distribution of Enterprises and Land Bases among Certified Organic Farms in Ontario, 2003[a]

Enterprise →	Vegetables	Herbs	Fruit and Nuts	Grain and Oilseed	Pasture and Hay	Clover, Green Manure, and Cover Crops
Farm Operations Reporting, No.	117	11	26+	765 (multiple crops per farm)	302	178
Landbase, ha	531	18	428	22,200 sown 1,000 in wild rice lakes	11,240	1,380
Comments	<1% of provincial total; mostly mixed vegetable or market gardens; 19% are potatoes	25% are Echinacea	1.6% of provincial total; 83% are apples	38% are spelt, 23% are soy, 5–7% each for mixed grain, wheat, and corn	n.a.	n.a.

Source: Adapted from Macey (2004).

[a] These are minimal figures, as not all certifying bodies reported all of the data sought.

We would conclude, therefore, that the primary constraint to sustainability for conventional operations is the lack of shared agreement on intent and necessity among stakeholders. The difficulty of eliciting voluntary participation even in a lengthy program suggests that most Ontario farmers do not prioritize either the need for or the cost of on-farm improvements to sustainability. For this reason, sustainability is also difficult to legislate or mandate, as will be shown below. In contrast, organic farmers are already in agreement with the need for sustainability and for absorbing its costs. The constraints they face to building ecological, social, and economic sustainability are discussed below, drawing from a three-year survey of 12 Ontario organic farmers by Clark and Maitland (2004), a one-year survey of 41 Ontario organic farmers by Sumner (2004, 2005), and a synthetic analysis by Martin (2004).

7.3.2 CONSTRAINTS ON ECOLOGICAL SUSTAINABILITY

Organic farmers are attempting to integrate the principles that sustain nature into remunerative agricultural systems, often without a clear body of technical evidence to guide on-farm practice. Following, in no particular order, are a sampling of the issues raised by surveyed farmers:

- *Need for better understanding of crop-nutrient interactions*—Producers often believe that soil nutrient balance affects crop vulnerability to weed, insect, and disease pests, as well as feed and food quality for livestock and human health. Cation balancing, as per Albrecht, is promoted by some consultants, but evidence in the literature is contradictory. A stronger understanding of soil nutrient impacts on pest dynamics and food and feed quality would give producers the tools they need to forestall pest buildup and promote health.
- *Difficulty accessing appropriate seeds*—Respondents identified several different problems related to seeds and genetics:
  - In the absence of breeding programs dedicated to organics, farmers are obliged to use varieties bred for conventional conditions, which may or may not be appropriate for organic systems. Farm groups or farmers seeking to breed their own organic cultivars need to be mindful of proposed changes to Canadian legislation bearing on ownership of their initial germplasm.
  - Because most field crop seed is routinely treated at the source with biocides, the need to use untreated seed often reduces choice to older or less popular varieties made available through organic suppliers.
  - Particularly for horticultural crops, seed identified as organically grown may be of inconsistent quality.
  - Because seed of non-genetically modified (GM) corn, soy, or canola is no longer guaranteed to be GM-free by the vendors, farmers may be inadvertently contravening the requirements of certified organic practice, which is to use non-GM seed.
  - Seed that is double certified—by both the Canadian Seed Growers Association and an organic certifying body—is expensive and largely unavailable, yet will soon be required to export grain to the EU.
- *Inappropriate and expensive machinery*—Machinery such as tractors and combines is increasingly costly to purchase new, yet used equipment can be unreliable. Martz (2004) reported that the farm input price index for combines and tractors rose by 74 and 61%, respectively, just between 1992 and 2003. These costs compare with a 37% increase for the overall input price index and a barely 10% increase in the product price index between 1981 and 2003 (Martz, 2004). There is need for both local repair shops able to work with older equipment and experiential opportunities to teach novice farmers the skills for both purchase and repair. Conventional machinery may also be ill-adapted to smaller-scale horticultural farms, necessitating either custom retooling or reliance on equipment dating from an earlier era.
- *Difficulty getting manure/compost from certified herds*—Organic horticultural and mixed grain–livestock producers rely on composted livestock manure as a soil amendment. Stockless horticultural farms, which are the norm, commonly import manure from neighbors. The requirement of some certifiers for manure from certified organic livestock is a logistic and economic challenge to farms distant from approved sources.
- *Limited options for pest control*—Pest control on organic farms is primarily prophylactic in nature. Options for control should pest outbreak occur

are often limited. Weed management through rotation and tillage is vulnerable to weather-induced delays. Delay in a planned operation can result in significant and uncontrollable weed pressure. Innovation is needed in implements and other practices, such as mulching or compost teas, to afford producers' fallback options.

- *Lack of options for worm control*—Controlling intestinal worms in sheep or young dairy cattle without using synthetic antihelminthics is a significant challenge. Shepherds consider this the single greatest limitation to organic production. Dairy producers routinely keep their young stock off of grass until six months of age, specifically to avoid worms. Few Ontario shepherds carry enough cattle to alternate pastures, as is the custom among New Zealand organic producers. Better understanding of how to strengthen animal response to the challenge of intestinal worms is urgently needed.
- *Inputs difficult to access*—Some organic farmers have difficulty accessing the products they need to comply with certification. Local farm supply depots may not carry organic inputs, which then must be sourced from farther away. Examples include potting media for greenhouse production and nutrient sources such as green sand or kelp.

Among the ecological concerns identified by Martin (2004) were the following:

- *Need for more systems thinking*—Acquiring the ability to adopt a true systems approach to farming instead of simply replacing synthetic with organic products is difficult, particularly for transitioning farmers.
- *Need for better access to approved products*—Biocontrol products, such as *Bacillus thuringiensis* (Bt), *Beauvaria bassiana*, *Gliocladium catenulatum*, neem oil, kaolin clay, acetic acid, corn gluten, garlic oil, and sulfur, may be beneficial for pest control, but have not yet been registered by the federal Pest Management Review Agency. Registration is an expensive and time-consuming process, and companies marketing these products have declined to pursue it due to small market size.

As may be judged from the foregoing, achieving ecological sustainability is constrained by everything from the novelty of mastering systems thinking to finding accessible suppliers of permitted products. Many of the issues confounding organic practitioners reflect the relative youth of the industry and may be resolved with experience and growth. However, some questions arising from organic systems may also be outside conventional experience, suggesting the need for targeted research and extension.

7.3.3   Constraints on Social Sustainability

The social hurdles that need to be overcome by organic farmers are shared by any group that challenges the status quo. However, many of these issues have become problematic only in recent years, as the scale and success of organic farming lifted it from a scattered fringe element to a vital component of rural communities.

- *Lack of support for organic agriculture*—In the province of Ontario, encouragement and even awareness is lacking at the level of institutions, communities, and banks.
  - Farm policies, whether for crop insurance or for nutrient management, do not yet acknowledge the distinct features of organic practice. Ontario has declined to provide any financial incentives for organic farmers, unlike the EU, the United States, or Quebec.
  - "Public good" research to address the unique needs of organic practitioners—from management practices to targeted crop and livestock breeding to market development—is just beginning to receive research funding. At the national level, funding for the Organic Agriculture Centre of Canada* (OACC) at the Nova Scotia Agricultural College accounts for most of what has been allocated to date. The Ontario Ministry of Agriculture and Food has designated an Organic Crop Production Program Lead to facilitate both research and extension efforts, and organics has begun to receive a small fraction of provincial research funding.
  - Only recently can organic farmers find educational opportunities focusing on their needs in Ontario colleges and universities. Ontario's University of Guelph was the first Canadian university to offer an academic major in organic agriculture.† Guelph's sister campuses at Kemptville and Alfred, which grant two-year diplomas, also offer organic courses.
  - On the level of rural communities, organic farmers may feel isolated and marginalized, and may also have difficulty buffering their crops from spray drift and GM pollen flow from neighboring farmers. Derisive labels, as "quacks," "hippies," or "hobby farmers," can be debilitating. Some have trouble overcoming their pride in clean, weed-free fields, afraid that others will judge them as poor farmers if weeds appear in their organic fields.
  - Bankers have not been sympathetic to loans for small or organic enterprises, which has retarded growth and expansion.
- *Lack of skills and training*—Novice farmers, as well as seasoned conventional farmers, may lack some technical skills important in organic farming. The absence of institutional support for organic training, coupled with the ongoing loss of farming skills through the industrialization of agriculture, can leave few options for those converting to organic.
- *High cost of certification*—Organic farmers are certified annually, necessitating not simply payment of a fee but keeping all the auditable information needed by the inspector. A common complaint is the prohibitive cost of maintaining the complicated paper trail needed for certification. Some experienced organic farmers are foregoing certification entirely, because their reputation

---

* http://www.organicagcentre.ca.
† www.organicag.uoguelph.ca.

is already established and the cost and complexity of certification is not worth the trouble. The efforts of the International Federation of Organic Agriculture Movements (IFOAM) in the EU to come up with a single, straightforward approach to certification are instructional for us as well.

In summarizing the state of the organic industry in Ontario, Martin (2004) identified several areas of social concern:

- Public confidence in the integrity of the term *organic* may be eroded by the confusing presence of competing labels, such as "pesticide free," "free range," "natural," etc., supporting the need for mandatory certification.
- Systematically collected public sector statistics and market intelligence are needed to formulate policy and to track price trends for the organic sector. Data currently available depend largely on tabulations by a volunteer at the Canadian Organic Growers.
- Focus is needed to build on existing value chain attributes of organic, and to capitalize on consumer-driven demand for "locally grown" product— despite the preference of retail and distributor links for imported product.

In summary, a range of social issues have arisen as organic farming approaches the mainstream. As the number of practitioners and the presence of organic foods in mainstream stores increase, so too, does the need for nuanced policies, informed bankers, and targeted research, teaching, and extension efforts. To respond effectively to this growing industry, government will require systematically collected data and thoughtful analysis. Both Canada and the United States now have a system of mandatory labeling in place to support consumer confidence in organic products.

7.3.4 Constraints on Economic Sustainability

Organic farmers face a variety of challenges to making their operations economically viable. Novice farmers may have a particularly difficult time of it because the high start-up costs related to land and equipment may be prohibitive. Both surveys found that while cash crop or mixed crop–livestock farmers tended to have lengthy experience in conventional agriculture before transitioning to organic, horticultural and pasture farmers often had no prior experience before entering organic farming. In general, both farmers without prior farming experience and those transitioning from conventional production express concern and uncertainty about farm financial issues. They confront the same high costs and low commodity prices as conventional farmers, but bear some additional burdens as well. Unlike most conventional producers, for example, organic farmers often have to take an active role in market creation, making marketing expertise essential to an economically viable organic operation. Yet it is difficult for many producers to master both marketing and production skills.

A major impediment for organic livestock producers is an incompatible infrastructure. Consolidation within the processing sector has not favored smaller, diversified, and decentralized livestock operations. As of April 2004, there were 33 federally inspected abattoirs in the province, which accounted for roughly 90% of the meat produced, with the remaining 10% from 191 smaller, provincially regulated abattoirs distributed in more remote communities (Haines, 2004). Provincially inspected abattoirs, which produce for consumption within the province, have declined by 28% just since 1998–1999. It is proving increasingly difficult for farmers to access abattoirs, egg-handling stations, bakers, or cheesemakers able to handle small-scale production within a realistic travel distance. Finding organically certified facilities further challenges those seeking to integrate livestock into their organic operation.

Perhaps the most serious—but rectifiable—issues are regulatory in nature. Agriculture and agrifood policies designed for conventional producers may inadvertently compromise the organic sector. Whether this is a social or an economic issue is arguable, but the outcome certainly has economic ramifications. Regulatory concerns from farmers ranged from the implications of the Nutrient Management Act of 2002 to restrictions on dairy and poultry production imposed under supply management. Supply management in Canada, which applies only to poultry and dairy production, obliges farmers to purchase a quota, which allows them to produce a specified fraction of total permitted production of a given commodity. While beneficial to medium- to large-scale operations, the price and scale restrictions of quota-based production are prohibitive to small-scale poultry or egg producers, encouraging consolidation and large size.

The Nutrient Management Act of 2002* is designed to manage nutrient-containing materials to enhance protection of the natural environment in Ontario. The impetus for this mandatory act was mounting societal concern about the ecological footprint of factory farming and the disposal of human waste. The fact that these problems had to be addressed with legislation is evidence for the ineffectiveness of voluntary agrienvironmental initiatives such as the EFP.

Mandatory regulations absorb administrative and enforcement staff to prepare or approve nutrient management plans and to issue permits and prohibitions, as well as to inspect for compliance. All aspects of manure management are regulated, from the size, capacity, siting, and construction of buildings intended to store nutrients or confine livestock, to standards for transport equipment and technologies for manure handling, including "temporary outdoor storage." To illustrate the impact of this legislation on organic farming, consider its effects on composting—a form of temporary outdoor storage (OMAF, 2003a).

Ten best management practices (BMPs) are required to calculate how many days of outdoor storage are permitted. In order to qualify for legal outdoor storage, farmers are required to maintain auditable records to demonstrate performance in enough of the 10 categories to support the desired duration of outdoor storage. Not all are required, as meeting all 10 BMP would represent 510 days, and the maximum permitted is 300 days. Each BMP is individually calibrated according to specific ranges, with preference (days permitted) given to drier manure and manure low in N and P content; to manure storage that is positioned at least 3 m from tilled land or more than 50 m from surface water, underlain by finer-textured soils, piled to ensure a perimeter of less than 100 m, fully covered by an anchored rain-shedding tarp from start to finish, at a site used no more than every three years; and to manure managed

---

* http://www.e-laws.gov.on.ca/DBLaws/Statutes/English/02n04_e.htm.

so that material is removed for application between August 15 and October 15 in any given year, and "turned so that every piece of material in the pile is displaced from its former position and mixed or inverted once weekly for the first three weeks, and once monthly after that" (which would be 12 times in the case of a 300-day windrow of compost).

The stated intent of this act is to promote sustainability. Whether the mandating of specific on-farm behaviors—in the absence of shared understanding—promotes or retards sustainability is arguable. As is occurring with abattoirs, millers, cheese-makers, and countless other local processors, the net effect of mandatory, auditable behavior is to impose costs that cascade into scale-dependent advantages for those large enough to absorb the added costs, further promoting consolidation.

McDonough and Braungart (2002) argued that "regulations are an indication of design failure," and that the more one needs to impose regulations "to keep people from killing each other too quickly," the more one should step back and reexamine the design of the system itself. In that regard, the excessive or unsafe manure application targeted by the act could be viewed as a symptom of a dysfunctional design rooted in high-density confinement production systems rather than as a problem in itself. By choosing to regulate the symptom rather than the cause, the Nutrient Management Act of 2002 may in fact be a diversion from what is actually needed to enhance sustainability. If the cost of compliance drives smaller operators out of business, the act will actually prolong and exacerbate the problems it purports to address.

In sum, organic farmers present several challenges to traditional farm economics. The considerable fraction of novices entering organic farming means that many of the financial skill sets and personal contacts presumed for conventional farmers are absent. Novice farmers will not necessarily be inheriting or buying a family farm, and will therefore be in a different financial position. Novice farmers may also be mid-career and bring with them considerable life skills and financial acumen. The scale of farming, reliance on management rather than on capital-intensive inputs, and the more direct producer-consumer linkages sought by many organic horticultural farmers will affirm different capabilities in farmers, as well as make different demands upon both the banking community and government policymakers/regulators.

7.4 EXPERIENCES OF CONVERSION TO ORGANIC PRACTICES

Conversion to organic agriculture is an important step, although not the only step, toward sustainability. While the process of conversion is often gradual, and essentially endless, there are definite catalysts that keep the process moving. Understanding these catalysts and ensuring that they continue can facilitate the conversion process.

Sumner (2004) found that 11 of 41 organic farmers had started out as organic farmers and the remaining 30 had converted from conventional to organic agriculture. The conversion experiences of these farmers provide a rich resource for understanding how fundamental change in farming practice occurs (see Table 7.3).

Only 4 of the 30 converting farmers reported that a particular event was pivotal in making up their minds to convert. The rest reported a web of personal and contextual episodes that eventually pushed them toward a process of conversion.

TABLE 7.3

Percent of Farmers Citing Various Reasons for Converting to Organic Farming

	Hort Farms (n = 2)	Mixed Farms (n = 16)	Pasture Farms (n = 12)	All Farms (n = 30)
Health and pesticide concerns	50	56	83	63
Influence of alternative agriculture advocates	50	44	75	56
Influence of family, friends, or experts	0	69	67	45
Environment, production, and soil concerns	50	38	42	43
Spiritual issues	50	6	0	19
Costs	0	38	17	18
Pivotal event	0	19	8	9

Source: Adapted from Sumner (2004).

The most cited catalyst was a concern about health and pesticides, with 63% of farmers reporting that this category was a factor in their decisions to convert. Regarding health concerns, for example, one dairy farmer explained that his father had died of cancer, and four other nearby dairy farmers had also died of cancer. "It makes you think," he said. A number of farmers reported feeling sick themselves, while some cited animal health problems. Pesticide concerns were sometimes directly linked to health concerns. One farmer said that his sister had died of DDT and lindane poisoning. Another experienced negative symptoms after using 2,4-D, while yet another reported seeing the farm foreman shaking after using fly spray. One farmer felt that some sprays could not be good for soil or people, and another said that he could smell pyrethrum on the vegetables. Other respondents simply expressed relief about no longer using pesticides. For example, one farmer reported being pleased that she no longer had to warn the children against playing with the pink seed, while another said that no chemicals were "a blessing."

The influence of advocates of alternative agriculture, including authors and organizations, ranked second overall as a catalyst to conversion. Fifty-six percent of farmers cited this category, referencing books, magazines, and newspapers, such as *Silent Spring* and *Acres USA*. Others had attended workshops and conferences, participated in seminars, or gone on farm tours. A number of farmers had attended EFAO meetings, while one farmer had taken a more formal route through courses and apprenticeship.

For 45% of respondents, people they knew or had met, including family members, friends and neighbors, current experts in the field, and local organic pioneers, provided the motivation to change. Some people were influenced by how their fathers or grandfathers farmed without chemicals. Others were first alerted by wives concerned with conventional farming practices, and one was influenced by his children.

Some talked to farmers who already farmed organically, and others were taken to EFAO meetings by neighbors. Local pioneers in the field were sources of inspiration for a number of the farmers.

Environmental protection and soil concerns were noted by 43% of farmers as a catalyst to conversion. While all the respondents clearly worked from an environmental consciousness, environmental concerns were not foremost in the minds of all when they recalled the conversion process. Likewise, while the importance of soil health was brought up by virtually every farmer at some time during the interview, not all farmers cited soil concerns as a factor influencing their decision to convert.

Spiritual and economic issues ranked fifth and sixth as catalysts to conversion, with 19 and 18%, respectively, citing these categories. While a number of respondents said that they realized a spiritual calling through the practice of organic agriculture, that calling was not necessarily reported as a catalyst to conversion. Similarly, only 8 out of 30 farmers cited the cost of conventional farming as a catalyst to conversion. This category was expressed in terms of the costs associated with conventional farming, including chemicals or fertilizers, and the problem of receiving less and less (or no) money for crops. One farmer reported, "The chemical companies and banks were taking just about everything I had."

In summary, for most farmers the decision to convert occurs over a period of time and comes as a result of a long process of being exposed to a variety of influences. Catalysts that keep the process moving include ongoing worries about health and pesticide use, exposure to information advocating organic methods or criticizing conventional practices, the influence of family and friends, concerns about the environment and the soil, links between farming practice and spirituality, and the high costs of conventional farming. These catalysts form a matrix of encouragement that needs to be understood in order to ensure the conversion process continues to be dynamic and ongoing.

7.5 LESSONS LEARNED

Farmers are generally conservative in nature, perhaps because of the uncontrollable elements they have to contend with, or because the repercussions of an error in judgment can be severe. Regardless, making fundamental changes toward more sustainable farming takes more courage and independence than might be appreciated by outside observers. A perceived threat or driver has to be powerful indeed to motivate a farm family to fundamentally change not simply their way of doing business, but their stature in the local community, their children's life experiences, and their very way of living. Friends and neighbors may turn away. Children may be ridiculed, creating tension at home. In a community where everyone knows not just everyone else but their parents and grandparents too, farmers risk more than just their livelihood when adopting novel production practices. That so many are doing just that speaks to the gravity of the issues facing farmers today.

It is noteworthy that neither institutions of higher learning nor government policies were mentioned as catalysts or stimulants for sustainability. When considering such a risky proposition, farmers listened to each other and to their own good counsel rather than to academics or government agents. This is a message with many implications. First, when it comes to changing on-farm practice, it must be recognized that change comes from within, and cannot simply be compelled from the outside. Second, it is tempting to ponder how much easier it would have been for farmers had institutional research and extension been willing to support their efforts. But at the same time, the obligation to depend on self and on neighbors for guidance created a resilient, self-reliant, and capable community of farmers. And finally, institutions can still be useful, but resources would be better allocated to develop, refine, and release new varieties, production practices, or system designs, than to convince farmers of the need to change.

The forces that motivated existing organic farmers to convert have not abated, and are arguably stronger with each passing year. Evidence of harm from biocides continues to mount. Organic food is receiving much more, and generally favorable, coverage in the media, and is available in mainstream retail chains as well as specialty stores. The EFAO has existed for more than 25 years, and the Guelph Organic Conference, which annually attracts an audience of 2,000, is approaching 30 years. The number of neighbors available for guidance or mentoring continues to grow, and organic courses are now available not simply at the University of Guelph campuses, but also online through the OACC.

Although economic issues ranked sixth among catalysts for conversion, the farm financial picture continues to worsen. In 2003, for the first time in history, net farm income across Canada was below zero (NFU, 2004). On a national basis, the fraction of farm family income coming from farming decreased from 47% in 1965 to 27% in 2000, with farm debt increasing by 50%, from $27 to $41 billion, just between 1996 and 2001 (Martz, 2004). For some, organic farming is seen as one of the few ways to remain financially viable.

As a key change agent for many organic and ecological farmers, the EFAO has always affirmed the need for farmers to decide for themselves. Rather than proselytizing or "spinning" a story, the EFAO embodies the low-key message "this works for me, and I'm happy to share it with you." Through workshops, farm tours, conferences, and kitchen table discussions, the EFAO affords farmers the space to come to the table in their own good time. This strategy—the opposite of the institutional, numbers-driven EFP—may be central to long-term success, given the many implications of the conversion decision and the necessity for a shared understanding of sustainability to enable genuine change.

Many of the concerns raised by organic farmers today will moderate over time, as the industry matures and government policy adjusts to the distinctive features of organic systems. Some production issues, such as achieving a better understanding of the interactions among crops, soil, livestock, and nutrients, or controlling intestinal parasites in sheep without using antihelminthics, will require new insights from *de novo* research. Learning to think in terms of whole systems rather than their components may take longer, both for researchers and for farmers. For example, while enterprise diversity was the norm on surveyed organic farms, the degree to which enterprises were integrated to effectively capture ecological as well as economic synergies was not clear. The preeminent focus on specialization in recent agricultural history may obscure the benefits from strategic enterprise combinations. Resolving other concerns, such as the pressure to consolidate and specialize, may depend on broader societal changes favoring full cost accounting to remove the chief advantage of large-scale agricultural production.

More problematic than accessing approved inputs or maintaining auditable records may be the site specificity of organic farming. Both field studies and farmer perceptions of problematic weed, insect, and disease pests suggest that different pests are selected for on different farms (Clark and Maitland, 2004). If confirmed on a wider range of farms, this would invalidate the one-size-fits-all "wide recommendation domain" research that supports inputs-driven farming. Site specificity arises because organic farming replaces purchased inputs with design and management decisions that take advantage of the specific location, e.g., its microclimates. Reductions in homogenizing inputs allow natural environmental heterogeneity to reexpress itself. Thus, fundamentally different roles both for farmers and for researchers may be needed to address researchable questions on organic farms.

Education for organic farming may also need to emphasize skill sets that are different from those needed for conventional agriculture, recognizing not simply the broader range of backgrounds and entry points or the substantive differences in farming practice, but also the ability to think holistically and integrate information into site-specific decision making. Novel marketing alternatives, whether direct producer-consumer sales or CSA or cooperatives, will also be needed in the educational portfolio.

In essence, experience to date suggests that a far-reaching matrix of off- and on-farm changes, involving not simply farmers but government and society at large, will be needed to support widespread conversion to organic farming, as well as sustainability. Those seeking to promote sustainability must acknowledge the impact of larger global issues—such as the export-based economy, globalization, and consolidation—on the viability of sustainable practices.

7.6   INDICATORS OF SUSTAINABILITY AMONG ORGANIC FARMS

As noted above, sustainability can be considered to have three main components: ecological, social, and economic. For each component, it is possible to develop sets of place-specific indicators—practices, actions, beliefs, and conditions—that point to the degree to which sustainability has been achieved. Once farms and farmers and farm communities have been assessed using these indicators, one can construct an index for that component that allows comparison of farms or farm types in terms of their progress toward achieving sustainability.

Research conducted during the previous several years has allowed us to develop such indicators for organic farms in Ontario. Field measurements and study of actual managerial practices on 12 farms (Clark and Maitland, 2004) produced data used to develop indicators of ecological sustainability and construct an ecological sustainability index (ESI). Interviews of 41 largely different farmers (Sumner, 2004, 2005) were used to create similar constructs for social and economic sustainability, and what we refer to as the sustainability commitment index (SCI).

In each area of sustainability, assessment involved the assigning of points for each indicator; a mean score on that indicator was calculated for all the farms in each farm category, expressed as a percentage of the possible points. The mean of the percentage scores for the farm category then became the score on the index for that category of farms.

7.6.1   ECOLOGICAL SUSTAINABILITY

The goal of ecological sustainability is to maintain or build biological capital, the living infrastructure from which life perenniates, evolves, and sustains a resilient and adaptive ecosystem. Biological capital in agriculture must acknowledge not just crops and livestock, but the ramifying food web, which cycles nutrients, maintains soil porosity and regulates pest community dynamics. Nature is the best and only standard of true ecological sustainability. The principles that sustain nature can be employed to design ecologically sustainable agriculture.

Nine indicators of ecological sustainability were developed: two for soils, three for biodiversity/biocontrol, two for nutrients, and two for energy use (Table 7.4). The 12 farms surveyed by Clark and Maitland (2004) were assessed using these indicators. Pasture farms proved to be the most ecologically sustainable, garnering an average of 96% of the possible points on the ESI; mixed farms were next at 58%; and horticulture (hort) farms were the least sustainable at 41% (Table 7.5). The trend in ESI was roughly consistent with documented trends in soil organic matter, which averaged 6.3% on pasture farms (n = 5), compared with 3.8 and 4.1% on field crop (n = 39) and hort crop (n = 30) farms, respectively (Clark and Maitland, 2004).

On pasture farms, crop rotations consisted predominantly if not wholly of perennial forages that included N-fixing species. The prevalence of N-fixing perennial forages and livestock on pasture farms supported scores at or near 100% in soil, biodiversity, and nutrient indicators. One tractor pass to spread compost or trim uneaten residue was assumed for the pastures. Pasture farms were least sustainable in travel distance for marketed commodities, which included provincial as well as local endpoints.

Mixed crop–livestock farms were able to achieve intermediate ESI scores because the prominence of perennial forages supported strength in soil, biodiversity, and nutrient indicators. Grain grown on organic farms is either used as on-farm feed or exported, and when exported, is sold either provincially or offshore, typically to Europe.

Although the lowest scoring among the three farm types, hort farms were comparatively strong in some biodiversity and energy use indicators, reflecting both smaller field size and entirely local or provincial sales.

Given that only one of the hort farms had livestock, the presence of forages and grains (e.g., living winter cover, perennial forages, and N-fixing crops) in hort crop rotations was noteworthy. The ecological necessity of livestock for sustainability was suggested by the willingness of even stockless hort farms to import and compost livestock manure and to include in their rotations crops for sale largely or solely as

TABLE 7.4
Indicators of Ecological Sustainability (ESI_c), Defined for Ontario Farms

Indicator	Measurement/Calculation	Rationale and Agronomic Implications
Soil		
1. Living winter cover	Percent of years in a rotation cycle in which winter soil is covered by a living forage or cereal crop	Unvegetated soil is a rarity in nature, where living soil cover is maintained even in winter, whether from the soil seed bank and seed rain or laterally encroaching species. Living winter soil cover resists erosion, absorbs the compactive kinetic energy of rainfall, promotes infiltration, and in both late fall and early spring, absorbs labile nutrients and suppresses weed seed germination.
2. Perenniality	Percent of years in a rotation cycle that is covered by perennial or biennial forages	Perennial and biennial forages withhold land from cultivation, add the OM that sustains soil biota and supports soil structure, promote water infiltration through continuous macropores, and maintain active nutrient absorption before, during, and after the growing season.
Biodiversity		
3. Rotation complexity	Scaled assuming more complex is better, e.g., 10 crops = 100%; 9 = 90%, ..., 1 = 10%	Biodiversity in time and space is the norm in nature. More complex rotations alter pest selection pressure, vary amount, timing, and location of resource demands, and leave different ecological signatures.
4. Surface access	Field edge:area ratio; linear field perimeter (m) divided by field area (ha); >300 = 100%; 250 to 300 = 80%; 150 to 250 = 60%; 125 to 150 = 40%; <125 = 20%	Smaller fields (higher edge:area ratio) facilitate movement of biocontrol agents from surrounding hedgerows and wild lands by maximizing exposure to unmanaged land and minimizing the spatial separation between predator and prey organisms.
5. Margin complexity	Each of the 4 sides of measured fields is scored as follows: wild/woods or ditches = 2; crop or lawn = 1; path or roadway = 0; ratings are summed over the 4 sides and divided by 8 to get percent of maximal biodiversity (8 reflects 4 wild edges = maximal biodiversity)	Unmanaged margins, hedgerows, woodlands, or wild lands adjacent to crop fields are a repository of biodiversity and potential biocontrol agents. Increased farming intensity would be reflected in fewer wild and more managed edges.

Continued

TABLE 7.4 (Continued)
Indicators of Ecological Sustainability, Defined for Ontario Farms

Indicator	Measurement/Calculation	Rationale and Agronomic Implications
	Nutrients	
6. N fixation	Percent of rotation years covered by an N-fixing species; forages and grain legumes (e.g., soy) scored as 1; underseedings of perennial or biennial forages counted as 0.5 y	N is the only macronutrient that can be generated on the farm, and is also a major, limiting nutrient that is exported in marketed commodities. Balanced replenishment, through N fixation, is needed to counterbalance losses.
7. Livestock integration	Three kinds of livestock impact are scored: (1) livestock on farm, (2) livestock compost applied (sourced on- or off-farm), and (3) perennial forages rotated with arable crops. Each is scored as 1, with farms scoring: $1 + 1 + 1 = 3 = 100\%$; 2 of $3 = 67\%$; 1 of $3 = 33\%$; 0 of $3 = 0\%$	Presence of livestock can benefit soil health and whole-farm management.[a] Livestock serve as tools of production, as well as sources of income. Livestock provide economic justification for growing—and receiving the benefits of—perennial forages, accelerate nutrient cycling from crop residue, control weeds, and convert weather-damaged crops to a marketable product. Grazing livestock displace machinery needed to harvest, store, and feed them, and to redistribute the manure, hence reducing dependence on fossil-fuel-powered machinery.
	Energy	
8. Tractor intensity	Inversely related, with 0 tractor passes (as on pasture or from horse-drawn equipment) = 100%; 1 pass = 90%; $2 = 80\%$, ..., 0 passes = 0%; field average = farm aggregate	With few exceptions, tractors are fueled by fossil fuels—a nonrenewable resource. Each tractor pass increases fossil-fuel loading to produce a crop. Frequency of use may also be an indirect indicator of design dysfunction, in that land working opens niches for weed proliferation as well as soil degradation and erosion.
9. Travel distance	Inversely related; grain for on-farm feeding or vegetables for local consumption (e.g., CSA) = 100%; provincial = 50%; export sales = 0; if both local and provincial = 75%; if both provincial and export = 25%	Lengthening food travel distance increases fossil-fuel loading on the marketed crop. Food purchased and consumed within a small foodshed places less demand on nonrenewable resources, releases less greenhouse gas, and retains both the nutrients/energy and profit within the local community.

Source: Adapted from Clark and Maitland (2004).

[a] It should also be acknowledged, however, that livestock-based enterprises can also be associated with higher N leaching losses, due to concentration of urine and manure.

TABLE 7.5

Mean Scores, as a Percentage of Possible Points, for Indicators of Ecological Sustainability and the Ecological Sustainability Index (ESI), by Farm Type

	Hort Farms (n = 4)	Mixed Crop–Livestock Farms (n = 5)	Pasture Farms (n = 3)
Living winter cover	35	66	100
Perenniality	17	39	95
Rotation complexity	70	64	100
Surface access	80	52	100
Margin complexity	46	56	100
N fixation	22	65	100
Livestock integration	33	80	100
Tractor intensity[a]	35	53	90
Travel distance	75	50	75
ESI (points out of 900 possible)	41	58	96
	(368)	(524)	(860)

Source: Adapted from Clark and Maitland (2004).

[a] On measured fields only; up to three hort (carrot, broccoli, and potato) and three grain crops (soy, spring cereal, and winter cereal) were monitored on each farm in each year.

livestock feed. Clark and Maitland (2004) found that surveyed hort farms allocated an average of 60% of their crop years to soil building/pest management crops, leaving just 4 years in 10 for economic returns from hort crops.

While organic farmers generally scored well in employing ecologically sustainable practices, this rating in and of itself does not constitute sustainable agriculture. Sustainable agriculture is more than a slate of allowed inputs and management practices. It encompasses larger ecological, social, and economic aspects that maintain or enhance life capital. To determine if farmers are truly committed to the goals and values that will continue to move them toward sustainability despite existing constraints, we need to ask deeper questions. Is the adoption of ecologically sustainable practices mirrored by engagement with larger ecological issues? Do farmers' concerns move beyond the ecological to include both social and economic sustainability? Do farmers follow the organic philosophy and build life capital or is organics just a way to maximize profit?

To create a framework for measuring these variables, seven indicators of commitment to ecological sustainability were developed. They addressed activities within the farm gate, such as following guidelines on waste and soil management, as well as activities beyond the farm gate, including marketing strategies and contributing to local environmental initiatives (Table 7.6). Together, the seven indicators were the basis for the construction of a sustainability commitment index (Table 7.7).

Applying the SCI to the survey of Ontario organic farmers, Sumner (2004, 2005) found that these farmers are indeed on the road to sustainable agriculture.

TABLE 7.6

Indicators of Commitment to Ecological Sustainability ESC$_c$, Defined for Ontario Organic Farms

Indicator	Rationale and Relation to Building Life Capital
	Within the Farm Gate
1. Follows guidelines on waste management	Controls livestock waste to avoid pollution of groundwater and aquifers (waste management is a serious concern in Ontario)
2. Follows guidelines on soil management	Manages to build healthy soil, which is at the heart of organic farming
	Beyond the Farm Gate
3. Belongs to or supports groups promoting environmental issues	Creates or supports environmental responsibility and promotes environmental action, thus building environmental consciousness and cooperative human agency
4. Sells produce locally	Avoids long distance transport of food, thus reducing pollution associated with air and road transportation
5. Supports local environmental initiatives	Encourages legitimate local environmental action, thus supporting cooperative human agency
6. Speaks to local groups on environmental concerns	Helps to spread environmental consciousness among local people, thus encouraging change to a life values perspective
7. Contacts elected representatives on environmental issues	Helps to convince elected representatives of the importance of environmental issues, thus opening spaces for increased environmental policy

Source: Adapted from Sumner (2004, 2005).

The results were generally consistent with the ratings on the ESI. Horticultural and pasture farms manifested nearly equal SCI ratings (86% of 700 possible points), and these were modestly higher than that of mixed crop–livestock farms (77%) (Table 7.7). Virtually all farmers within all three farm types showed commitment in their on-farm practices, and were generally also strong in showing commitment in their off-farm activities. Between 71 and 100% of farmers were committed to selling locally and supporting local environmental initiatives, and somewhat lower percentages, 24 to 89%, showed commitment in the arenas of speaking up publicly or contacting politicians on environmental issues. It is noteworthy that commitment did not vary significantly with farm type, even though some farm types (e.g., pasture-based systems) appear to lend themselves more readily to ecologically sustainable farm practices.

7.6.2 Social Sustainability

Promoting social sustainability by building social capital can mean many things, as discussed by Wall et al. (1998) and Fine (1999). This chapter allies social capital with socially produced life capital or life wealth. Social capital sustains and adds value to

TABLE 7.7

Mean Scores, as a Percentage of Possible Points, for Indicators of Commitment to Ecological Sustainability and the Sustainability Commitment Index (SCI), by Farm Type

	Hort Farms (n = 9)	Mixed Crop–Livestock Farms (n = 17)	Pasture Farms (n = 15)
Follows guidelines on waste management	100	100	100
Follows guidelines on soil management	100	100	93
Belongs to or supports groups promoting environmental issues	78	100	100
Sells produce locally	89	88	87
Supports local environmental initiatives	78	71	87
Speaks to local groups regarding environmental issues	67	59	80
Contacts elected representatives on environmental issues	89	24	53
Mean SCI (points out of 700 possible)	86	77	86
	(601)	(542)	(600)

Source: Adapted from Sumner (2004, 2005).

life capital by providing more and better life goods. McMurtry (2003) adds to this new meaning when he argues that social capital "produces more social capital in each cycle of reproduction if it is not made stagnant by repressive custom or depleted by a competitive market which selects against cooperative agency" (p. 1).

In the present context, social sustainability is conceived as a condition in which farming is an integral part of a local and regional system that reproduces social capital. Accordingly, movement toward this condition is indicated by farmers engaging in a variety of business, cultural, social, and political activities that build social capital. Eight such activities were identified for Ontario organic farmers (Table 7.8), ranging from supporting local businesses to participating in local elections. These indicators were used to construct a social sustainability index (SSI) (Table 7.9).

Scores on the SSI were similar over farm types, with horticultural farms (73% of total points) scoring moderately better than pasture farms (68%), and pasture farms doing slightly better than mixed farms (66%). Across all farm types, involvement was strong or strongest in support for local businesses and farmers, and weakest for personal participation in panels or protests and for local elections (other than voting). All other indicators reflected roughly similar and intermediate levels of engagement. This pattern of expression roughly paralleled that for the indicators of commitment to ecological sustainability, where farmers were seen to engage most readily in cultural and community activities and to be most reticent to participate in public panels, protests, or elections.

Mean scores for the SSI were slightly lower than for the SCI (Table 7.7); we can interpret this to mean that fewer farmers are engaged in (or committed to) building

TABLE 7.8
Indicators of Social Sustainability, Defined for Ontario Farms

Indicator	Rationale and Relation to Building Life Capital
1. Supports local businesses and farmers	Makes an effort to buy locally, for both farm and household necessities, thus building viable communities
2. Volunteers in the community	Supports those in need, thus increasing access to life goods
3. Supports local cultural events and institutions	Helps to maintain the social fabric, thus building viable communities
4. Engages with local government	Encourages participatory democracy, thus supporting increased accountability to life values
5. Is a member of a local club or organization	Maintains the social fabric, thus building viable communities
6. Demonstrates neighborliness	Builds bonds of mutual help, thus increasing access to life goods
7. Participates in local roundtables, panels, protests	Encourages participatory democracy, thus increasing accountability to life values
8. Participates in local elections	Encourages civic engagement, thus increasing cooperative human agency

Source: Adapted from Sumner (2004, 2005).

TABLE 7.9
Mean Scores, as a Percentage of Possible Points, for Indicators of Social Sustainability and the Social Sustainability Index (SSI), by Farm Type

	Hort Farms (n = 9)	Mixed Crop–Livestock Farms (n = 17)	Pasture Farms (n = 15)
Supports local businesses and farmers	78	100	100
Volunteers in the community	67	82	80
Supports local cultural events and institutions	78	82	67
Engages with local government	89	65	60
Is a member of a local club or organization	78	53	80
Demonstrates neighborliness	89	66	53
Participates in local roundtables, panels, protests	78	53	53
Participates in local elections	11	29	47
Mean SSI (points out of 700 possible)	73 (568)	66 (530)	68 (540)

Source: Adapted from Sumner (2004, 2005).

social capital than are engaged in building ecological capital. It is noteworthy, too, that similar to the findings on commitment to ecological sustainability, there was little in the findings on social sustainability engagement to distinguish horticultural from mixed or pasture farmers.

7.6.3 ECONOMIC SUSTAINABILITY

In its current fetishized form, economic capital (also known as money capital or financial capital) is an unaccountable demand system that increasingly depletes life capital in order to multiply money demand (McMurtry, 2003). For economic sustainability to occur, economic capital must instead be recognized as just one part of the real economy, and it must contribute to the reproduction of life capital. In other words, economic capital must be steered or regulated to produce and distribute goods that enable rather than disable life systems.

In agriculture, moving toward economic sustainability means engaging in economic activities that help build life capital. Indicators of this movement include supporting the local economy over the export economy (i.e., selling within the local or regional foodshed), cutting out the middleman through direct sales, and becoming involved in alternative business and transaction models that retain value within the local community. The seven indicators of economic sustainability developed for Ontario organic farms (Table 7.10) were used to construct an economic sustainability index (EcSI) (Table 7.11).

Scores on the EcSI were similar among farm types, with pasture farms and mixed crop–livestock farms roughly equal at 33 and 34% of possible points, respectively, and hort farms not far behind at 24%. Farm types differed somewhat in areas of strength, however. Mixed farms did the most bartering (65% of possible points), pasture farms were strongest in farm gate sales (40% of possible points), and CSA involvement was most common with hort farms (56%). Among all three farm types, direct sales to local businesses were emphasized over sales at farmers' markets, which were often viewed as being too time consuming.

TABLE 7.10

Indicators of Economic Sustainability, Defined for Ontario Farms

Indicator	Rationale and Relation to Building Life Capital
1. Sells directly to local businesses	Builds local economy
2. Barters produce or services	Offers an alternative economic form
3. Sells at farm gate, farm store, or produce stand	Direct sales of fresh, local produce
4. Owns share in or sells to a cooperative	Offers an alternative economic form
5. Sells to family, friends, or local farmers	Direct sales outside the market
6. Is involved in CSA	Offers an alternative economic form
7. Sells at farmers' market	Direct sales to build the local economy

Source: Adapted from Sumner (2004, 2005).

TABLE 7.11

Mean Scores, as a Percentage of Possible Points, for Indicators of Economic Sustainability and the Economic Sustainability Index (EcSI), by Farm Type

	Hort Farms (n = 9)	Mixed Crop–Livestock Farms (n = 17)	Pasture Farms (n = 15)
Sells directly to local businesses	44	71	53
Barters produce or services	22	65	40
Sells at farm gate, farm store, or produce stand	11	29	40
Owns share in or sells to a cooperative	0	24	53
Sells to family, friends, or local farmers	11	41	27
Is involved in CSA	56	6	0
Sells at farmers' market	22	0	20
Mean EcSI (points out of 700 possible)	24	34	33
	(166)	(236)	(233)

Source: Adapted from Sumner (2004, 2005).

The relatively low scores on the EcSI show that Ontario farmers were less engaged in moving toward economic sustainability than they were in moving toward either ecological or social sustainability. This may suggest that farmers have more readily adopted practices and behaviors conducive to ecological and social sustainability, and that greater effort is needed in market development and transactions that support the economy of the local community. It may also reflect market dominance by large-scale distributors, processors, and retailers.

7.7   CONCLUSIONS

Are Ontario organic farmers on the path to sustainability, or are they being diverted? The multifaceted analysis presented above suggests considerable room for optimism in the areas of ecological and social sustainability, although economic trends are less clear.

Chuck Francis of Nebraska has argued that because *sustainability* is such a good word it has been readily co-opted. The term is often employed to rational-ize decisions favoring specific beneficiaries, such as lobbying for subsidized growth in ethanol plants to give an advantage to corn producers. Aligning sustainability with McMurtry's concept of life capital helps avoid this problem because life capital embodies not growth per se but growth equitably distributed for societal benefit. In the same context, McDonough and Braungart (2002) consider that simply sustaining something is an unworthy goal. They aspire to more uplifting, inspiring, and animat-ing goals, not inconsistent with McMurtry's life capital.

Sustaining and building the life capital of Ontario agriculture is contingent upon several equally important requisites. First, as demonstrated by comparing the insti-tutional EFP program and the grassroots EFAO organization, fundamental change must come from within, from a shared acceptance of both the meaning of sustainability and the cost of making needed changes. Institutional input—in the form of informed policies and regulations, economic incentives, or the commitment of societal resources to conduct research, teach, and extend information—could facilitate the process. However, organics is a consumer-driven movement in Ontario, and it has gotten to where it is today on its own terms. Therefore, it would be prudent for well-intentioned academics and government agencies to cultivate a life capital perspective and adopt the overarching principle "first, do no harm."

Of equal importance will be refocusing such societal interventions as are actually needed, whether in the form of regulations or policies, on actual problems rather than on symptoms of problems. The recurrent tendency to "fix" symptoms rather than problems necessarily prolongs and exacerbates causal problems, benefiting not society, farmers, or the environment, but rather, the purveyors of the fixes. Rather than acknowledging that high-density confinement feeding is an ecologically dysfunctional way to manage livestock, for example, we opt instead to allocate scarce societal resources to produce the transgenic Enviropig™, which excretes less phosphorus (Golovan et al., 2001), and mandate costly and cumbersome nutrient regulations that exert scale-dependent pressures favoring consolidation. If Monsanto's Roundup™ worked at the causal end of weed control, it would put Monsanto out of business in a year. Publicly funded institutions that genuinely seek to build the life capital of Ontario agriculture must channel their interventions to benefit their actual clients—farmers, society, and the environment that sustains us.

The final requisite is societal willingness to employ full cost accounting when setting the price on goods and services. Part of the reason for the higher price of organic produce in the marketplace is that organic farmers intentionally internalize costs of production. Organic horticultural farms sacrifice return from high-value vegetable crops when they integrate lesser-value grain and forage crops into their rotations to build soils and regulate pests. In so doing, they make no demand on society to cover the cost of either degraded soil or biocide-related illness or lost pollinators. Organic dairy producers may be satisfied with a rolling herd average of 8,000 liters, knowing that to push for 10,000 liters or more would invite mastitis and other productivity-related ailments. As such, they do not ask society to absorb the risk of antibiotic-resistant bacteria carrying over into the human food chain. Direct producer-consumer links, as through farmers markets or CSA, reduce food travel distance to tens instead of thousands of kilometers (Pirog and Benjamin, 2003). The CSA farmer does not impose an involuntary bill upon society for the adverse human health implications of the effluents released by routine, long-distance transportation of food and other products* (Davis, 2002; Townsend et al., 2003).

Unless and until full cost accounting becomes the norm, organic farmers will be subject to the same market domination and consolidation pressures that have destroyed not just rural communities but communities in general. In 1991, Member of Parliament Ralph Ferguson stated that "current prices for farm commodities do not allow for sustainable agriculture" in Canada (quoted in Martz, 2004). Fourteen years later, net farm income for Canadian agriculture as a whole is below zero (NFU,

---

* See also http://www.epa.gov/air/urbanair/ozone/hlth.html.

2004), not because farmers are inefficient or incompetent but because no money is left in primary production agriculture (NFU, 2003).

Producing food in ecologically and socially sustainable ways is within the grasp of organic farmers themselves. And as shown by both the field studies and interviews reported above, organic farmers in Ontario have a pretty good handle on it. It is in the sphere of economics that the least progress has been made, and where the greatest threat to sustainability remains.

ACKNOWLEDGMENTS

We take pleasure in applauding the willingness of the many Ontario producers who participated in our surveys. Their patience and forthrightness were much appreciated. The poise and technical expertise of Karen Maitland in collecting and processing the resultant mountain of data are gratefully acknowledged. Appreciation is extended to Hugh Martin, Organic Crop Production Program Lead, Ontario Ministry of Agriculture and Food, for reviewing parts of the manuscript. We are grateful to Dr. John McMurtry for inspiring discussions on sustainability as life capital.

REFERENCES

Clark, E.A., and K. Maitland. 2004. On-farm survey of organic farm practice in Ontario, 2001–2003. Unpublished. Accessible from eaclark@uoguelph.ca.

Cummings, H., G. Kora, and D. Murray. 1999. Farmers' markets in Ontario and their economic impact, 1998. Available at http://www.ofa.on.ca.

Davis, D. 2002. *When smoke ran like water.* New York: Basic Books.

Fine, B. 1999. The development state is dead—Long live social capital? *Development and Change* 30:1–19.

Gelinas, J. 2001. You try living down here. *Globe and Mail,* Toronto, ON, Canada, October 3.

Golovan, S.P., R.G. Meidinger, A. Ajakaiye, M. Cottrill, M.Z. Wiederkehr, D. Barney, C. Plante, J. Pollard, M.Z. Fan, M.A. Hayes, J. Laursen, J.P. Hjorth, R.R. Hacker, J.P. Phillips, and C.W. Forsberg. 2001. Pigs expressing salivary phytase produce low phosphorus manure. *Nature Biotechnology* 19:741–45.

Grudens-Schuck, N. 2004. *The mainstream environmentalist: Learning through participatory education.* Westport, CT: Greenwood Publishing Group.

Guthman, J. 2004. *Agrarian dreams: The paradox of organic farming in California.* Berkeley: University of California Press.

Haines, R.J. 2004. *Farm to fork. A strategy for meat safety in Ontario.* Report of the Meat Regulatory and Inspection Review. Toronto: Queen's Printer for Ontario.

Hall, A., and V. Mogyorody. 2001. Organic farmers in Ontario: An examination of the conventionalization argument. *Sociologia Ruralis* 41:399–422.

Higgins, E. 1998. *Whole farm planning. A survey of North American experiments.* Henry Wallace Institute for Alternative Agriculture Policy Studies Report 9, Greenbelt, MD.

Hill, S.B., and R.J. MacRae. 1992. Organic farming in Canada. *Agriculture, Ecosystems and Environment* 39:71–84.

Lampkin, N.H. 1994. Organic farming: Sustainable agriculture in practice. In *The economics of organic farming: An international perspective,* pp. 3–9, ed. N.H. Lampkin and S. Padel. Wallingford, UK: CAB International.

Lewis, J. 2004. Experience, networks, practice: The form and function of networks among community shared agriculture (CSA) projects in Ontario. Paper presented at the First Annual Conference for Social Research in Organic Agriculture, University of Guelph, Guelph, Ontario, Canada, January 23.

Macey, A. 2004. *Certified organic: The Status of the Canadian organic market in 2003.* Report to Agriculture and Agri-Food Canada Contract 01B6830423 (rev. May 2004). http://www.ota.com/pics /documents/Organic%20Stats %20Report%20r evised%20 May%20 2004.pdf.

Martin, H. 2004. *Report of the Ontario Organic Research Advisory Committee 2004.* Ontario Ministry of Agriculture and Food. Accessible from hugh.martin@omaf.gov.on.ca.

Martz, D.J.F. 2004. *The farmers' share: Compare the share 2004.* St. Peter's College, Muenster, SK: Center for Rural Studies and Enrichment. http://www.stpeterscollege.ca/crse/crse.html

McCallum, C. 2003. Identifying barriers to participation in agri-environmental programs in Ontario. http://www.christianf armers.org/sub2_spec ial_reports/pub_ special_report. html#cmc specreport2003.

McDonough, W., and M. Braungart. 2002. *Cradle to cradle.* New York: North Point Press.

McMurtry, J. 2003. The life capital calculus. Paper presented at the Canadian Association for Ecological Economics Conference (CANSEE), Jasper, Alberta, October 18.

NFU (National Farmers Union). 2003. The farm crisis, bigger farms, and the myths of "competition" and "efficiency." Saskatoon, SK. http://www.nfu.ca/briefs/Myths_PREP_ PDF_TWO.bri.pdf.

NFU (National Farmers Union). 2004. Black Friday for Canadian farmers. Saskatoon, SK. www.learningcha nnel.org/article/view/78794/1/.

OMAF (Ontario Ministry of Agriculture and Food). 2002. Ontario census farms with sales of $2,500 or more, classified by major product type, 1991, 1996, and 2001 (Bill McGee). www.gov.on.ca/OMAFRA/english/stats/census/product.html.

OMAF (Ontario Ministry of Agriculture and Food). 2003a. Construction and siting protocol NSTS-08 temporary field nutrient storage sites for solid manure and other prescribed materials. http://www.gov.o n.ca/OMAFRA/en glish/nm/regs/conpro /conpro08.htm.

OMAF (Ontario Ministry of Agriculture and Food). 2003b. Organic farming in Ontario (Hugh Martin). Agdex 100/10. http://www.gov.on.ca/OMAFRA/english/crops/facts/03-063. htm#growth.

OMAF (Ontario Ministry of Agriculture and Food). 2004a. Farm cash receipts from farming operations, Ontario 1997–2003 ($'000) (John Cummings). http://www.gov.on.ca/ OMAFRA/english/s tats/finance/receipts.html.

OMAF (Ontario Ministry of Agriculture and Food). 2004b. Milk shipments by county of origin to milk processing plants in Ontario, 1998–2003 (kilolitres) (Bill McGee). www. gov.on.ca/OMAFRA/english/stats/dairy.chip03.html.

OMAF (Ontario Ministry of Agriculture and Food). 2004c. Number of cattle, Ontario by county, July 1, 2003 (John Cumming). www.gov.on.ca/OMAFRA/english/stats/livestock/ctycattle03.html.

OMAF (Ontario Ministry of Agriculture and Food). 2004d. Number of pigs, Ontario by county, July 1, 2003 (John Cumming). www.gov.on.ca/OMAFRA/english/stats/livestock/ctypigs03.html.

OMAF (Ontario Ministry of Agriculture and Food). 2004e. Ontario egg production and value, 1998–2003 (John Cumming). www.gov.on.ca/OMAFRA/english/stats/livestock/eggs.html.

Pirog, R., and A. Benjamin, 2003. *Checking the food odometer: Comparing food miles for local vs. conventional produce sales to Iowa institutions.* Leopold Center for Sustainable Agriculture. http://www.le opold.iastate.edu/p ubs/staff/files/food_tra vel072103.pdf.

Sumner, J. 2004. The contributions of organic farmers to rural community sustainability in southwestern Ontario. Unpublished manuscript.

Sumner, J. 2005. Organic farmers and rural development: A research report on the links between organic farmers and community sustainability in southwestern Ontario. www.organicagcentre.ca/ResearchDatabase/res_social_science.html.

Townsend, A.R., R.W. Howarth, F.A. Bazzaz, M.S. Booth, C.C. Cleveland, S.K. Collinge, A.P. Dobson, P.R. Epstein, E.A. Holland, D.R. Keeney, M.A. Mallin, P. Wayne, and A. Wolfe. 2003. Human health effects of a changing global nitrogen cycle. *Frontiers in Ecology and the Environment* 1:240–46.

Wall, E., G. Ferrazzi, and F. Schryer. 1998. Getting the goods on social capital. *Rural Sociology* 63:300–22.

8 Mexico
Perspectives on Organic Production

María del Rocío Romero Lima

CONTENTS

8.1 INTRODUCTION

In Mexico, as elsewhere, it is difficult to point to a particular type of agricultural production system and say with confidence that it is sustainable. The existing production systems generally referred to as *organic*, however, have many of the characteristics required for sustainable systems. Organic agriculture in Mexico has experienced a rapid growth in recent years, which indicates that farmers are actively converting conventional systems (and systems that mix conventional and traditional practices) into systems that are the closest approximations we have of sustainable systems.

Since this chapter will explore the status, history, and future of organic agriculture in Mexico, it is worthwhile to first discuss what farmers, officials, and organizations mean when they talk about organic agriculture. In a general sense, the term *organic agriculture* refers to a form of production that excludes the use of synthetic agricultural chemicals and makes use of a series of practices that favor the recycling of nutrients, the use of natural and biological controls of pests and diseases, and the conservation of natural resources. There are many other related ways of referring to this form of production, such as biodynamic agriculture, natural agriculture, biointensive agriculture, regenerative agriculture, biological agriculture, ecological agriculture, and permaculture. The term *organic* is more common in America, with *ecological* and *biological* being more frequently used in Europe. In addition to referring to a system of production, each of these terms also connotes a certain philosophical orientation toward production and the market, as indicated by Guzmán et al. (2000).

The International Federation of Organic Agriculture Movements (IFOAM), a coordinating body at the world level in organic production, defines organic agriculture as "all farming systems that promote safe and secure production of food and textile fibers from environmental, social, and economic points of view. These systems are grounded in soil fertility as the foundation for good production. They respect the natural demands and capacities of plants, animals, and the landscape, while trying to optimize the quality of agriculture and the environment in all of their aspects." In Cuba, where there has been governmental support and a specific orientation toward food self-sufficiency, organic agriculture is defined as a "productive system with its foundations in agroecology, and which has as a principal proposition the production of safe food, protection of the environment and human health, and the intensification of biological interactions and natural processes" (Pérez, 2003, p. 9).

A definition in Mexico, which was adapted from the Food and Agriculture Organization (FAO) by Goméz et al. (2005) and considers standards as well as some of the aspirations of IFOAM, states that organic agriculture is a

> holistic system for the promotion of food production that foments and improves agro-ecosystem health, and in particular biodiversity, biological cycles, and soil biological activity, using practices that avoid the use of synthetic chemical products such as fertilizers, insecticides, herbicides, hormones, plant and animal growth regulators, as well as genetically modified organisms, sewage water, and synthetic food dyes and preservatives in processed products. In synthesis, it has as a primary objective to obtain safe food with high nutritional quality. Organic production systems are based on precise and specific production standards whose final goal is to achieve optimum agroecosystems that are sustainable from social, ecological, and economic points of view. (p. 11)

In Mexico, the defining of organic agriculture is associated with a perceived need to connect it with standards and certification. For example the Mexican Association of Ecological Farmers (AMAE), founded in 1992, defines organic agriculture as "the art and the science employed to obtain safe farm products using techniques that favor the use of natural sources of soil fertility without the use of contaminating agrichemicals, using a preestablished program of ecological management, which can be certified for all processing steps from the selection of seed through the sale of the product." Similarly, the Standards Commission of ECOMEX (Campesinos y Indígenas Ecológicos de México) calls organic agriculture "the practice and art employed in the production of food that is safe and highly nutritious, using a sustainable management of natural resources, where the production process takes advantage of ecological cycles, free of synthetic pesticides and fertilizers. This agriculture responds to the standards of production and quality, through which they differ from traditional and conventional agriculture."

In February 2006, after a broad movement of diverse actors in the organic sector, the law of organic products was officially published in Mexico; it defines *organic production* as "the system of production and processing of food, animal products and subproducts, vegetables, and other foodstuffs, with the controlled use of external inputs, restricting and in its case prohibiting the use of synthetically produced chemicals" (Diario Oficial, 2006). This definition, although a bit formal, was the product of an agreement between parties with distinct visions of organic agriculture who recognized the necessity that legal boundaries be established that positioned this activity in the official context of agriculture in Mexico, and therefore would be taken into account by the governmental sector.

8.2 STATUS OF ORGANIC PRODUCTION IN MEXICO

Globally, Mexico ranks as the country with the greatest number of organic producers. In addition, Mexico is in first place internationally in the production of organic coffee and the production of organic tropical fruit (Willer and Yussef, 2007). Within the context of Mexican agriculture, the organic sector represents only about 2% of the total agricultural land area, but it is currently the most dynamic sector, having grown considerably in the last several years in area, number of producers, and value generated (see Table 8.1).

The growth of organic production in Mexico has been achieved despite a large number of barriers, which makes its achievements all the more impressive. These barriers (Granados and Lopez, 1996) include the following:

- Higher costs due to the increased need for labor,
- The limited availability of proper organic fertilizers,
- Restricted access to credit among poor farmers,
- Insufficient access to reliable sources of farming information; a lack of specialized research information,
- Limited internal markets due to a lack of recognition in Mexico of organic products and their benefits for health and the environment,
- Problems with marketing approaches and difficulty in getting consumers to overcome their distrust of organic products and their reticence to pay the higher prices, and
- Lack of any state policy that provides incentives for organic production.

TABLE 8.1
Dynamics of Organic Agriculture in Mexico, 1996–2007

	1996	1998	2000	2002	2004–2005	2007
Land area (ha)	23,265	54,457	102,802	215,843	307,692	545,000
Number of producers	13,176	27,914	33,587	53,577	83,174	126,000
Value generated (US$ × 1,000)	34,293	72,000	139,404	215,000	270,503	430,000

Source: Gómez et al. (2005); Schwentesius (2007). (With permission.)

One of the first statistical documentations of organic agriculture in Mexico was presented by Laura Gómez in 1996 in her undergraduate thesis in agroecology at Universidad Autónoma de Chapingo (UACh). Later a group of researchers from CIESTAAM (Centro de Investigaciones Económicas, Sociales y Tecnológicas de la Agroindustria y la Agricultura Mundial; located at UACh) followed up with national statistics for organic agriculture (Gómez et al., 2001, 2005). Some of the data presented in this chapter were taken from their work.

8.2.1 SCALE OF PRODUCTION

To examine trends in the scale of organic production in Mexico, it is helpful to distinguish between small-scale producers and medium-scale producers. Small-scale producers are those growing on less than 30 ha; medium-scale producers farm an area of land 30 or more hectares in size. (Very large-scale producers—generally those with more than 200 ha—are not included in this scheme.) An important characteristic of organic production in Mexico is that it is dominated by small-scale producers. These are mostly indigenous and peasant farmers who work in some organized manner. As can be seen in Table 8.2, the number of small-scale producers has increased dramatically since 1996. During the same period, the number of medium-scale farmers has seen a small net increase, but these farmers have greatly increased the area they have in organic production, as can be seen in Table 8.3.

TABLE 8.2
Number of Certified Organic Growers
in Mexico, 1996 to 2004–2005

Type of Producer	1996	2000	2004–2005
Small	12,847	33,117	80,319
Medium	329	470	345
Total	13,176	33,587	80,664

Source: Gómez et al. (2005). (With permission.)

TABLE 8.3
Organic Area in Mexico by Type of
Farmer, 1996 to 2004–2005 (ha)

Type of Producer	1996	2000	2004–2005
Small	20,705	86,507	233,967
Medium	2,559	16,299	58,491
Total	23,265	102,802	292,459

Source: Gómez et al. (2005). (With permission.)

TABLE 8.4

Surface Area of the Principal Certified Organic Crops in Mexico in 2004–2005 (ha)

Crop	Area	Crop	Area
Coffee	147,136	Maguey (tequila and mezcal agave)	5,943
Vegetables (22 species)	33,416	Prickly pear (pads and fruit)	5,039
Aromatic and medicinal herbs	30,166	Corn	4,530
Cacao (chocolate)	17,313	Avocado	2,652
Grapes	12,032	Sesame	2,497
Coconut	8,400	Mango	2,132

Source: Gómez et al. (2005). (With permission.)

In this differentiation between small- and medium-sized certified organic producers, small producers represent almost 98% of the total number of producers. In 1996 small producers accounted for 89% of the organic area, and for 2004–2005, it is estimated that this figure dropped to about 80%. This shows a tendency for organic production to become more concentrated in the hands of medium-sized certified organic growers. Between 2000 and 2004–2005, the average land area for a medium-sized operation increased from 34.7 to 169.5 ha, whereas the average land area for small growers increased from 2.6 to 2.9 ha.

Medium-sized organic producers are commonly involved in cattle production, which requires a more extensive land area, or with the production of specialty products for export, such as tropical fruit and winter vegetables; in these systems, some production practices are very similar to those in conventional farming because they have mostly employed an input substitution strategy.

8.2.2 CERTIFIED ORGANIC PRODUCTS

Considering the products that are grown, coffee is the most important, with 51.3% of the organic area cultivated. In spite of the dominance of coffee, however, there are more than 200 Mexican organic products. The crops with greatest area are presented in Table 8.4.

Among other crops, there are sábila (aloe vera), citrus, olives, sugar cane, safflower, guava, vanilla, cassava, button and seta mushrooms, apples, pineapple, cashews, bamboo, and small fruits. Some certified organic processed products include honey, milk, cheese, candies, and some cosmetic items.

8.2.3 MARKET AND CERTIFICATION

Eighty-five percent of the organic production in Mexico is exported, with only 15% going to the domestic market. The principal export destinations are the United States, Germany, the Netherlands, Japan, the United Kingdom, and Switzerland.

For the domestic market, stores with organic products and organic areas in super-markets in cities of greater importance are becoming important. Since 2003 local organic markets are appearing in response to an urban demand for safe products and organic markets. There are now more than a dozen local markets in cities in the center and south of Mexico (www.chapingo.mx/ciestaam/to, www.mercadosorganicos.org).

Geographically, organic agriculture—which began in the south of the country—has now spread to the center and north parts of the country. Organic production sites are now found distributed in all parts of the country. Of the producing states, Chiapas, Oaxaca, Michoacan, and Veracruz concentrate 61.7% of the organic area (Gomez et al., 2005).

Organic certification itself is provided by private companies, with 19 companies recognized by Gómez et al. (2005); only one of them is a Mexican business—the Mexican Certifier of Ecological Products and Processes (CERTIMEX, S.C.). This company holds the lead in the number of certified production units and second place in area certified. Other companies that have considerable presence in Mexico are Bioagricert, Organic Crop Improvement Association (OCIA), IMO Control, Naturland, Quality Assurance International (QAI), and Oregon Tilth Certified Organic (OTCO).

8.3   TRANSITION TO ORGANIC PRODUCTION

This chapter begins to explore several important questions about organic production in Mexico: Why are farmers in Mexico undergoing the conversion process? What happens during the process? How far have organic systems come from the point of view of sustainability? To fully answer these questions, of course, more detailed analysis of emerging data will eventually need to be done.

When current experienced organic producers are asked the first of these questions (the why of conversion) they often respond by pointing to the better price they receive for organic products, coupled with their commitment to the environment. Some remark that they are motivated by the better income that comes with a more diversified production system, while others indicate the desire to be more connected to mother earth. There are also those who converted their farms as a response to sickness or poisoning from farm chemicals, and others who shifted due to training or exchange with other farmers.

Overall, farmers have been converting to organic production because of growing recognition of the many advantages of this form of agriculture over conventional production. Among its many benefits, organic production

- Allows farmers to receive higher prices for their products.
- Favors the conservation of water and soil resources.
- Generates safer and healthier food.
- Creates a safer, pesticide-free work environment.
- Generates more and better employment for the community.
- Contributes to local organizational development that provides better access to resources, inputs, and markets.
- Reduces and sometimes eliminates the need for market intermediaries.
- Helps to build a more equitable market with a humanitarian conscience.
- Reduces costs by moving away from synthetic agrochemical inputs.

- Revalidates traditions that form an integral part of the culture of many communities.
- Counteracts the economic forces that have resulted in increased permanent and seasonal migration away from rural areas, which results in strengthening of family relationships and community.
- Offers hope for reaching a dignified quality of life in the present and the future.

8.3.1 History of Organic Production in Mexico

The combination of economic, social, and philosophical motivations discussed above matches what we know about the history of organic production in Mexico. It has been stated that organic agriculture came about in the developed countries of Europe and North America as mostly a response to consumer demand for safe food and a better environment, whereas in Latin America organic agriculture arose as more of a survival strategy for peasant agriculture. In contrast to the emergence and development of organic agriculture in industrialized countries, in Mexico organic agriculture has been more associated with traditional, peasant, and indigenous production.

The economic crisis facing small growers, especially coffee growers, pushed them to diversify their production and look for better markets for their products. They found organic production to be a good niche market, and as they began to make inroads into this type of production and looked for sources of employment and for markets that showed stronger social commitment, personal conviction became an important motivating force as well.

The pioneers in Mexican organic agriculture were the coffee growers in the southeast of the country, who now place Mexico in the top position among producers of organic coffee in the world (Martínez and Peters, 1995). The first organic certification in Mexico—of a private family coffee-producing business in the Soconusco region of Chiapas called Finca Irlanda—occurred in 1967.

Organic production of products other than coffee also has its roots in the south of Mexico, specifically in the states of Oaxaca and Chiapas, where small-scale indigenous growers sought alternatives in the face of official abandonment by the state, cultural segregation, hunger, and the exploitation of their natural resources. In the face of such conditions, social organization and organic production gave them an opportunity to get ahead. From these origins in Chiapas and Oaxaca, organic agriculture in Mexico progressed to the center of the country and then to the north.

Since its beginnings a few decades ago, several important trends have characterized organic agriculture in Mexico. These trends, which reinforce each other, include the following:

- Isolated, individual producers have formed collectives and networks.
- Local-level commitments and organizations have grown to a national scale, a trend that culminated in the passing of the national law of organic products in 2006.
- The initial focus on organic coffee has spread to hundreds of other crops.

- Although the export market still dominates, production for domestic consumption has increased significantly.
- Private certification has increasingly yielded to participatory certification.
- Farming practices once based primarily on traditional methods have increasingly incorporated practices based on scientific research, as farmers have moved from sharing information solely among themselves to actively soliciting information and advice from scientists.
- Conversion based on input substitution has been largely replaced by holistic practices, diversified production, and other system redesign strategies. Today, few organic farmers in Mexico have production systems based solely on input substitution.

Examples are presented below to illustrate these shifts in organic production and show how actual farmers and organizations have confronted the challenges that face them. The production systems incorporated in these examples are in the "agroecological consolidation" phase, as indicated by Gliessman (2002). In this phase, agroecosystems are minimally dependent on external inputs, manage pests and diseases using internal regulation mechanisms, and are capable of recovering from the disturbances caused by cultivation. Their managers have increased the efficiency of cropping practices, used alternative practices, and redesigned their management so that they function on the foundation of ecological processes that go beyond just ecological sustainability. This means that they have influenced regional, national, and even international public policy as demonstrated by the modification of the criteria used for certifying groups of producers by international certifying agencies. They have made inroads in work with solidarity certification, fair trade, participatory research, and the creation of alliances and networks.

8.3.2 The Union of Indigenous Communities of the Isthmus Region

The best known example of how small-scale indigenous organic producers have organized to advance organic agriculture and their own interests is the Union of Indigenous Communities of the Isthmus Region (UCIRI) in Oaxaca. The members of UCIRI have developed both the technical means of organic production and strategies for social organization, commercialization, and promotion of better living conditions for the members of their organization (Santoyo et al., 1995, Hernández, 2001; Ramirez, 2003). The roots of UCIRI go back to 1981, when indigenous peasants of the region, together with a group of priests from the Diocese of Tehuantepec, began to meet regularly to analyze their problems. They decided to form an organization to push for better markets for their coffee, for which they had been receiving an unfair price.

UCIRI was constituted in 1983, with a membership made up of various indigenous groups (*zapotecos*, *mazatecos*, *mixes*, *chontales*, and *chatinos*). They created their own center for peasant education in order to train peasant promoters of organic agriculture and initiate certification with Naturland. In 1988 UCIRI joined with a solidarity group in Holland (Solidaridad) to create the first fair trade seal, Max Havelaar. As an organization committed to sustainability, UCIRI has promoted a range of projects apart from organic production, including a hardware store (Lachinavani, SA de SV), women's projects, technical assistance projects, a savings and credit fund (FAC), a clothes-making venture (Xhiiña Guidxi SCL), and various efforts related to community health, education, communal work, and transport. They focus, however, on the national and international marketing of their organic products, which include coffee, maracuya (passion fruit), and marmalades. Coffee is produced in several roasts and presentations, and is exported to Sweden, Italy, Holland, France, Switzerland, Germany, Canada, the United States, and Japan; they also market their products domestically in Mexico. At the current time, UCIRI has 2,547 members working a total of 11,592 ha.

The organization and work developed by UCIRI has had various impacts, with their organizational model serving as an example for other groups of peasant and indigenous small producers in Mexico. They developed the foundations for forming such national organizations as the National Coordinator of Coffee Grower Organizations (CNOC) in 1989, ECOMEX in 1994, CERTIMEX in 1997, and Fair Trade Mexico (Comercio Justo México) in 1999. As stated by one of their advisors (Vanderhoff, 2005), UCIRI has succeeded in developing a national and international market for organic products, collaborating in environmental improvement, improving the living conditions of the people in the countryside, and most importantly, organizing indigenous producers. Although this has all taken a lot of work, they continue to forge ahead.

8.3.3 OTHER EXAMPLES

Another example of a successful collective in southern Mexico is the group Indigenous People of the Sierra Madre de Motozintla (ISMAM) in Chiapas. ISMAM began the production of organic coffee in the middle of the 1980s with the help of UCIRI. In 1990, Roberto Sánchez López published the *Practical Manual for the Biological Cultivation of Organic Coffee*, one of the first documents about organic agriculture in Mexico, where he describes how ISMAM enabled peasants and technicians to cooperate in the production and commercialization of coffee in ways that overcame the limitations of infrastructure and capital.*

Another organization that deserves mention is the Center for Agroecology of Saint Francis of Asis (CASFA, S.A.), which is also in Chiapas. It came about in 1986 as an outgrowth of the indigenous and peasant movement in the Sierra region of the state. The founders, part of the Commission of Cooperatives of the Diocese of Tapachula, were searching for integrated development alternatives. In 1991 they formally constituted themselves with the goal of stimulating social processes through which local people could develop a consciousness of their current situation and create alternatives that could allow them to overcome the challenges they faced. Applying the principles of cooperativism adapted to the local situation, they formed more than 500 communal work groups under the motto "For organized communal work!" In

---

* Sánchez also participated in the publication of the book *Scientific Fundamentals of Mexican Agroecology*, in which he emphasizes the importance of social organization, teamwork, and technological consolidation of the productive process backed up by science (González et al., 1995).

2003, in cooperation with other organizations, they developed the Maya Network of Organic Organizations, which promotes integrated development for small-scale indigenous farming communities through the practice of agroecology, certification of quality, the incorporation of value, and integration with the market. By prioritizing geographic localization, local crops, and interests of the members, they have created seven programs: coffee production, tropical crops, Mayan vegetables, apiculture, rural tourism, technology, and Bio-Mexico (see www.redmayacasfa.com). The network's motto is *Por un desarrollo integral con raíces profundas* ("For an integrated development with deep roots").

Producers of crops other than coffee have also created successful organizations. In 1986, growers of herbs, spices, vegetables, and fruits in southern Baja California established the Society of Social Solidarity of Organic Producers of Cabo. Seeking to raise the level of family livelihoods of nearly 1,500 peasant families in the region, strengthen their agricultural communities, protect and improve the environment, and promote organic agriculture, they switched to organic production practices on their small family parcels, diversified their systems, and sent their products to the international market (Martínez, 1997; Ceseña, 1997).

There have also been successful efforts to promote organic production for subsistence and domestic consumption. Using methodologies developed by Ecology Action in California, the Mexican Institute of Social Security (IMMS) and later Ecology and Population (ECOPOL) have promoted, since 1984, the development of biointensive organic agriculture (Jeavons, 2002). This kind of agriculture, which is not certified, uses small production units and a high diversity of crops (in association as well as in rotation). Initially begun as a program for home gardens in communities in the northeast of Mexico, biointensive methodology and practices have extended across the country, principally among small growers and for subsistence production. It has made it possible for families to produce food for themselves, their animals, and the soil (see www.growbiointensive.org).

Another successful organization that has followed the scheme of biointensive organic agriculture is the Las Cañadas cooperative in the cloud forest of Veracruz, a civil association constituted in 1999. In addition to running programs in agroecology and biointensive agriculture, this nonprofit manages an ecological reserve and works on capacity building for peasants and environmental education. Their work has diffused to the regional and national levels, and they work with Cosecha Sana, a network that focuses on the exchange of organic vegetable seed. In order to not lose the equilibrium that they have achieved in their work, they have learned to impose clear limits to the growth of each of the activities they carry out (see www.bosquedeniebla.com.mx).

Finally, another successful example of medium-scale organic production is Pro-Organico, a family business that formally began operations in the middle of 2003 once it received organic certification from Oregon Tilth for a small orchard of oranges in the citrus growing region of Nuevo Léon in the northeast of Mexico. Pro-Organico's founders left professional careers to begin an operation that would offer urban consumers safe and fresh food of high quality, and at the same time be an attractive and economically viable model for small-scale rural farmers. They have placed their products in Mexico, the United States, and Canada. They have a staff of full-time workers and professionals in management positions. They are in the process of opening a store to direct market their products with the idea of personalizing service to customers through direct contact. They are currently producing vegetables, lamb, milk, eggs, fruit, forage crops, grains, organic fertilizers, and seedlings. Everything is certified as organic (Elizondo, 2007; see www.pro-organico.com).

This brief survey of organic agriculture in Mexico shows that it has developed on the basis of the experiences of small-scale farmers, as discussed by González Jácome in Chapter 9. Traditional farming practices have persisted with these farmers, and they have adapted to organic production based on a strong understanding of soil, water, and biotic resources; the management of soil fertility, pest, diseases, and weeds with nonchemical methods; the value of spatial and temporal diversity; and the management of genetic resources *in situ*. Such a situation largely indicates that it was not technological packages that led to this type of production, but rather the adaptation and adjustment of the practices that these peasant farmers have been developing as a part of local culture over time.

8.4 CONCLUSIONS

Past experience has demonstrated that we should not adhere to a type of agriculture that uses synthetic fertilizers and pesticides to increase short-term yields but in the long-term reduces productive capacity (Rodale, 1973). In a relatively brief time we have seen that resources are limited, and that with organic methods it is possible to use practices that control pests and diseases biologically, reduce contamination of the environment, grow safe food of high nutritional quality, improve the soil, conserve energy, and use as resources substances that would otherwise be considered waste.

There is a consensus in Mexico (Torres, 1999) that organic agriculture is not just a matter of changing the technological aspects of agricultural production, but that it also implies a questioning of the role that agriculture plays in society and what kind of model of development is desired. It is also related to food security, the creation of more equitable relationships between the rural and the urban, between agriculture and industry, and between economy and energy, and greater participation of the peasant sector in the definition of agricultural and food policy. Organic agriculture in Mexico builds on the belief that the elements of a new paradigm for agriculture can be found in peasant and indigenous systems.

Although at the current time organic agriculture based on input substitution is growing significantly in Mexico, and is being supported by a public policy that favors the development of international markets more than an internal national market, there is also much evidence showing the development of a different kind of organic agriculture—one grounded in the principles indicated by IFOAM, and which sustains and strengthens the health of the soil, plants, animals, human beings, and the planet as a whole system. This is an agriculture based on ecological cycles and relationships that ensure equity, managed in a way that protects the health and quality of life of present and future generations. Mexican organic producers have shown that it is possible to generate, little by little, the changes in public policy, attitude, and consumption that will be necessary for this truly sustainable organic agriculture to flourish.

REFERENCES

Ceseña, B.A.S. 1997. Productores orgánicos del Cabo experiencias en agricultura orgánica. In *Memoria Segundo Foro Nacional sobre Agricultura Orgánica*, 1–4. La Paz, San José del Cabo, Baja California Sur.

Diario Oficial. Martes 7 de febrero de 2006. Secretaria de Agricultura, Ganadería, Desarrollo Rural, Pesca y Alimentación. *Ley de Productos Orgánicos*, Primera sección, pp. 45–53.

Elizondo, L. 2007. Pro Orgánico. Experiencia en la producción de hortalizas y ganado orgánicos. In *México Orgánico. Experiencias Reflexiones, propuestas*, ed. R.R.M.A. Schentesius, C. y H. Gómez, and B. Blas, 91–95. Chapingo, México: UACh. CIESTAAM.

Gliessman, S.R. 2002. *Agroecología. Procesos Ecológicos en Agricultura Sostenible*. Turrialba, Costa Rica: CATIE.

Gómez, C.M.A., R. Schwentesius R., L. Gómez T., I. Arce C., Y. Morán V., and M. Quiterio M. 2001. *Agricultura Orgánica de México Datos Basicos*. Chapingo, México: SAGARPA-CEA, UACH-CIESTAAM.

Gómez, C.M.A., R. Schwentesius R., M.A. Meraz A., A.J. Lobato G., and L. Gómez T. 2005. *Agricultura, Apicultura y ganadería orgánicas de México-2005. Situación, Retos y Tendencias*. Chapingo, México: CONACYT, SAGARPA, CEDERSSA, UACH, CIESTAAM, PIAI.

Gómez, T. L. 1996. La agricultura orgánica de México: una opción viable para los agricultores de escasos recursos. Tesis professional, Agroecología UACH, Chapingo, Méx.

González, S.A., R. Sánchez L., and E. San Martín de la C. 1995. *Fundamentos científicos de la Agroecología mexicana (La Agroecología como alternativa para el desarrollo rural sostenible)*. Motozintla, Chiapas, México: Unión de Ejidos Prof. Otilio Montaño.

Granados, S.D., and G.F. López R. 1996. *Agroecología*. Chapingo, México: UACH.

Guzman, C.G., M.M. González, and E. Sevilla Guzman. 2000. *Introducción a la agroecología como desarrollo rural sostenible*. Sevilla, España: Ediciones Mundi Prensa.

Hernández, C.R.A. 2001. *La otra frontera. Identidades múltiples del Chiapas poscolonial*. México: Ed. CIESAS.

Jeavons, J. 1991. *Cultivo biointensivo de alimentos*. Willits, CA: Ecology Action. EEUU.

Martínez, E.C. 1997. Dimensión social de la agricultura orgánica: un enfoque integral. In *Memoria Segundo Foro Nacional sobre Agricultura Orgánica*, 11–15. La Paz, San José del Cabo, Baja California Sur.

Martínez, T.E., and W. Peters G. 1995. Cafeticultura orgánico/biodinánica en la Sierra Madre de Chiapas, México, 1963–1993. In *Memorias Conferencia Internacional sobre café orgánico*, 13–26. México: AMAE, IFOAM, UACH.

Pérez, C.N. 2003. *Agricultura Orgánica: bases para el manejo ecológico de plagas*. La Habana, Cuba: ACTAF, CEDAR.

Ramírez, G.J.A. 2003. La experiencia de UCIRI en México. In *Producción, comercialización y certificación de la agricultura orgánica en América Latina*, 119–131. Chapingo, México: UACH.

Rodale, R. 1973. The basics of organic farming. *Crops and Soils* 26:5–7, 32.

Sánchez, L.R. 1990. *El cultivo biológico del café orgánico. Indígenas de la Sierra Madre de Motozintla, San Isidro Labrador, Sociedad de Solidaridad Social*. Chiapas, México: Motozintla.

Santoyo, C.V.H., S. Díaz C., and B. Rodríguez P. 1995. *Sistema Agroindustrial café en México. Diagnóstico, problemática y alternativas*. Chapingo, México: CIESTAAM-SARH.

Schwentesius, R., and M.A. Gómez C. 2007. México en el mundo orgánico. In *México Orgánico: Experiencias, Reflexiones, Propuestas*, 23–28. Chapingo, México: UACH. CIESTAAM.

Torres, T.F. 1999. Recuperación de viejos paradigmas para la agricultura del tercer milenio. In *1er. Simposio Internacional de Agricultura Sostenible y Orgánica. "La Huasteca hacia el Tercer Milenio,"* pp. 34–56. Pachuca, Hgo: Fundación Produce Hidalgo.

Vanderhoff, B.F. 2005. *Excluidos hoy, protagonistas mañana*. México.

Willer, H., M. Yussefi, and N. Sorensen. 2007. *The world of organic agriculture. Statistics and emerging trends*. Frick, Switzerland: IFOAM, FiBl.

9 Mexico
Traditional Agriculture as a Foundation for Sustainability

Alba González Jácome

CONTENTS

9.1 INTRODUCTION

Across the Mexican countryside, approximately 5,654,000 small-scale farmers—including *ejido* landholders, communal landholders, and peasants in possession of some amount of agricultural land (*posesionarios*) in the ejido properties—manage a total of about 3,392,000 plots in and around 31,518 communities (INEGI, 2007, 2008). These farmers represent about 5.4% of the total population in Mexico, which in 2005 stood at about 103,263,388 inhabitants (INEGI, 2000, 2005).

The diverse, small-scale agricultural systems managed by these rural Mexican peasants represent, collectively, a broad type of agroecosystem with ancient origins that has been called the "Mexican model of agriculture" (Palerm, 1968). This model is considered a prime example of "traditional agriculture" by many ecologists, agroecologists, and social scientists (Gliessman, 2001).

Mexican traditional agriculture has a long history of endogenous development, with roots in the pre-Columbian systems that flourished as early as 7,000 to 9,000 years ago (Benz, 2001; Iltis, 2006). As an ancient, locally adapted model based on nontechnological inputs, it exhibits many of the characteristics that are required of sustainable systems. It integrates with and supports (or at least does not harm) local biodiversity; it does not require imported or purchased inputs; it is able to satisfy (at least partially) the food needs of both the rural population and nearby urban centers; it relies on time-tested methods of ecological agriculture that use nutrient cycling and biological interactions to maintain fertility and control pests; it incorporates practical methods of risk management; and it is intimately connected with, and supportive of, rural culture and urban society.

Throughout Mexico, traditional agriculture has served in various ways as a foundation for the development of sustainable systems of "organic agriculture" (see Chapter 8). Although these systems may differ in scale, practices, crops, and other aspects from traditional systems, the imprint of their source in traditional agriculture is unmistakable. Traditional Mexican agriculture, therefore, has already proven itself to have an important role in the ongoing effort to create sustainable agricultural systems in the country.

Mexican traditional agriculture, however, faces many threats, mostly linked to external, large-scale forces such as industrialization, modernization, urbanization, globalization, national agricultural policy, and demographic trends. These threats, which include labor shortages, migration to urban centers, dietary changes, and devaluation of traditional knowledge and culture, are so serious that we may classify traditional agriculture as endangered. Nevertheless, there are examples all over Mexico of successful responses to these threats. In every case, they involve modifying traditional systems or practices to fit the modern context. Viewed through the lens of history, this is just the latest in a series of adaptations to imposed conditions that rural Mexicans* have made since the beginning of corn domestication and cultivation (Benz, 2001; Iltis, 2006). These adaptations have much to teach us about moving toward sustainability.

9.2   HISTORY OF TRADITIONAL AGRICULTURE IN MEXICO

Ancient land use patterns are basic to understanding the rise of contemporary Mexican agricultural systems, their relationship with natural ecosystems, and the current discussions about sustainability. The diverse systems of traditional agriculture that exist today in the Mexican countryside are the product of a long history of indigenous and rural people adapting ancient agricultural techniques and land use patterns to a series of disruptions and appropriations of land that began with Spanish

---

* "Rural Mexicans" include indigenous and nonindigenous people, peasants, small-scale farmers, small-scale private cattle ranchers, and communal organizations. In some contexts, these are important distinctions, but for the purposes of this chapter, they add an unnecessary layer of complexity that can obscure the more general dynamics being discussed. I will use the terms farmers, agriculturalists, and peasants in a similar broad manner.

colonization; continued with the spread of estates (*haciendas*), development of large-scale production systems for export, the Mexican Revolution, and agricultural "modernization"; and persist into the present day with globalization, economic crises, and seasonal agricultural work migration to the United States and Canada.

When the Spanish arrived in Mexico in the early 1500s, the agricultural systems they encountered were already very old—probably the oldest in the Americas. These systems had been productive enough to produce food surpluses and permit the rise of urban civilizations such as those of the Maya, Zapotec, Nahua, and Totonac. As the Spanish colonized Mexico, they introduced many new crops, animals, technologies, farming practices, and land use patterns. Wheat, barley, sugar cane, citrus, peaches, pears, apples, grapes, watermelons, horses, and cattle from the Old World joined the corn, beans, squash, chilies, and turkeys native to Mesoamerica (Dunmire, 2005). These imported agricultural elements were often in conflict with indigenous elements, but over a long period of time, hybrid systems were created with both Old and New World elements.

The combining of European and indigenous agriculture was seldom peaceful, and it had devastating consequences for many indigenous people. At the beginning of the colonial era in the 1500s and 1600s, the relationship between agriculture, land tenure, and the economic goals of the new society produced conflicts that resulted in rebellions of Indians against Spaniards. These conflicts were accentuated by the introduction of cattle ranching and new models for land tenure and irrigation (Hoekstra, 1992). Cattle and irrigation competed for the same land and water resources that the Indians had used for intensive agricultural systems before the Spanish arrival in New Spain (Chevalier, 1975; Gibson, 1975).

The pre-Hispanic intensive *chinampa* agricultural system in the Valley of Mexico suffered from reduction of available water. Agricultural lands on the plains were converted from irrigated corn cultivation into seasonal rainfall corn cultivation as the available irrigation water was directed to wheat crops and applied to vegetables and fruits of Old World origin (González, 2009; Olivares, 2007; Quiñones, 2005). As indigenous people adapted to the introduction of new technologies and plants, many of the management practices, such as weeding, that had characterized systems like the *milpa* were modified or abandoned (González, 2004).

Depletion of natural areas in several regions of New Spain was a common practice (Melville, 1994, 1997). Several regions in the Viceroyalty, including forested areas and semiarid subtropical areas, began to be used for dryland agriculture and subsequently suffered degradation. Many ecologically rich wetlands and lakes—like the ones located in the Valley of Mexico—were drained to use the land for farming and urban expansion purposes. Moreover, after the destruction of the Indian water control systems in the Valley of Mexico, annual floods and heavy rains during the summer increased the necessity of draining the old lakes and wetland areas (Gurría, 1978; Palerm, 1973).

In the central part of the country, indigenous people responded by developing new corn seed varieties adapted to the lack of water through the cultivation cycle. In the high-altitude Valley of Toluca, where hailstorms and frosts were a problem, farmers developed the *Palomero toluqueño* corn variety, which was very well adapted to the harsh conditions. In the tropics, farmers developed a corn variety called *marceño* that was adapted to an excess of water and annual flooding (Gliessman, 1999; Orozco, 1999). As Indians lost their lands to the Spaniards—who built large estates (*haciendas*) to grow commercial crops such as sugar cane, wheat, and vegetables— indigenous agricultural systems were refocused on family subsistence and payment of the taxes imposed by the Spanish Crown.

In the three centuries between the Spanish conquest (1519) and Mexican independence (1821), the Indian population decreased drastically, Spaniards expanded their control of empty lands, crop monocultures for export came to dominate the agricultural economy, cattle ranching expanded, and urban centers grew rapidly— all of which radically altered the agrarian landscapes of Mexico (Aguirre Beltrán, 1991; Siemens, 1983, 1990). At the end of this period, traditional agriculture had incorporated new plants, animals, ideologies, and farming practices, as well as new agricultural systems, such as the *solar* in the Mayan communities, in which stone walls (*albarradas*) divided home gardens, restricting the movement of the animals that had formerly been hunted for food (Mariaca et al., 2007; Vanderwaker, 2006). There is some historical evidence that the milpa system in tropical zones survived with reduced crop diversity (Blanco, 2003, 2006).

During the 1800s, Indian communities were subordinated to the newly independent Mexican state, its institutions and economic programs. Programs directed to the development of new industries were organized in different parts of the country and new crops—most notably coffee—were introduced into commercial agriculture to meet the demands of the industrialized countries (González, 1996, 2004; Sartorius, 1961). Indian communities practiced subsistence agriculture and supplied seasonal labor on the neighboring estates. They obtained wood, charcoal, plants, animals, mushrooms, and many other supplementary resources from the surrounding natural areas (González, 2008a).

Beginning in the 1800s, and then more importantly during the 1900s, a series of local and regional programs were organized throughout the country to develop water control systems. Some lakes, lagoons, and rivers were drained or diverted to expand agricultural lands over their old basins; there are good examples in the Lerma River basin near Toluca and also in the southwest of Tlaxcala (Albores, 1995; González, 1992, 1999, 2003a, 2008b). The land was divided among large estates, ranches, and Indian and peasant communities, which had communal organizations of property with respect to natural resources such as forests, mines, ravines, lakes, and rivers (Bilbao, 1989; Blanco, 2006; Servín, 2000). Commercial agriculture was mainly concentrated on private properties, and water was controlled for irrigation purposes; rural communities mainly focused on rain-fed seasonal subsistence agriculture. Corn was the basic crop for the peasants, and it was cultivated mainly for subsistence purposes.

Modernization of industry and large-scale agriculture proceeded throughout the country from 1830 onward. At the end of the 1800s, estate owners introduced British agricultural machinery (Nickel, 1996).* From the 1930s on, development programs

---

* Lucas Alamán was the ideologist for the industrialization of the country during the first years of the nineteenth century. The need to modernize Mexican agriculture is found at least in the eighteenth century with the Bourbon reforms of the New Spain economy. The idea grew during the nineteenth century; it also was publicly expressed by several politicians and estate owners during the Porfiriato. Around the 1880s a process of mechanization can be documented for some regions of Mexico, as in southwest Tlaxcala; it was related to the lack of human labor from the towns around the estates (Nickel, 1996).

were organized along the lines of U.S. agricultural and technological models. Formal education reinforced the ideology of progress: agriculture had to be modernized. This included the idea of having spaces without natural vegetation around cultivated plots, which act against natural diversity. Agronomists and politicians could not appreciate the value of many traditional practices, which included managing unweeded, overgrown natural spaces near cropping areas.

In the sparsely populated regions of the southeast of the country, forested areas became important sources of raw materials for foreign companies. European and American companies cut timber in the tropical forest of Marqués de Comillas in Chiapas (Mariaca, 2002). Timber was taken by Canadian companies in the temperate woods of the Sierra de Juarez in Oaxaca (Guhs, 1992), and American companies obtained chicle from sapote trees in the tropical forests of Los Chenes in Campeche (Morales, 2004).

Land tenure in Mexico underwent radical changes as a consequence of the Mexican Revolution (1910–1921). The peasantry took control of some of the land in the hands of estate owners. Starting in 1916, a collective system of land tenure called *ejido* was imposed by the political leaders. Small-scale farmers had usufructuary rights but were forbidden to rent or sell the plots they had been allocated. Various natural areas were held in common. Isolated communities all over the country were able to use natural areas for the development of new agricultural lands for obtaining other resources, or even the creation of new agricultural systems.

One example is the *banquetera,** established in the ravine slopes of the Xopilapa town in central Veracruz, where coffee and mango production was directed toward the regional markets and the local fauna was not affected by agriculture (Servín, 2000, 2001, 2002). Another strategy was developed by the Popoluca people of Soteapan in the Tuxtlas region of southern Veracruz, where coffee trees were intermixed with the natural forest (Blanco, 2003, 2006, 2007). In both cases the new agricultural systems were located in the subtropical forest; they were dedicated to the cultivation of commercial products while staples were grown in the plots located near the houses (*milpa*). A combination of self-sufficiency and commercial production was the result of these attempts to integrate local society with regional and national sociopolitical developments. Staple crops remained the same through the years while commercial crops changed to adapt to market necessities.

The developmental policies of the Mexican government from 1940 to 1970 were mainly directed toward the industrialization and modernization of the nation, and this included agriculture. Their objective was to impose the American model of agriculture. This model mainly involved the use of agricultural machinery instead of human labor, the intensive application of agrochemicals, the use of hybrid seeds for the cultivation of corn, and the cultivation of monoculture commercial crops instead of the basic staples. In addition, earth and plant borders and the natural areas located around cultivated plots were leveled to create terrains that could be worked with machinery (Márquez, 2007; Martínez and Gándara, 2007; Palerm, 1968).

---

* The *banquetera* is an agricultural system in which small, perched, triangle-shaped plots on hillsides are intermixed with the subtropical forest and are cultivated with coffee and mango trees in such a way that the vegetation looks undisturbed (Servín, 2000, 2001).

In recent decades, government policy has favored large-scale conventional agriculture and paid little attention to the needs of small-scale farmers in the countryside. The *Procampo* governmental program mainly helped medium- and large-scale farmers. At the same time, however, social programs like *Progresa* and *Oportunidades* were used by small-scale farmers as a way to improve their agriculture-based family economies (Márquez, 2007). Neoliberal economic policy in Mexico started with the Miguel de la Madrid government in 1982; it was reinforced with the new agrarian laws in 1992 and the NAFTA (TLCAN) agreement in 1993. The government of Vicente Fox did not make any fundamental changes in agrarian policy. The NAFTA agreement with the United States and Canada forbids direct aid to agriculture, so peasants get assistance for converting subsistence agriculture into commercial enterprises indirectly, by using governmental social policies aimed at improving the life of poor rural families.

The recent financial crisis in the world economy has impacted the jobs of Mexican migrants and the remittances they send home, causing changes in the agrarian policies of the current Calderon government. At the beginning of 2008, the government approved a new program of economic support to large- and medium-scale agricultural enterprises, with only a little more than 300 pesos each going to the poorest farmers in rural areas. However, on November 13, 2008, the Mexican Congress responded to the rise in food prices by approving more than 230,000 million pesos for the 2009 annual national agricultural budget; a part of this amount will go to support agricultural producers. This could signal a change in the perspective of Mexican politicians about agriculture.

Despite the powerful forces arrayed behind it, the modern agricultural model has not completely displaced traditional agriculture. Traditional practices survive in some form, recognizable as the basis for the various distinctive agricultural practices that are common to most mestizo and Indian rural communities in Mexico. These practices include the following:

1. Combining agriculture with the collection of plants and animals from the surrounding natural areas.
2. Focusing agricultural effort on both family subsistence and the selling of food and products such as mushrooms, medicinal plants, wood, and charcoal to local and regional markets.
3. Managing diverse crops in the traditional agroecosystems such as the *milpa*.
4. Creating agroecosystems that involve growing crops in partially modified natural systems, such as managing commercial tree crops within forests.
5. Modifying ancient agricultural practices as needed to maintain soil fertility (adding more green and animal manure to the fields, for example).
6. Leaving forested areas around the cultivated fields to protect crops from excess of sun and wind (such as is done in the *tolché* in the Mayan area of Yucatán).
7. Keeping alive a cultural context for farming that is shaped by a syncretic system of myths and rituals both old and new (Albores and Broda, 2003; Alcorn, 2006; Blanco, 2006; Ellis and Porter, 2007).

9.3   THREATS TO TRADITIONAL AGRICULTURE

There are many forces acting against Mexican traditional agriculture and its sustainable agricultural practices. Some of these forces, such as NAFTA and emigration to the United States, are closely related to economic policies at the national and international global scale. Their impacts on agriculture and rural areas located in the central and southern regions of Mexico are unmistakable, even if our understanding of them could benefit from further study.

9.3.1   Dietary Changes

During the last 30 years, the diets of people in rural communities have changed dramatically, becoming more like those of urban people. Hunting and fishing for food is a thing of the past in the majority of the Mexican rural areas, and the consumption of traditional plant foods prepared at home in traditional ways has declined. The preferred foods are often processed products imported over long distances; even though they are more expensive, people prefer them because they are related to the idea of modern life. Traditional food, in contrast, is tied to old people and the past. To increasing numbers of rural people, it is irrelevant that traditional cooking may improve nutrition and health, or that it is part of a valuable cultural heritage.

Changes in the consumption of beverages are indicative of what is happening with diet generally. The consumption of manufactured soft drinks has increased dramatically; Mexico is first in the consumption of such drinks in the world (Coca-Cola Company, 2001). At the same time, many traditional, local beverages have nearly disappeared. These include *posol* (corn and cacao), *posole* (corn meal mixed with water, chile, and salt, or corn meal mixed with coconut pulp), *balché* (balche tree bark mixed with water, honey, and anise), *aguamiel* (nonfermented *Agave* juice), *pinole* (toasted and milled corn mixed with water), flavored water (pulped fruits mixed with water), *tepache* (pineapple peel mixed with brown sugar and water), *atole* (corn milled and mixed with water, sugar, and sometimes the flavor of some fruit such as *Prunus capulli*), chocolate (cacao mixed with water, sugar, and cinnamon), and many others (González et al., 2007).

These dietary changes have many negative consequences. Among them are public health effects. The rise in soft drink consumption, for example, is tied directly to a sharp rise in diabetes, which is now the fourth cause of mortality in the country (Tecontero, 2005).* The current health minister, through the media, is advising people to monitor their food intake and reduce consumption of sweetened commercial beverages. More relevant to the current discussion, however, are the effects of dietary changes on agriculture and sustainability.

As rural people shift toward processed, industrialized foods, it undermines the main reason for engaging in traditional agriculture—to grow food for one's family. At the same time, it motivates farmers to increase production for sale to urban markets so that there is more cash available for the purchase of the processed or distantly

---

* Diabetes Atlas, 2000, p. 10. The estimated prevalence (for the 20 to 79 age group) for Mexico is 14.2%. Only three countries had a higher incidence: Papua New Guinea (15.5%), Mauritius (15.0%), and Bahrain (14.8%).

grown food. As a result, there is a decline in the cultivation of corn and bean varieties and other crops for subsistence—and this is one of the foundations of traditional agriculture. There are fewer reasons to maintain all the sustainable practices that characterize traditional agriculture: maintaining a diverse home garden made up of vegetables, fruits, aromatic and medicinal plants, and flowers; practicing intercropping; managing surrounding natural areas; and integrating livestock animals in ways that facilitate closed nutrient cycling (Gliessman, 2001).

Dietary changes are also linked to another factor that threatens traditional agriculture: the decline of the oral tradition. The oral tradition in rural societies is an important means of preserving local culture and traditional agriculture. Older members of the family teach children about the local history of the community, the family's relationship with other community members, and the stories about the mythical figures related to forested areas, canyons, ravines, mountains, lakes, lagoons, and rivers. The rural kitchen was an important place to discuss family business, to learn local histories, and to cook, but it was also where much of agricultural knowledge was transmitted from old to young. Thus, as dietary changes have diminished the prevalence of traditional cooking, so too have they removed from daily life an important means for the transmission of traditional knowledge.

9.3.2 LOSS OF PEASANT KNOWLEDGE AND DEVALUATION OF CULTURAL HERITAGE

Traditional agriculture depends fundamentally on traditional knowledge, which includes local practices about all aspects of food and sustenance: cooking, food processing, collection of foods and medicinal plants from natural areas, cultivation of crops, pest and weed management, soil, local irrigation, nutrient management, animal husbandry, weather and climate, and risk management (Del Amo, 2001, 2007, 2008; González, 2007). This practical knowledge, in turn, is intimately related to traditional culture, which includes not only ways of life but also beliefs, values, ethics, and worldview (Albores and Broda, 2003; López Austin, 2007).

Much traditional knowledge is not easily detected by the outside observer because it is taken for granted by the traditional farmer. For example, farmers in former times based their decisions on what to plant in particular locations based on local classifications of different soil types; these classification systems have only recently been recovered in anthropological studies (Juan, 2003, 2007; López Montes, 2008).

The crops of peasant farmers are threatened by weather and climatic events. By careful observation of weather patterns, peasants are able to predict with some accuracy when these events may occur and adjust their strategies accordingly. For example, peasants may have several plots of corn located at different altitudes or in different microenvironments. During dry years, flood-prone areas produce the best crops, whereas well-drained areas are most valuable for obtaining crops during rainy years (González, 2003b). Another strategy is to plant different varieties of corn in mixed plantings (Blanco, 2007); the resulting diversity in maturation time, drought resistance, and other characteristics ensures the success of at least some of the crop. Corn and broad bean are cultivated together in places where frost is a common climatic event. If the year is not so harsh in frosts, the peasants will be able to obtain the two crops. If the year is a bad one, they will at least be able to harvest broad beans

(Mariaca, personal communication, 2003). Farmers may also plant staple crops in narrow strips within a plot of a commercial crop so that if the commercial crop fails the plot produces at least some food (Wilken, 1969).

The knowledge involved in managing the germplasm of the multitudes of traditional varieties of crops is prodigious. Peasants know exactly which varieties are best adapted to higher altitudes, lack of water, an excess of water, and other environmental variables. Continuous experimentation is the key to continuing genetic modification (the pots and cans with plants located near to the houses in Tepeyanco home gardens, and the *kanché* in Yucatán, with new plants growing in pots over a rustic table made with tree trunks are good examples). Peasants know how to reproduce the varieties so that they remain distinct. And, of course, it was traditional knowledge that allowed the development of the different crop varieties in the first place.

This all-important traditional local knowledge is being eroded in two major ways. First, the practical knowledge itself is not being transmitted from generation to generation as it used to be. This is happening for several reasons. Cultural and economic change has weakened the old socialization systems through which older generations trained the younger. Outmigration and international migration (see below) have taken young people out of their rural communities for considerable periods of time, reducing the opportunities for transmission of knowledge. Finally, even when young people are present in their home communities, they tend to be engaged in nonagricultural work, in which traditional knowledge has no role.

The second way in which traditional knowledge is eroded is less direct; it is through the general devaluing of cultural heritage and tradition. In modern Mexican society, there is a general feeling that rural culture and peasant farmers are backward. Agronomists, economists, rural developers, and urban people in general consider rural people illiterate and incapable of adapting to modernization and change. They look at peasants' prediction of weather events and see superstition; they look at traditional crop varieties and see inferior productive capacity. The attitudes underlying this deprecation of peasants' farming abilities have been around for a long time. During the Porfirian government (1880–1910) many Italian and French peasant families were brought into the country with the specific task of teaching Mexican peasants how to cultivate the land with modern techniques (Alfaro, 2001).

The ideologies of progress and modernism are so pervasive that many rural people have convinced themselves that they are indeed backward. When this happens, they deprecate their own valuable knowledge, and traditional practices become even more vulnerable to displacement with conventional, industrialized agricultural practices. Hybrid corn varieties replace traditional varieties, for example, and peasants lose control over their crop germplasm and the exquisitely tuned adaptation between crop and microclimate in exchange for the possibility of (temporarily) higher yields.

9.3.3 INTERNAL AND INTERNATIONAL MIGRATION

There is a long history of Mexican peasants migrating to urban centers and agricultural areas in the United States and Canada to obtain the monetary income they need for meeting their financial needs and those of their families. This kind of migration, usually seasonal and temporary, began as early as the late 1800s. It is very well known that migrants worked in constructing railroads between Mexico and the United States in the 1880s and 1890s. It was said that 60% of the men who worked the western railways were Mexicans (*El Paso Times*, 1963). The typical pattern before the 1970s was for the residents of a community to establish a connection with a particular city and a specific type of job. In this way, a yearly migration pattern to that place was organized. For example, Mexican migrants from Santa Barbara, Guanajuato, all traveled to Fort Worth, Texas, to work in the slaughterhouses and for frigorific companies (Andrade, 2004).

After the 1970s, farmers and their families began to develop new strategies for acquiring monetary resources. It became common for rural people to migrate to cities inside Mexico, such as Mexico City, Puebla, Acapulco, Guadalajara, and Cuernavaca, where construction and industrial jobs were available; later, during the 1980s, the number of places to work increased to include services in tourist zones such as Playa del Carmen, Cancun, Huatulco, and Los Cabos. From the middle of the 1970s to date, legal seasonal migration to agricultural jobs in Alberta and Montreal (Canada) has also been an important source of monetary income (Caloca, 1999). Beginning in the 1990s, legal and illegal international outmigration to the United States became another very common way to obtain monetary income and send remittances back home to one's family.

In recent years, internal seasonal migration to the cities has remained an important phenomenon. For example, in Naranjales, a town in the mountainous Totonac region of northern Veracruz, young men go out to work in seasonal masonry employment in cities such as Mexico City, Zacatecas, and Monterrey (Moctezuma, 2008). Similarly, in Ocotal Chico, in the Tuxtlas region, many young men leave to work in salaried jobs in Chiapas, Veracruz, Yucatán, Quintana Roo, and even Mexico City (Blanco, 2006). This internal migration has combined with international migration to create significant changes in the economic and social structures of rural areas all over Mexico.

The major consequence for traditional agriculture is an acute labor shortage. Families in rural communities do not have sufficient nonpaid human labor to apply to growing crops. This is particularly true for the complex intercropped agricultural systems characteristic of traditional agriculture, which requires significant human labor for sowing, weeding, cultivation, pruning, harvesting, seed collection, and other activities (González et al., 2007). Some agricultural systems, such as the *milpa* mixed system, are totally eroded by the lack of family labor (Blanco, 2007). In the states of Mexico, Tabasco, and Yucatán in particular, many towns are crowded with old people with no strength to work in their *milpa* and home garden agroecosystems (González, 2008; Ramírez, 2007; Robles, 2008).

The increasing tendency of the family workforce to find employment in nonagricultural activities is related to the abandonment of traditional agricultural practices. Among the practices that are already abandoned in some areas, it is possible to cite repeated manual weeding (González, 2003a), the four- to five-year process for the creation of manure compost, the local ways of soil erosion control (González, 1992, 2003a; Mountjoy, 1985), growing tree barriers to protect plants in windy zones, rotating cultivated crops with natural vegetation to maintain fertility in the soils, and rotating cereals with legumes and alfalfa. These agricultural practices need a nonpaid workforce to exist because payment for them would be very costly. Market prices paid to producers are so low that it is impossible to use salaried labor, and they are only able to compete using nonpaid mutual aid labor to reduce crop cultivation costs. Where there are labor shortages due to migration, there is a tendency for traditional agroecosystems to become simplified and move toward corn monoculture, which is less labor intensive. This trend is clearly related to the lack of human labor, although other factors are involved.

Labor shortages in rural areas are exacerbated by the increasing tendency for young people to work in jobs outside the agricultural sector, such as construction, services, teaching, administration, and commercial labor. Employed in these ways, they do not learn farming skills and are less likely to want to engage in agricultural labor when they return (if at all) to their home communities. Furthermore, agroecosystems located next to the family house are being depleted because the land is currently being used to build houses for family members—many of them recently married—who are working outside the town and sending money to their families. Old home gardens in places like Yaxcabab and Chan Kom, in the Yucatán Peninsula, are being converted into habitation spaces (González, 2008).

International migration is having another kind of impact in several rural areas of central Mexico: people are selling plots of land to obtain the money the migrant will need to pay to the person who helps him or her cross the American border. Even though the risk is great and the cost high, migrants from rural areas like Soteapan and many towns in Oaxaca will take the risk involved and go so far as to sell the land on which their subsistence has depended in order to leave poverty in Mexico (Blanco, 2006; Caloca, 1999; Guhs, 1992). Migration is an important process and more studies are needed to understand its many effects on Mexican rural communities. However, its impact on the agricultural labor force in the Mexican countryside is one of the most important issues at this point.

9.3.4 LOSS OF BIODIVERSITY

Natural diversity is recognized as a basic component for the maintenance of ecosystems around the world; ecosystems, in turn, provide the ecosystem services—such as nutrient cycling, water purification, and pollination—that support human life and human activities like agriculture. However, biodiversity is in rapid decline in many rural societies, especially as population numbers continue to grow and densities continue to increase. This is certainly the case in Mexico, where natural systems are being reduced in total area and the number of species they support diminish (Del Amo, 2001, 2007).

Natural systems have been depleted to create more agricultural land, to accommodate the needs of modern agricultural machinery, and to build access roads for the trucks that are needed to move produce rapidly (González, 2008a). Natural systems are also under pressure from rural communities themselves; because of internal and external constraints, such as the scarcity of capital for investment in agriculture, the loss of nonpaid family labor due to outmigration, and low market prices for agricultural produce and other factors, natural areas are exploited for marketable products beyond their carrying capacity (Del Amo, 2007; Molina, 2003; Velasco, 2002).

Natural systems have also been exploited by herb sellers and national and foreign firms seeking valuable commodities such as plants containing medically useful compounds. From 1940 to 1960, international corporations such as Syntex, Ciba-Geigy, Beisa, and Diosynth exploited natural vegetation mainly in tropical regions of the country to obtain medicinal plants such as *barbasco* (*Dioscorea composita*), which was used for birth control. Depletion of natural vegetation was a result of this activity. In 1960, the Mexican government had to declare some plants like *Dioscorea* in danger of extinction and established legal mechanisms to protect them (Rodríguez, 2003). New synthetic substances were developed in the 1960s and 1970s to replace Mexican *barbasco* in international markets, but the growth of traditional medicine continued the exploitation of natural vegetation by increasing the demand for medicinal plants, which were sold in local markets by small-scale herb sellers* (Rodríguez, 2003; Robles, 2008).

The existence of functioning, diverse natural ecosystems in areas surrounding human-cultivated plots is important to Mexican traditional agriculture (Wilken, 1987). They permit the biological control of diseases, support useful plants, and provide fodder for domestic animals. The plant species in these natural systems are also a source of genetic variety for crop breeding. Plants and animals from natural areas such as ponds, riverbanks, ravines, and forests are used by peasants to fulfill basic necessities and generate some monetary income (Juan, 2003). It has been shown that communities that lose the biodiversity and natural resources around them tend toward monoculture of maize and experience heavy outmigration (Blanco, 2003, 2006).

9.3.5 EXPANSION OF CONVENTIONAL PRACTICES

The green revolution ideas underlying the agricultural modernization that occurred in the country from 1940 to 1970 gained a foothold all over Mexico and influenced the ways many farmers grew food (Martínez and Gándara, 2007). Modern agricultural methods and technology increased internal national production but demanded dependence on technological inputs. In a short period of time, machinery and agrochemicals were introduced into traditional agriculture because there was a strong temptation to abandon traditional practices in favor of modern practices that promised to increase productivity or reduce labor needs. Today, conventional practices retain their allure.

Many farmers have adopted modern agricultural technology because it reduces the amount of labor needed for some activities, such as the preparation of soil, sowing, and harvesting. As a consequence, they have abandoned the use of animals for traction, resulting in an overall decline in the number of animals and teams used for agricultural purposes (Caloca, 1999; Cruz and Martínez, 2001). Fewer animals, however, means a reduction in the quantity of manure available to maintain fertility in the plot and an increased reliance on purchased, inorganic fertilizer. Another

---

* Medicinal plant sellers send their produce to markets called *tianguis*. In Mexico City there is a large
  market (Mercado de Sonora) that specializes in medicinal plants, sorcery, and witchcraft, and that also
  sets the prices for noncultivated medicinal plants in Mexico.

trend with serious implications is the increasing use of hybrid corn varieties all over Mexico, and the more recent introduction of genetically engineered seeds.

In the long-term, the gradual adoption of the modern agricultural model has caused a drop in agrodiversity, contamination of soil and water resources, and loss of local varieties of seeds that were adapted to specific environmental conditions (Brush et al., 2003; Gliessman, 2001, 2002).

9.3.6 OTHER THREATS

There are many other threats to traditional agriculture. One important one is the way in which land is passed from generation to generation. Inheritance systems vary, but many can have negative effects. When land is divided among the sons or all the children of a family, this division can result in economically untenable overdivision of fields (González, 1996). This also generates problems in obtaining credit for agriculture because the owners of several small plots do not qualify for credit. On the other hand, inheritance systems that pass land to only one son may have the effect of encouraging urban outmigration. Another threat worth mentioning is that many of the natural areas on which rural communities depend for firewood, supplemental food products, grazing areas, and collection of medicinal plants are becoming increasingly less accessible (Velasco, 2002).

9.4 SUSTAINABILITY FROM TRADITIONAL FOUNDATIONS: POSITIVE STEPS

As noted above, traditional Mexican agriculture—because of its locally adapted, nontechnological, and biologically based principles and practices—has many characteristics that should be considered ecologically and socially sustainable. Realizing the goal of developing agroecosystems in Mexico that are sustainable both ecologically and socially depends on incorporating many of traditional agriculture's sustainable characteristics into the new systems. This is true for both the conversion of conventional systems to organic production and the modification of traditional systems into agroecosystems that can operate in a market context.

The small scale of traditional systems is an especially important characteristic because it is only on a relatively small scale that many of the other sustainable features can function; small-scale systems also make sense in rural settings where there is a shortage of agricultural land (Wilken, 1987). A focus on maximizing soil fertility in the broadest sense is another critical feature; this is what allows the highest productivity per unit area of land. In the relatively arid areas of Mexico, efficient use of water resources is essential; many traditional agroecosystems across the country rely on sophisticated irrigation systems that make the best use of the available rainwater or soil moisture and do not require long-distance transport or large-scale storage of water (Del Ángel, 1988; Martínez, 2006; Rebolledo, 2007).

A key feature of all traditional systems is diversity, which is manifested in a variety of ways. Biodiversity—which includes diverse crops, diverse genomes, and use of natural vegetation—is what makes it possible for fertility and pest management to come largely from biological interactions; home gardens and *banqueteras* are good examples of this feature (Allison, 1983; Blanco, 2006; González, 1985; Servín, 2001, 2002). Physical or spatial diversity is also important: this occurs when annual crops are planted with taller fruit trees, when crops are grown under a loose canopy of natural vegetation, or when small-scale plots containing different crops are arrayed over a landscape that includes areas of natural vegetation (González, 2003b; Wilken, 1969). Lastly, cultural diversity helps support the variety of different ways farmers manage biodiversity, conserve agricultural and natural resources, and grow food for both subsistence and commercial sales.

Regardless of what constitutes a sustainable practice from an ecological viewpoint, the agroecosystems that will actually survive in rural Mexico are those that fulfill the needs of rural families. A farming family in rural Mexico has at least five basic requirements for maintaining an adequate quality of life. It must (1) have control of a plot for the cultivation of staple crops (a *milpa* or home garden); (2) have an economic activity with monetary earnings; (3) have access to natural zones for obtaining basic materials and supplemental food (firewood, timber, medicinal plants, mushrooms, animals, and so on); (4) be part of social networks (family, extended family, mutual aid network) that together can contribute the equivalent of one person's full-time labor to the cultivation of staple crops and the maintenance of the home garden; and (5) engage in activities that maintain the natural resources belonging to the community (González, 2007; Mariaca, personal communication, 2003; Reyes, 2003).

Despite the many forces acting against them, some communities have managed to find ways to provide their members with these necessities. In many cases, their solutions—if not actually sustainable themselves—point the way toward sustainability. They allow the basic principles and social contexts of traditional agriculture to survive and inform the transition to new models and systems.

9.4.1 Solving the Labor Problem

In former times, the labor needs of rural communities were satisfied in two relatively distinct ways. Family members carried out the sowing, cultivation, weeding, harvesting, and other activities involved in the management of traditional agricultural systems, and community members—organized into cooperative networks or mutual aid systems—attended to the community's noneconomic needs by building new houses, participating in social and religious holidays (Korsback, 1996), paving roads,* and repairing the infrastructure of communal services (Caloca, 1999).†

Internal and international migration, however, have in the last few decades drawn so many family members away from rural areas for long periods of time that families

---

* From the 1970s to the 1990s government offices were using communal nonpaid labor to repair and construct the roads connecting the towns with regional cities. The people involved in these projects received food or some money to work for several weeks.
† During the 1970s the towns on the slopes of La Malinche, in Tlaxcala, were using money from migrants and a workforce of peasants to repair churches and public buildings and to introduce services like electricity and potable water into their towns. Sometimes the community paid with labor half of the cost for the construction of the school building.

no longer have enough members on hand to cover the labor needs of farming. In some communities, this labor shortage has been filled, at least partially, by the mutual aid systems.

One community in which this has occurred is Progreso Hidalgo, a town in the south of the state of Mexico. Here the mutual aid system is called *macoa* (Juan, 2003). The *macoa* is formed by male members of the families of the kin who are related to a family head. The system is seldom used for sowing and harvest activities in the cultivation of corn; more commonly, this type of labor is applied to the cultivation of commercial crops during the one or two months of the year when money is scarce and cultivation has to be done. There are other communities in Mexico in which mutual aid systems are used mainly to support traditional agriculture (Reyes, 2003).

Research shows that rural families may be able to mitigate labor shortage problems by controlling the number of children they have. A small family—one with less than four members—is generally unable to maintain an intensive agricultural system without hiring labor (Robichaux, personal communication, 2003). On the other hand, a family with more than seven members also has difficulty because of the larger quantity of food required and because such a family is more likely to have more of its children work outside the farm in paid activities (Robichaux, personal communication, 2003). Medium-size rural families—those with five to six members—seem to be the best adapted for the development of an intensive and sustainable agricultural system (Robichaux, personal communication, 2003). Similarly, it has been shown that a rural family combining subsistence agriculture with commercial crops needs to have from five to six members in order to be successful (Reyes, 2003).

9.4.2 Generating Cash Income

Many small rural families in both indigenous and mestizo communities have difficulty surviving without a means of generating cash income. It has been mentioned above that this need is often filled by finding paid work in the cities or outside of Mexico, but this solution, rational from the perspective of the individual or family, contributes to the labor shortages and economic problems that undermine traditional agriculture. A different solution is to grow crops that can be sold in local, regional, or even international markets, while maintaining plots of subsistence crops also. This strategy is being tried in many parts of Mexico with success. A variety of different commercial crops are being grown in small-scale, sustainable systems, but one very promising example is medicinal plants.

Mexican people have used medicinal plants since pre-Hispanic times (González, 2009). Since 1990, the economic importance of medicinal plants in Mexico has grown as traditional medicine and alternative therapies have increased in popularity. As a result, growing medicinal plants has become an important commercial enterprise. There are now several towns dedicated to the cultivation of medicinal plants (Robles, 2008). There are also communal organizations in Oaxaca,* Chiapas,† Morelos, and

---

* Sociedad de Solidaridad Social Bejarano, Tuxtepec, Oaxaca.
† Political problems affect the market, but there is a program in some Indian communities for the cultivation of medicinal plants.

Guerrero* that facilitate the conversion of land to organic cultivation of medicinal plants. Universities have also become involved. In 1997 the University of Tlaxcala began a program dedicated to the study and production of medicinal plants in small communities and towns such as San Francisco Tepeyanco, a community very well known for its home gardens (Allison, 1983; González, 2003b).

An increase in monetary earnings is generally good from the standpoint of local economic development, but it can have negative consequences. The accumulation of monetary earnings may permit families to pay for higher education for the offspring who will not inherit the land; these community members become well educated, but they are also much more likely to move to urban areas for salaried jobs, leaving their communities without their expertise (González, 2003b).

When the attractiveness of earning a monetary income growing for markets is combined with certain gender biases or gendered divisions of labor, traditional agricultural systems can suffer. For example, during the 1990s, males of the Otomi people in Temoaya became very involved in selling produce in cities such as Toluca and Mexico City, leaving the women in charge of all the agricultural work (Acle, 2000). The result was a drop in cultivated land area and a conversion from mixed cropping to corn monoculture.

9.4.3 Modifying Traditional Systems

In many cases, the crops that can be marketed commercially are not traditional crops, and the somewhat larger scale of their cultivation requires techniques different from those that have existed in purely traditional systems. Therefore, both new crops and new techniques have been introduced into the small-scale systems of rural Mexico as peasants have moved toward commercial production.

Frequently, younger people are the ones who have introduced the new crops (Reyes, 2003). Communities such as Santiago Yeché and Progreso Hidalgo (Juan, 2003; Reyes, 2003) began growing commercial crops after some of their youth went to work outside of their towns and learned about the cultivation of these other crops.

The small-scale agriculturalists from Progreso Hidalgo learned how to grow strawberries when they were temporary workers in Watsonville, California, and today they are growing California strawberry varieties in Mexico. In this case traditional agriculture is combined with commercial crop cultivation, and together the crops are giving people the food and monetary income they need for daily life. This combination of Mexican traditional agriculture and modern Californian agriculture is only one example of how people can learn to manage two different agricultural systems with different goals and turn this into a way of making a living.

In the Atoyac and Zahuapan river basin, the last 20 years have seen a conversion from corn cultivation to commercial vegetable production. Using family labor alone, the small-scale farmers here are growing cold-season vegetables such as green tomatoes, spinach, chili, carrots, and white beets on nearly 3,000 hectares (Martinez, personal communication, 2003). The huge and very important markets in nearby Puebla and Mexico City are easily absorbing the production. In addition, it is possible to

---

* Coalición de Ejidos Productores de Plantas Medicinales en Guerrero.

grow tree crops all year in these areas, and there are some towns in which cultivation of commercial peaches is giving monetary income to the peasants (Márquez, 2005; Montero, personal communication, 2008).

In many ways, the innovation represented by the addition of new crops is the opposite of the conservatism of traditional agriculture. Yet these examples show that the two can coexist very well. To generate enough income to maintain their families without emigration, farmers can combine commercial crops with subsistence crops grown using traditional agricultural systems (Cortés, 2007).

9.4.4 Investing in Development of Sustainable Systems

A considerable amount of money flows into rural areas of Mexico from migrants working across the border and in Mexican cities. Workers typically send money to family members as regular remittances. These remittances are used in a variety of ways; some uses of the money support sustainable farming while others may undermine it.

In many extremely poor communities, remittances are directed toward survival and subsistence, and nothing is left over for other purposes. More commonly, remittances are invested in the improvement or expansion of local agricultural systems. A typical path is to use the additional capital to convert traditional systems to larger-scale conventional systems, which then provide income to pay for the salaries of agricultural workers, seeds, electricity for water pumps, fertilizers, herbicides, and other types of agrochemicals.

For example, there are rural communities, such as San Lucas Tecopilco, where legal seasonal migration to Canada has permitted the development of conventional monocultural systems. With the remittances from their work in Canada, community members have purchased agricultural machinery like tractors, harvesters, and combines, which helps make up for the shortage of labor. The flow of money has also allowed the community to pay for the introduction of electricity and potable water systems, and for the paving of the streets and roads in their towns (Caloca, 1999). Although standards of living have improved dramatically, it has been at the expense of traditional agriculture. During the last 30 years, the traditional corn, squash, and bean systems have been converted into barley monoculture. Furthermore, the semi-terraced hillsides that formerly checked soil erosion and created channels for water movement have been leveled to allow the use of the agricultural machinery.

However, there are a few places where remittances have financed conversion to organic agriculture or other more sustainable systems that offer the possibility of monetary income while preserving traditional practices. Furthermore, there are also places (such as Xalapa and Soteapan in central and southern Veracruz) where organic markets are being organized for the producers' direct selling of corn, beans, vegetables, and fruits (Blanco and García Rañó, personal communication, 2008).

It is very well known that during the Second World War, the migration that occurred in response to the demand for agricultural workers in the United States provided many Mexican rural families with money that they could invest in local agriculture. There are good data that show how money from migration to the United States during the war favored the development of agriculture in some rural communities of

Mexico. Some towns in southwest Tlaxcala, for example, improved their traditional agriculture through the introduction of technology for moving irrigation water, as well as buying agricultural machinery and trucks to move agricultural produce from their towns to the markets (González, 2003b).

9.4.5 USING NATURAL SYSTEMS SUSTAINABLY

Economic necessities have put enormous pressures on Mexico's natural systems, especially those in close proximity to rural farming communities. As noted above, natural systems around rural communities are important sources of firewood, construction materials, charcoal, medicinal plants, textile fibers, and food for animals and people. It is very easy for exploitation of these resources to go beyond the carrying capacity of their ecosystems, resulting in loss of biodiversity, erosion, and deforestation (Del Amo, 2007). On the other hand, some collection and use of wild resources can be beneficial to natural ecosystems because it can ensure the survival of certain sensitive species, reduce the accumulation of fire-prone fuels, and replace the ecological roles of native species (such as carnivores and grazers) that have been lost from the ecosystems.

It is important to restore or re-create the mentality of taking care of the surrounding natural systems. Some communities have done this by constructing ideological mechanisms such as myths—stories about imaginary beings that take care of the natural areas and impose some type of punishment on people who destroy natural resources (Servín, 2000). In the canyons of central Veracruz, for example, Juan Del Monte and the siren take care of the forest and the river, controlling the use of wild fauna and fishes and preventing depletion. People who violate the rules may suffer mysterious illnesses or become lost in the forest for several days (Servín, 2000, 2001).

Regulations and other legal means are often necessary when ideological mechanisms fail. Unfortunately, state regulation of resource use is mostly ineffective because the authorities in charge of taking care of the natural environment and natural resources are often corrupt (Molina, 2003; Velasco, 2002). Regulations created at the local level with the approval of communities work much better than laws imposed from outside. In Flor del Marqués, Chiapas, the *ejidatarios* control the use of natural resources through rules obeyed by the members of the community (Mariaca, 2002). Local authorities control the use of resources such as aquatic and terrestrial animals, the months of the year these animals can be caught, hunted, fished, or collected, as well as the number of animals to be collected by any one person. In Progreso Hidalgo, the natural resources of the canyons may only be used during the part of the year when there is no monetary income. The rest of the year people live on income obtained from irrigated commercial agriculture and staple produce obtained from seasonal agriculture (Juan, 2003).

Developing a more or less harmonious relationship between agricultural activity and the natural environment has benefits that go beyond ecological concerns. For example, combining subsistence agriculture and the management of natural resources in a seasonal regime can reduce the necessity for industrial foods. In Progreso Hidalgo, the rural population has developed a diet that includes collected natural resources like purslain, *epazote*, wild spinach, the flowers of *zompantli*

(*Erythrina* spp.), watercress, and pods from wild legumes like the so-called *guajes* (*Leucaena esculenta*). These wild foods are combined with commercial products like eggs, meat, chicken, and pork. Fish collected from natural water catchments are very important from June to October when money is scarce (Juan, 2003).

9.5 PRESERVING TRADITIONAL AGRICULTURE IN MEXICO

Because many social and economic trends create an environment that is decidedly unfavorable, Mexican traditional agriculture may not survive without some assistance from institutions and organizations such as universities, federal and state governments, nongovernmental organizations based in Mexico and elsewhere, and local cooperative and economic development groups. Any efforts to preserve traditional agriculture and use its basic principles and practices to create small-scale, sustainable commercial systems, however, must be based on accurate knowledge of how Mexican rural communities and their agroecosystems actually function. In addition, these efforts—whether they involve government policies or nongovernmental development initiatives—must have certain priorities if they are going to make a difference (Turrent, 2007).

Although we already know a good deal about traditional agriculture and the ways peasants have converted traditional systems into much more economically productive systems, there are still many holes to fill in our knowledge. There are wide geographic variations in culture, climate, and techniques, and a broad array of factors have to be taken into account when developing ecologically sound and profitable agricultural systems able to meet the multiple needs of rural communities. For urban and educated people, it is difficult to understand how rural culture operates, because it involves many interactions among different aspects of nature, society, and culture. Managing interrelationships among these factors is a characteristic of rural societies. When factors are examined independently, removed from their context, complex interactions can be easily missed, and what is actually complicated may appear simple.

The study of rural societies needs to apply complex models in which interdisciplinary approaches are basic. This has not occurred to an adequate extent because the natural and social sciences usually have different languages and different approaches to research problems. Since the 1950s, however, some models have been developed to work in this direction. The University of Yucatán has been developing an important self-help program for tropical areas.* Academics and research consultants have played important roles in the development of these programs. The most important lesson from their experiences is that local development must be based on the people's decision to select and participate in programs.

In addition to this basic principle, it is important to understand the specifics of local culture. This includes values, mores, gender roles, inheritance patterns, relationships between architecture and land use, use of traditional crops, and more. Often, what may seem like a small detail to an urban researcher can have huge impacts. For example, the importance of certain social and religious holidays and

---

* PROTROPICO, which is directed by Dr. Juan José Jimenez-Osornio.

observations (such as the birthday of the head of the family, graduation from elementary school, and religious family holidays) extends beyond the social realm into the economic: staging these events in the proper manner may require considerable monetary resources, and rural people consider these expenditures as important as providing food, clothing, and shelter.

Another important factor that could be easily overlooked is how the construction, architecture, orientation, and location of dwellings and other structures relate to various traditional agricultural practices. For example, there is a community in Tlaxcala (Juan Isidro Buen Suceso) in which the traditional sweat bath (*temascal*) plays an agricultural role: ashes from the bath are used to improve the quality of home garden soils, while the insects in the garden zone are controlled by the fire and smoke of the sweat bath when it is used every weekend by the family members (Romero, 1998). It is not an accident that sweat baths are constructed near the granaries and lined up with the prevailing wind during the year. In a similar way, the location of stables, pig sties, and yards in peasant communities is related to the use of manure for maintaining soil fertility (González, 2003b).

The manner in which land is passed from generation to generation is an aspect of social organization with obvious and important links to agriculture. Inheritance systems vary across Mexico, and so the particulars of the local system and its effects must be understood. The old Spanish system (*mayorazgo*) did not divide land because the oldest son in the family became the owner of all properties when the father died. The original native inheritance system (*xocoyotazgo*)* also did not divide the land, but put it in the hands of the younger son. Contemporary inheritance systems in rural Mexico may be based on either system, or the land may be divided among all the sons or all the children, depending on ethnic background and also on the quantity of land owned by the family. There are places where women inherit at least a plot, but there are other places where women do not have access to family land. In the rural community of Xiloxoxtla in Tlaxcala, for example, women are not permitted access to family land unless the family has no man who can assume care of the land (González, 2003b).

The success or failure of an initiative or program can hinge on an understanding of such cultural foundations as family social organization or the gendered division of labor. For example, an external cooperative organization introduced a program in Santiago Yeché in which married women were encouraged to grow organic tomatoes for markets in Mexico City, but the program failed because the personnel of the program did not know that married women in Santiago are not involved in commercial agriculture; the male farmers in the community are the ones who grow green tomatoes for the markets in Mexico City (Reyes, 2003).

---

* The *xocoyotazgo* includes the duty of taking care of the parents when they get old. There are severe relationship problems between the mother-in-law and the son's wife when the mother-in-law is a younger person. This fact has to do with the age of marriage that is socially accepted by the community. It seems that indigenous communities are more tied to xocoyotazgo than mestizo communities.

9.6 CONCLUSIONS

For rural development to be sustainable, alternative livelihood strategies need to be developed that build on local knowledge and create local opportunity. Traditional agricultural systems are an ecologically sustainable and locally adapted foundation for this kind of development, and everything possible should be done to preserve these systems and facilitate their adaptation and integration into the national food system (Calva, 2007).

The necessary balance between use and conservation of wild resources can be achieved. Historically, traditional agriculture in Mexico promoted this balance because it was part of a set of diverse management practices related to the maintenance of resources. Peasants developed agricultural systems with close and relatively harmonious connections to the surrounding natural systems. Human-maintained areas connected with food production—such as home gardens, water catchments and canals, and diverse farming plots—could serve as wildlife habitats, while use of surrounding natural areas was limited by an ecological resource management mentality.

Many goals must be pursued at the same time if traditional agriculture in Mexico is to survive. Some of the most important are listed below:

- Preserve subsistence-oriented corn cultivation as an important practice for the diet and economy of local communities.
- Reinforce the importance of a plot near the house that can be used to grow vegetables, fruit trees, aromatic and medicinal plants, and staples (Cortés, 2007).
- Reintroduce mixed cropping practices in areas of the country where they have been abandoned. Such practices are a fundamental aspect of sustainability and are important for maintaining a balanced and healthy diet in rural societies.
- Recover and preserve ancient knowledge about soil, water sources,* cultivated and noncultivated plants, animals, agricultural practices in relation to manure use, and management of seeds.†
- Recover and preserve traditional knowledge surrounding the understanding of nature, climatic events, and the traditional processes of social and cultural adaptation (Albores and Broda, 2003).
- Direct formal education toward filling the holes left by the disintegration of traditional modes of intergenerational transmission of traditional knowledge. At present, formal schooling does not teach local history and it does not teach people how to deal with local knowledge in a written form.

---

* There are current projects related to local irrigation systems in different parts of Mexico. One of them is directed by Jacinta Palerm Viqueira and Tomás Martínez Saldaña at the Colegio de Postgraduados (CP) in Texcoco. Another such research project is located in the north of Mexico and it is directed by Casey Walsh at the University of California in Santa Barbara. There is also a network of people doing research on this basic aspect of rural development (Web Net RISSA).

† Efraím Hernández Xolocotzi organized an important research program at the Postgraduate College (CP) on tecnología agrícola tradicional. The project functioned from 1976 to 1991, when Hernández X. died. The regions studied were the Bajío of Guanajuato, the north of the Puebla Mountains, the central valleys of Oaxaca, and the highlands of Chiapas, Yucatán, and Tlaxcala.

- Stress the importance of rural communities conserving and enhancing the natural systems surrounding them.

In addition to gaining a better understanding of traditional agriculture and rural communities, it is important to investigate the impacts on rural society of the changes occurring around them. Many social, political, and educational policies in the country are affecting rural areas, and we know very little about the specifics of these impacts and the variables that influence them.

ACKNOWLEDGMENTS

This work is based on the results of five research projects organized to study several rural communities in central and southern Mexico. The research was focused on the anthropological study of traditional agricultural systems and their relationship with environment, society, and culture; cultural ecology was the methodology. The results of these studies are written and discussed in many dissertations that were presented to obtain master and doctoral degrees in the graduate program in social anthropology at the Iberoamericana University in Mexico City, from 1975 to date. The studies were done in central, south, and southern Mexico, and they include several rural communities in the Valley of Mexico and also in the states of Mexico, Guanajuato, Tlaxcala, Tabasco, Veracruz, Campeche, Quintana Roo, Chiapas, Oaxaca, and Yucatán. During the development of these research projects the participation of several specialists in ecology, geography, and rural development was very important. Thanks to Stephen R. Gliessman, Alfred H. Siemens, and Tomás Martínez Saldaña for their advice and help over all these years. Thanks to Ramón Mariaca and William L. Crothers for their comments on this chapter.

REFERENCES

Acle Tomasini, Guadalupe. 2000. Gentes de Razón: Educación y Cultura en Temoaya. PhD, dissertation in social anthropology, Iberoamericana University, Mexico City.

Aguirre Beltrán, Gonzalo. 1991. *Regiones de Refugio*. Vol. IX. México: FCE, Colección Obra Antropológica.

Albores Zárate, Beatriz. 1995. *Tules y Sirenas. El Impacto Ecológico y Cultural de la Industrialización en el Alto Lerma*. México: El Colegio Mexiquense AC.

Albores Zárate, Beatriz, and Johanna Broda, eds. 2003. *Graniceros: Cosmovisión y Meteorología Indígenas de Mesoamérica*. México: El Colegio Mexiquense AC, UNAM.

Alcorn, Janis B., Barbara Edmonson, and Cándido Hernández Vidales. 2006. Thipaak and the origins of maize in northern Mesoamerica. In *Histories of maize: Multidisciplinary approaches to the prehistory, linguistics, biogeography, domestication, and evolution of maize*, ed. John E., Staller, Robert H. Tykot, and Bruce Benz, 600–11. Amsterdam: Elsevier, Academic Press.

Alfaro Telpalo, Guillermo. 2001. Mare Nostrum, Terra Nostra. Acercamiento Etnohistórico a los Proyectos de Colonización en México durante el Siglo XIX, Italianos y franceses en el norte de Veracruz. MA thesis in social anthropology,Iberoamericana University, Mexico City.

Allison, J. 1983. An ecological analysis of home gardens (huertos familiares) in two Mexican villages. MA thesis, University of California, Santa Cruz.

Andrade Torres, Juan. 2004. La otra cara de la migración: Santa Bárbara, Guanajuato, una comunidad que se extiende a la Ciudad de Forth Worth, Texas. PhD dissertation in social anthropology, Iberoamericana University, Mexico.

Benz, F. Bruce. 2001. Archaeological evidence of tesinte domestication from Guilá Naquitz, Oaxaca. *Proceedings of the National Academy of Sciences* 98:2104–6.

Bilbao, Jon Ander. 1989. *Fieldwork report on Huexoyucan, Tlaxcala.* Mexico: Iberoamericana University.

Blanco Rosas, José Luis. 2003. *La Erosión de la Agrodiversidad en la milpa de los zoque-popoluca de Soteapan.* Mexico City: Iberoamericana University.

Blanco Rosas, José Luis. 2006. La erosión de la agrodiversidad en la milpa de los zoque-popoluca de Soteapan: Xutuchincon y Aktevet. PhD dissertation in social anthropology, Iberoamericana University, México.

Blanco Rosas, José Luis. 2007. Los cambios en el sistema milpero de los zoque-popoluca del sur de Veracruz: el manejo de la agrodiversidad. In *Los nuevos caminos de la agricultura: procesos de conversión perspectives*, 197–99. México: Plaza y Valdés, Universidad Iberoamericana, PROAFT.

Brush, Stephen B., Dawit Tadesse, and Eric Van Dusen. 2003. Crop diversity in peasant and industrialized agriculture: México and California. *Society and Natural Resources* 16:123–41.

Caloca Rivas, Rigoberto. 1999. Migración y Desarrollo Autogestivo en San Lucas Tecopilco, Tlaxcala. PhD dissertation in social anthropology, Iberoamericana University, Mexico City.

Calva, José Luis, ed. 2007. *Desarrollo Agropecuario, forestal y pesquero.* Agenda para el desarrollo, Vol. 9. México: Cámara de Diputados, Miguel Ángel Porrúa y UNAM.

Chevalier, François. 1975. *La Formación de los Latifundios en México.* México: Fondo de Cultura Económica.

Coca-Cola Co. 2001. *Annual report.*

Cortés Flores, José I. 2007. La milpa intercalada con árboles frutales (MIAF), una tecnología multi objetivo para las pequeñas unidades de producción. In *Desarrollo Agropecuario, forestal y pesquero*, 100–16. México: Cámara de Diputados, Miguel Ángel Porrúa y UNAM.

Cruz León, Artemio, and Tomás Martínez Saldaña. 2001. *La Tradición Tecnológica de la Tracción Animal.* México: Universidad Autónoma de Chapingo.

Del Amo Rodríguez, Silvia. 2001. *Lecciones del Programa de Acción Forestal Tropical.* México: Plaza y Valdés Editores.

Del Amo Rodríguez, Silvia. 2007. La acción del hombre en las selvas tropicales: panorama general y soluciones posibles. In *Los nuevos caminos de la agricultura: procesos de conversión y perspectives*, 97–146. México: UIA, Plaza y Valdés, PROAFT.

Del Amo Rodríguez, Silvia. 2008. Paisaje y memoria totonaca: la relación entre ecología cultural y el manejo permanente de los recursos. In *Nuevas rutas para el desarrollo en América Latina. Experiencias globales y locales*, ed. Juan Maestre Alfonso, Ángel María Casas Grajea, and Alba González Jácome, 263–302. México: Universidad Iberoamericana.

Del Ángel Pérez, Ana Lid. 1988. *Fieldwork report in Santa Inés Tecuexcómac, Tlaxcala.* Mexico: Iberoamericana University.

Diabetes Atlas 2000. 2000. Brussels, Belgium: International Diabetes Federation.

Dunmire, William W. 2005. *Gardens of New Spain. How Mediterranean plants and foods changed America.* Austin: University of Texas Press.

El Paso Times. The Bracero program. May 30, 1963.

Ellis, Edward A., and Luciana Porter-Bolland. 2007. Agroforestería en la selva maya: antiguas tradiciones y nuevos retos. In *Los nuevos caminos de la agricultura: procesos de conversión y perspectives*, 213–42. México: UIA, Plaza y Valdés, PROAFT.

Gibson, Charles. 1975. *Los Aztecas Bajo el Dominio Español.* México: Siglo XXI.

Gliessman, Stephen R. 1999. Un enfoque agroecológico en el estudio de la agricultura tradicional. In *Agricultura y Sociedad en México: Diversidad, Enfoques, Estudios de Caso,* 25–31. México: Plaza y Valdés, UIA, PROAFT.

Gliessman, Stephen R. 2001. *Agroecosystem sustainability. Developing practical strategies.* Washington, DC: CRC Press.

Gliessman, Stephen R. 2002. *Agroecología. Procesos Ecológicos en Agricultura Sostenible.* Turrialba, Costa Rica: CATIE.

González Jácome, Alba. 1985. Home gardens in Central Mexico. In *Prehistoric agriculture in the tropics,* ed. I.S. Farrington, 521–37. BAR International Series 232. Manchester, Great Britain: BAR International.

González Jácome, Alba. 1992. Manejo del Agua en Condiciones de Secano en Tlaxcala, México. *TERRA, Orstom* 10:494–502.

González Jácome, Alba. 1999. El paisaje lacustre y los procesos de desecación en Tlaxcala, México. In *Estudios Sobre Historia y Ambiente en América,* ed. Bernardo García Martínez y Alba González Jácome, 191–218. Vol. 1. El Colegio de México, México, D.F. IPGH.

González Jácome, Alba. 2003a. Paisajes del Pasado. In *Estudios Sobre Historia y Ambiente en América,* ed. Bernardo García Martínez and Rosario Prieto, 200–20. Vol. II. El Colegio de México, IPGH.

González Jácome, Alba. 2003b. *Cultura y Agricultura. Transformaciones en el Agro mexicano.* México: Universidad Iberoamericana, México, D.F.

González Jácome, Alba. 2004. The ecological bases of the indigenous Nahua agriculture in the sixteenth century. *Journal of the Agriculture, Food and Human Values Society* 21:221–31.

González Jácome, Alba. 2007. Conversión social y cultural. De los agroecosistemas tradicionales a los alternativos en México. In *Los Nuevos Caminos de la Agricultura: procesos de conversión y perspectivas,* 59–95. México: UIA, Plaza y Valdés.

González Jácome, Alba. 2008a. *Humedales en el Suroeste de Tlaxcala. Agua y Agricultura en el Siglo XX.* México: UIA, Colegio de Historia de Tlaxcala.

González Jácome, Alba. 2008b. Unpublished fieldwork notes. Yucatán.

González Jácome, Alba. 2009. *Historias Varias. Caminando con los Agricultores Mexicanos.* México: UIA.

González Jácome, Alba, Silvia del Amo Rodríguez, and Francisco Gurri García, eds. 2007. *Los Nuevos Caminos de la Agricultura: procesos de conversión y perspectivas.* México: UIA, Plaza y Valdés.

Guhs, Bernard. 1992. *Fieldwork report on the Cajono River Basin in the Sierra de Oaxaca, Mexico.* México: UIA.

Gurría Lacroix, Jorge. 1978. *El desagüe del valle de México durante la época novohispana.* México: UNAM.

Hoekstra, Rik. 1992. Profit from the wastelands: Social change and the formation of haciendas in the Valley of Puebla, 1570–1640. *European Review of Latin American and Caribbean Studies* 52:91–123.

Iltis, Hugh H. 2006. Origin of Polystichy in Maize. In *Histories of maize: Multidisciplinary approaches to the prehistory, linguistics, biogeography, domestication and evolution of maize,* 22–53. Amsterdam: Elsevier, Academic Press.

INEGI. 2000. *Population and housing national census.*

INEGI. 2005. *Population and housing national census.* Addenda.

INEGI. 2007. *Censo Agropecuario 2007.* Aguascalientes, Aguascalientes.

INEGI. 2008. *Censo Ejidal.* Aguascalientes, Aguascalientes.

Juan Pérez, José Isabel. 2003. Tiempo con dinero y tiempo sin dinero: de la agricultura tradicional a la comercial en Progreso Hidalgo, Estado de México. PhD dissertation in social anthropology, Iberoamericana University, Mexico.

Juan Pérez, José Isabel. 2007. Huertos, ambiente y cultura en el ecotono sur del Estado de México. In *Los nuevos caminos de la agricultura: procesos de conversión y perspectives*, 261–80. México: UIA, Plaza y Valdés.

Korsback, Leif. 1996. *Introducción al Sistema de Cargos*. México: Universidad Autónoma del Estado de México.

López Austin, Alfredo. 2007. Cuatro mitos mesoamericanos del maíz. In *Sin maíz no hay país*, pp. 29–35, ed. Gustavo Esteva and Catherine Marielle. México: Culturas Populares de México.

López Montes, Genaro. 2008. Desecación de las Lagunas del Alto Lerma y Agricultura en la Zona Lacustre de Santa Maria Jajalpa, Estado De México. Manuscript.

Mariaca Méndez, Ramón. 2002. Marqués de Comillas, Chiapas: Procesos de Inmigración y Adaptabilidad en el Trópico Cálido Húmedo de México. PhD dissertation in social anthropology, Iberoamericana University, Mexico.

Mariaca Méndez Ramón, L. Arias, and A. González. 2007. El Huerto Familiar en México. Un Agroecosistema Antiguo que Puede Ser Sustentable. In *Avances en agroecología y ambiente*, Vol. 1, pp. 119–138, eds. Jesús Francisco López Olguín et al. México: Universidad Autónoma de Chapingo y BUAP.

Márquez Mireles, Leonardo Ernesto. 2005. De la Agricultura Tradicional a la Convencional: ahorro y capital en Cruz de Piedra, Estado de México. PhD thesis in social anthropology, Iberoamericana University, Mexico.

Márquez Mireles, Leonardo Ernesto. 2007. De la agricultura tradicional a la convencional en Cruz de Piedra, Estado de México. In *Los nuevos caminos de la agricultura: procesos de conversión y perspectivas*, 351–71. México: UIA, Plaza y Valdés.

Martínez Saldaña, Tomás. 2006. Los Rituales del Agua en el Río Bravo. *Anduli* 6:193–201.

Martínez Saldaña, Tomás, and Leticia Gándara Mendoza. 2007. La agricultura sustentable: una opción de desarrollo para una dimensión social de la agricultura. In *Los nuevos caminos de la agricultura: procesos de conversión y perspectivas*, 147–60. México: Plaza y Valdés, Universidad Iberoamericana AC, PROAFT.

Melville, G.K. Elinor. 1994. *A plague of sheep: Environmental consequences of the conquest of Mexico*. Cambridge, UK: Cambridge University Press.

Melville, G.K. Elinor. 1997. Environmental change and social change in the Mezquital Valley, Mexico 1521–1600. In *Agriculture, resource exploitation, and environmental change*, ed. H. Wheatley, 69–98. Brookfield, VT: Variorum.

Moctezuma Pérez, Sergio. 2008. Ambiente, caficultura y migración: los indígenas totonacos de Naranjales, Mecatlán, Veracruz. MA thesis in social anthropology, Iberoamericana University, Mexico.

Molina Hampshire, Daniel Alejandro. 2003. El Entorno Limitado. Una comunidad en el Área de la Reserva de la Mariposa Monarca. MA thesis in social anthropology, Iberoamericana University, México.

Morales Valderrama, María del Carmen. 2004. Población y sistemas de cultivo de maíz en Los Chenes, Campeche durante el Siglo XX. Unpublished manuscript.

Mountjoy, Daniel C. 1985. Adaptation and change in a local agroecosystem of Tlaxcala, Mexico. MA thesis, University of California, Santa Cruz.

Nickel, Herbert J. 1996. *Morfología Social de la Hacienda Mexicana*. México: FCE.

Olivares Rodríguez, Felipe. 2007. Agricultura Campesina Cambio y Permanencia: el Caso de Míxquic. PhD dissertation in social anthropology, Iberoamericana University, Mexico.

Orozco Segovia, Alma D.L. 1999. *El marceño en las zonas inundables de Tabasco*. Agricultura y Sociedad en México. Diversidad, Enfoques, Estudios de Caso. México: Plaza y Valdés, Universidad Iberoamericana AC, PROAFT.

Palerm, Ángel. 1968. *Productividad Agrícola: Un Estudio Sobre México*. México: Centro Nacional de Productividad.

Palerm, Ángel. 1973. *Obras hidráulicas prehispánicas en el Sistema Lacustre del Valle de México*. México: SEPINAH.

Quiñónez, Columba. 2005. Chinampas y chinamperos: los horticultores de San Juan Tezompa. PhD dissertation in social anthropology, Iberoamericana University, Mexico.

Ramírez Martínez, Miguel Ángel. 2007. Ambiente, cultura y sociedad: los productores de cacao de pequeña escala de José María Pino Suárez, Comalcalco, Tabasco. PhD dissertation in social anthropology, Iberoamericana University, Mexico.

Rebolledo Recéndiz, Nicanor. 2007. Jefaturas del Agua. Ambiente y política en el sureste de Tlaxcala. Mesa Redonda Agua y Agricultura, INAH-DEAS, Seminario de Historia, Filosofía y Sociología de la Antropología Mexicana, Noviembre 30.

Reyes Montes, Laura. 2003. Adaptación sociocultural en Santiago Yeché: un estudio de ecología cultural en México. PhD dissertation in social anthropology, Iberoamericana University, Mexico.

Robles Cervantes, Maribel. 2008. Purificación Tepetitla: Organización familiar para la producción de plantas medicinales en los huertos del Somontano. MA thesis in social anthropology, Iberoamericana University, Mexico.

Rodríguez, M.D. 2003. Personal communication.

Romero Contreras, Alejandro Tonatiuh. 1998. *Los Temascales de San Isidro Suceso. Cultura, Medicina y Tradición de un Pueblo Tlaxcalteca*. Tlaxcala: Tlaxcallan Ediciones del Gobierno del Estado.

Sartorius, Carl. 1961. *Mexico about 1850*. Stuttgart: Brockhaus.

Servín Segovia, Jorge Aníbal. 2000. Sistemas de cultivo en una barranca: el caso de Xopilapa en Veracruz Central. MA thesis in social anthropology, Iberoamericana University, Mexico.

Servín Segovia, Jorge Aníbal. 2001. La globalización en una "Economía Enemiga," el Caso de Xopilapa en el Veracruz Central. *Actas Latinoamericanas de Varsovia* 24:165–77.

Servín Segovia, Jorge Aníbal. 2002. Banqueteras: un sistema agrícola Mesoamericano. *Suplemento Antropológico* XXVII:285–321.

Siemens, Alfred H. 1983. Modeling prehispanic hydro agriculture on levees back slopes in northern Veracruz, Mexico. In *Drained field agriculture in Central and South America*, pp. 27–54, ed. J.P. Darch. BAR International Series 189. Oxford: BAR International.

Siemens, Alfred H. 1990. *Between the summit and the sea. Central Veracruz in the nineteenth century*. Canada: University of British Columbia Press.

Tecontero Tlacopanco, Margarita. 2005. Alimentación y Salud en Almoloya del Río, Estado de México. MA thesis in social anthropology, Department of Anthropology, University of the State of Mexico (UAEM), Toluca, México.

Turrent Fernández, Antonio. 2007. Plan estratégico para expandir la producción de granos a niveles superiores a la demanda. In *Desarrollo agropecuario, forestal y pesquero*, 179–98. México: Cámara de Diputados, Miguel Ángel Porrúa y UNAM.

Vanderwarker, Amber M. 2006. *Farming, hunting and fishing in the Olmec world*. Austin: University of Texas Press.

Velasco Orozco, Juan Jesús. 2002. *Subsistencia Campesina y Desarrollo Sustentable en la Región Monarca*. México: Universidad Autónoma del Estado de México.

Wilken, Gene C. 1969. Drained field agriculture. An intensive farming system in Tlaxcala, Mexico. *Geographical Review* 59:215–41.

Wilken, Gene C. 1987. *Good farmers. Traditional agricultural resource management in Mexico and Central America*. Berkley: University of California Press.

10 Cuba

A National-Level Experiment in Conversion

Fernando R. Funes-Monzote

CONTENTS

10.1 INTRODUCTION

Cuba has a long tradition as an exporter of agricultural crops produced under conditions of monoculture and natural resource extraction (Le Riverend, 1970; Moreno Fraginals, 1978; Marrero, 1974-1984). Practiced over approximately four centuries, these agricultural patterns have generated a dependence on imported inputs and caused an enormous negative environmental impact on soils, biodiversity, and forest cover (CITMA, 1997; Funes-Monzote, 2004). During the last 15 years, however, agricultural development has been reoriented (Rosset and Benjamin, 1994; Funes et al., 2002; Wright, 2005). Today, agricultural production in Cuba is concerned, as never before, with food self-sufficiency and environmental protection. In 1994, the National Programme for Environment and Development (the Cuban adoption of the United Nations Division for Sustainable Development's Agenda 21) was instituted, and two years later the National Environmental Strategy was approved (CITMA, 1997; Urquiza and Gutiérrez, 2003). In 1997 the Cuban law of environment became the environmental protection policy of the state (*Gaceta Oficial*, 1997). Although environmental protection is still not practiced as fully as it might be, government support for preserving the environment has helped put Cuban agriculture on a more sustainable course.

A principal goal of the revolution of 1959 was to resolve what were perceived as long-standing problems of Cuban agriculture, mainly national and foreign (basically North American) ownership of large farms and lack of agricultural diversification (Anon, 1960; Valdés, 2003). However, the rapid industrialization of state-controlled agriculture based on conventional methods after the revolution tended to concentrate land in large state enterprises, and consequently resulted in environmental problems similar to those caused by the old *latifundios*. Although on one hand, this model successfully increased both levels of production and rural well-being owing to the social goals of the political system, on the other hand it produced negative economic, ecological, and social consequences that cannot be ignored.

The excessive application of externally produced agrochemical inputs (i.e., produced outside the country), the implementing of monocultural, large-scale production systems, the concentration of farmers in the cities or rural towns, and the dependence on few exports conferred a high vulnerability to the nationally established conventional agricultural model. This vulnerability became evident at the beginning of the 1990s with the disintegration of socialist Eastern Europe and the USSR, when the majority of the favorably priced inputs, both material and financial, disappeared. Cuban agriculture, along with the other branches of the national economy, entered into its greatest crisis in recent history; at the same time, however, these factors provided exceptional conditions for the construction of an alternative—and far more sustainable—agricultural model at a national scale.

The transformation that occurred in the Cuban countryside during the last decade of the twentieth century is an example of a large-scale agricultural conversion—from a highly specialized, conventional, industrialized agriculture, dependent on external inputs, to an alternative input substitution model based on principles of agroecology and organic agriculture (Altieri, 1993; Rosset and Benjamin, 1994; Funes et al., 2002). Numerous studies of this conversion attribute its success to both the form of social organization employed and the development of environmentally sound technologies (Rosset and Benjamin, 1994; Deere, 1997; Pérez Rojas et al., 1999; Sinclair and Thompson, 2001; Funes et al., 2002; Wright, 2005).

Unlike the isolated sustainable agriculture movements that have developed in most countries, Cuba developed a massive movement with wide, popular participation, where agrarian production was seen as key to food security for the population. Still in its early stages, the transformation of agricultural systems in Cuba has mainly consisted of the substitution of biological inputs for chemicals, and the more efficient use of local resources. Through these strategies, numerous objectives of agricultural sustainability have been serendipitously reached. The persistent shortage of external inputs and the surviving practices of diverse production systems have favored the proliferation of innovative agroecological practices throughout the country.

Under current conditions, however, with about 5,000 enterprises and cooperatives and nearly 400,000 individual producers (*Granma*, 2006b), neither the conventional model nor that of input substitution will be versatile enough to cover the technological demands of such a heterogeneous and diverse agriculture. Consequently, the author believes it is necessary to develop a more integrated, participatory, long-term agroecological focus and to more strongly combine the economic, ecological, and sociopolitical dimensions of agricultural production. A mixed farming systems approach is presented here as the next step toward sustainable agriculture, one that can address these needs at a national scale.

10.2 GEOGRAPHIC AND BIOPHYSICAL BACKGROUND

Cuba, the biggest of the Caribbean islands, is strategically located between the two Americas, allowing it to play an important role for the Spaniards in their conquest of the New World. Cuba is approximately three times the size of the Netherlands, and half the size of Minnesota, the 12th largest state in the United States. With a total area of 110,860 km^2, the country is dominated by expansive plains (occupying about 80% of the total) and three well-defined mountain ranges.

Cuba may even be considered a micro-continent, owing to the highly diverse nature of its natural biodiversity, soil types, geographic landscapes, geological ages, and microclimates (Rivero Glean, 2005). The country comprises 48 well-defined natural regions, each with specific characteristics of climate, vegetation, and landscape, ranging from rainforest to semidesert (Gutiérrez Domenech and Rivero Glean, 1999). Such heterogeneity favors a high natural biodiversity: the island supports 19,631 known plant and animal species, of which 42.7% are endemic (ONE, 2004).

TABLE 10.1

Demographic, Physiographic, and Climatic Features of Cuba

			Climate	
			Season	
General Data			Wet	Dry
Length of country, km	1,250	Rainfall, mm	1,104	316
Area, km²	110,860	Mean temperature, °C	26.9	23.2
Highest elevation, m	1,974			
Total population (millions)	11.3			

Source: ONE (2004).

Cuba is long (1,250 km) and thin (the average width is less than 100 km, with a maximum of 191 km and minimum of 31 km). This physiography facilitates sea transport. The most important cities, connected by some 5,700 km of railway, are located an average of less than 40 km from the coast, with its more than 200 bays and coves.

According to the climate classification system recognized by the Food and Agriculture Organization (FAO) (Koppen, 1907), Cuba's climate is tropical savannah. However, it is also considered to have a tropical oceanic climate (Alisov and Paltaraus, 1974). These and other general classification criteria have been adapted in various forms to heterogeneous Cuban conditions (Lecha et al., 1994). Except for some specific areas, the whole island is influenced by the ocean.

Near to the Tropic of Cancer and the Gulf Stream, the island receives the destructive effects of tropical storms and hurricanes (with winds of 150 to 200 km/hour and more) as well as severe droughts that directly affect agricultural activity and the infrastructure in general. The climate is characterized by a wet season, with high temperatures and heavy rains, between May and October (70% of the total annual rainfall) and a dry season from November to April with low rainfall and cooler temperatures (Table 10.1).

Although Havana is the main economic center, each of the country's 14 provinces is important agriculturally, culturally, and economically. Population density is higher in Cuba (101.7 inhabitants/km²) than in Mexico (50), Central America (68), and South America (17), but lower than the average for the Caribbean region (139) (FAOSTAT, 2004). More importantly, Cuba has a high percentage of arable land, so that each arable hectare only needs to feed less than two people per year. Whereas agricultural land accounts for about 34% of the total land area in Latin America as a whole, in Cuba approximately 60% of the land is appropriate for agriculture (ONE, 2004; FAOSTAT, 2004). However, according to the last national census, currently less than 25% of the Cuban people live in rural settlements, only 11% work in the agricultural sector, and probably less than 6% are directly linked to farming activities (ONE, 2004).

Soils in Cuba are heterogeneous. Soil fertility, as based on available nutrients and classified as a percentage of the total arable land, is 15% high fertility, 24% fair fertility, 45% low fertility, and 14% very poor fertility (CITMA, 1998; ONE, 2004;

TABLE 10.2
Principal Limiting Factors of Cuban Soils

Factor	Affected Agricultural Area	
	(million ha)	(percent of total)
Salinity and sodicity	1.0	14.9
Erosion (very strong to medium)	2.9	43.3
Poor drainage	2.7	40.3
Low fertility	3.0	44.8
Natural compaction	1.6	23.9
Acidity	2.1	31.8
Very low organic matter content	4.7	69.6
Low moisture retention	2.5	37.3
Stony and rocky areas	0.8	11.9

Source: CITMA (1998), ONE (2004).

Treto et al. 2002). According to these sources, Cuban soils are predominantly oxisols and ultisols (68%), and the remaining areas are mostly inceptisols and vertisols. The primary limiting factors of soils used for agricultural activities are low organic matter content, low fertility, erosion, and poor drainage (Table 10.2).

Despite these limitations, Cuba possesses an exceptional natural environment for agriculture. Due to its continuous growing season and diversity of plants and animals used for agricultural purposes, crop cultivation and raising animals in open air are possible throughout the year. The ample infrastructure of roads and railroads with access to the sea, the existence of high-water reservoir capacity for irrigation, electrification of the countryside, and high investment in agricultural facilities are all valuable preconditions for greater agricultural production in Cuba. In addition, the extensive network of scientific institutions is a considerable asset in carrying out agricultural changes. However, these resources are not being efficiently used for several reasons, including a lack of maintenance of the agricultural infrastructure, continued specialized organization of agriculture, a scarcity of agricultural labor, and the high cost (or lack of availability) of necessary inputs for production.

10.3 BRIEF HISTORY OF CUBAN AGRICULTURE

10.3.1 MIGRATORY ABORIGINAL GROUPS

The first inhabitants of Cuba arrived about 10,000 years ago from North America through the Mississippi River watershed, via Florida and the Bahamas (Torres-Cuevas and Loyola, 2001). Called *Guanahatabeyes,* these groups were hunters, fishers, and gatherers. The second migratory stream came from South America about 4,500 years ago. Known as *Ciboneyes,* they were also fishers and gatherers, but introduced a variety of more advanced instruments for hunting and food processing. Some 1,500 years ago, a third group of people called *Taínos* came to the island. Part of the South American aboriginal family known as *Arawaks*, they were advanced hunters and fishers, but they also practiced agriculture (Le Riverend, 1970). They were the most numerous and dominant Native Americans when the Spanish arrived on the island in 1492. One of their most productive agricultural systems utilized raised beds, called *camellones*, which were planted mounds of earth and organic matter. These communities applied the system of small-scale slash and burn for the cultivation of crops, especially cassava and corn, and those used in their rituals, such as tobacco and cotton.

10.3.2   Spanish Colonization of Cuba

At the time of the Spanish arrival, an estimated 60 to 90% of Cuba was covered with forest (Del Risco, 1995). Initially the conquerors resettled indigenous people in *vecindades* or reserves. In these reserves, most inhabitants continued using traditional agricultural methods. As colonists, the Spanish became landholders, employing predominantly mixed crop–livestock systems called *estancias* with a high proportion of crops (Le Riverend, 1970). The transition from indigeous agriculture to the new form implemented by the Spanish may be considered the first major step in the process of conversion to European agricultural practices.

The small population of Spaniards focused on cattle raising as their principal economic activity. To this end, they distributed lands in extensive circular areas called *hatos* and *corrales*. At the same time, around their population centers they established less extensive areas of crop cultivation (Le Riverend, 1992). In the middle of the 1500s, increasing demand for wood for ship construction, swelling populations in the main villages of the island, and the growing external market for agricultural products led to an expansion in timber extraction and sugar and tobacco production and processing. These activities extended into the interior of the cattle ranches, transforming the original Spanish agrarian structure.

Beginning in the early 1600s, commercial agriculture experienced more rapid development with the advent of sugar cane and tobacco production in the *estancias* (Le Riverend, 1992; Marrero, 1974–1984; Funes-Monzote, 2004).

The outbreak of the Haitian slave revolt in 1791 gave Cuba the opening it needed to begin competing with the French colonies as the principal producer and exporter of sugar worldwide. The consequent establishment of sugar processing plants in the Cuban countryside meant a radical transformation in the structure of agriculture and a definitive jump in the economy of colonial Cuba. The great expanses of land dedicated to cattle ranching, interspersed with forest and grassland, were subdivided into smaller properties. The increased scale of production and the specialization in sugar cane accentuated the social and environmental impacts in the countryside that had accompanied the industry from the beginning. Early criticism of the system was based on damage to the natural resource base, specifically forest destruction and the abandonment of "tired," unproductive lands (De la Sagra, 1831; Reynoso, 1862).

10.3.3   Neocolonial Agricultural Patterns and Their Consequences

Concentration and centralization of sugar production continued into the 1900s. After Cuba achieved independence from Spain in 1898, North American capital flowed into the country, helping to establish giant sugar *latifundios* on the eastern half of the island, which until this time had been the area least affected by agriculture. During the first two decades of the twentieth century, the planting of sugar cane produced the most intense deforestation in Cuba's history. By around 1925, most of the extensive plains of Cuba had been planted with sugar cane. The largest ranches, both foreign and nationally owned, were predominantly sugar cane and cattle, and these occupied 70% of the agricultural land. A little more than 1% of the landowners owned 50% of the land, while 71% held only 11% (Valdés, 2003).

However, the lands managed by the *latifundios* were inefficient at producing food, and many of these large farms (around 40%) were gradually abandoned. Meanwhile, the *campesino* sector, which practiced a diversified agriculture with traditional mixed farming strategies, was having a considerable impact on the agrarian economy. According to the agricultural census of 1946, almost 90% of the farms were diversified. These 5 to 75 ha farms, with their mixed crop–livestock production and better organizational efficiency, generated about 50% of the country's total agricultural production but occupied only 25% of the total agricultural area (Censo Agrícola Nacional, 1951).

Despite the existence of many diversified small farms, the structure of land tenure and the export-oriented economic model combined to create an agriculture sector that as a whole specialized in only a few agricultural crops. Rural Cuba was characterized by an economic and political dependency on the United States, a scarcity of subsistence foods, social inequity, and a high rate of unemployment during the "dead period" (months where there was no sugar processing). This unstable situation greatly influenced the emergence of the Cuban Revolution of 1959, which was grassroots, agrarian-based and anti-imperialist. During the 46 years since the revolution, unprecedented events have taken place with arguable relevance to the future of world agriculture.

10.4   POSTREVOLUTION SCENARIO

10.4.1   AGRARIAN REFORMS

The revolutionary government adopted two agrarian reform laws that passed ownership of rented lands to the peasants who had worked them. This considerably reduced farm size. First, in May 1959, the maximum land holding was reduced to about 400 ha. Later, in 1963, a Second Agrarian Reform established an upper limit of 67 ha in order to eliminate the landed social class and thus the exploitation of farmers (Anon, 1960; Valdés, 2003). In the first stage, 40% of arable land was expropriated from foreign companies and large landholders and passed into the hands of the state. In the second stage, another 30% of the land became state owned (Valdés, 2003).

At that point, there were four prioritized objectives for the transformation of Cuban agriculture: (1) to meet the growing food requirements of the population, (2) to generate monetary funds through the exportation of products, (3) to obtain raw materials for the food processing industry, and (4) to eradicate poverty from the countryside (Anon., 1960). A number of educational, cultural, and economic approaches were developed, including literacy campaigns, the development of planned rural communities to supply social and health care services to farmers, the building of thousands of kilometers of new roads, and the extension of electricity to rural areas (Anon., 1987). The government's will to change was reflected clearly in the first decree of the first law of agrarian reform: "The progress of Cuba is based on the growth and diversification of industry to take more efficient advantage of its natural and human resources, as well as the elimination of the deep dependency on monocultural agriculture that is a symptom of our inadequate economic development" (*Gaceta Oficial*, 1959).

10.4.2   THE CONVENTIONAL AGRICULTURE MODEL

Although the government expressed its official desire for diversification, its actual on-the-ground administration of agriculture supported large-scale monoculture. The commitments to export primary materials such as sugar, citrus, coffee, tobacco, etc., to the countries of the Council for Mutual Economic Assistance (COMECON)—the economic block of the former socialist countries—forced Cuba to fulfill five-year plans at high environmental costs. Consequently, the dependency on processed food imported from Eastern Europe reached unprecedented levels (Espinosa, 1992).

The application of green revolution concepts was facilitated by Cuba's strong relationship with the socialist countries of Eastern Europe, particularly with the Soviet Union (USSR). As a national policy, Cuba adopted the world trend of substituting capital for labor in order to increase productivity. This method was characterized by the physical and agrochemical management of agricultural processes—specifically large-scale, mechanized production with a high application of external inputs to a monocultural crop. The application of the industrialized model of agriculture, along with the 10-fold increase in food imports over a 30-year period (1958–1988), was successful in achieving increases in per-capita calorie consumption from 2,552 kcal/day in 1965 to 2,845 kcal/day in 1989. Protein consumption per capita also increased in the same period from 66.4 g/day to 76.5 g/day. In spite of this progress, however, per-capita consumption rates still fell short of the calculated nutritional needs of 2,972 kcal/day for calories and 86.3 g/day for protein (Pérez Marín and Muñoz, 1991).

These improvements were achieved and sustained through a model that relied on high external inputs, a few export crops, and trade with the socialist countries of Eastern Europe. Throughout the 1980s, 87% of external trade was undertaken at favorable prices with socialist countries, and only 13% at world market prices with other countries (Lage, 1992). In 1988, Cuba sent 81.7% of its total exports to the socialist bloc of Eastern Europe, while 83.8% of its total imports came from those countries (Pérez Marín and Muñoz, 1991). The COMECON agreement allowed Cuba to sell its goods in the socialist market of Eastern Europe at high prices while imports were purchased from them at low cost.

Consequently, the dependency of the agricultural economy on a few export products was impressive, and the land dedicated to these crops was enormous. Three of the principal export crops—sugar, tobacco, and citrus—covered 50% of agricultural land. Importing energy (petroleum), machinery, and diverse raw materials in large amounts was favorable for Cuba in economic terms, but not for its food self-sufficiency. Under these conditions the country imported 57% of its protein requirements and more than 50% of its energy, edible oil, dairy products and meats, fertilizers, herbicides, and livestock feed concentrates (PNAN, 1994).

As early as the 1970s, Cuban scientific research institutions had become aware of the concepts of low external inputs and input substitution. Policies and research began to focus on the economic implications of substituting local raw materials for imported. Nevertheless, at the end of the 1980s, Cuban agriculture was characterized by a high concentration of state-owned land (80% of total land area was in the state sector), high levels of mechanization (one tractor for every 125 ha of farming land), crop specialization, and high input usage (13 million tons diesel, 1.3 million tons fertilizers, US$80 million in pesticides, and 1.6 million tons livestock feed concentrates applied per year) (Lage, 1992).

10.4.3 CONSEQUENCES AND COLLAPSE

The continued application of this agricultural model resulted in several economic, ecological, and social consequences. Among the most important were soil salinization (1 million ha affected), an increased frequency of moderate to severe soil erosion, soil compaction with its resultant soil infertility, loss of biodiversity, and deforestation of agricultural land (CITMA, 1997). From 1956 to 1989, an accelerated rural population exodus to urban areas caused a drop in the rural population from 56% to 28%, and then to less than 20% by the mid-1990s (Funes et al., 2002).

As result of this situation, at the end of the 1980s crop and livestock yields and subsequent economic efficiency started to decrease (Pérez Marín and Muñoz, 1991). The conventional agricultural model, which had been applied for about 25 years, demanded higher amounts of chemical inputs and capital to keep yields stable. The depression of agricultural production provoked a shortage of goods in the agricultural markets. To counter this situation, an ambitious food program was initiated in order to recuperate the infrastructure and subsequent volume of production and cover internal demand (ANPP, 1991). This program essentially carried on the conventional high-input focus because it could count on abundant externally derived inputs. Even when the disintegration of Eastern European and Soviet socialism resulted in the loss of these inputs, the government decided "to continue developing the Food Program despite whatever difficult conditions might have to be faced" (ANPP, 1991, p. 7). Without the expected aid, however, it would be necessary to seriously adjust the technology and structure of production.

10.5 SITUATION AFTER THE COLLAPSE OF THE SOCIALIST BLOC

Today Cuba faces the most difficult challenge in its history … in addition to the worsening blockade exercised for more than 30 years by the United States, it now has to resist the effects of a second blockade provoked by changes in the international order.

—**Fidel Castro, 1992**

The unexpected collapse of the socialist countries of Eastern Europe and the USSR fully highlighted the contradictions and vulnerabilities of the agricultural model that Cuba had developed. The island lost the principal markets and guarantees that these countries had provided in the past. Foreign purchase capacity was drastically reduced from US$8,100 million in 1989 to US$1,700 million by 1993, a decrease of almost 80%. In that year, some US$750 million was required solely for the purchase of fuel for the national economy and US$440 million for basic foods (Lage, 1992; PNAN, 1994).

Cuba's reduced foreign exchange greatly affected its ability to obtain necessary agrochemical inputs, leading to a drastic reduction in production. This shortage was most severely felt by the large state farm enterprises that were dependent on high inputs to maintain their monoculture systems. In fact, all farmers suffered under the difficult situation, but small- and medium-size farmers were less affected due to their more locally oriented agricultural strategies, the practice of a more diversified agriculture, greater control of farm management, and lower dependence on external inputs.

Although small- and medium-scale traditional farming exhibited higher resilience to the crisis, in 1989 this sector of agricultural production represented only 12% of the total agricultural land area. The remaining agricultural lands, which were being managed using high-input, industrialized, and large-scale methods, dramatically collapsed. This led to the drastic reduction of each citizen's food ration, which seriously affected food security. One of the first effects was caloric deficiency, and consequently, widespread weight loss among the population. In addition, many diseases started to appear as a result of low intake of certain nutrients (PAHO, 2002) (Table 10.3). For example, epidemic neuropathy, caused by vitamin B deficiency, affected the vision of more than 50,000 people (Arnaud et al., 2001). The consequences of the food security crisis would have been far more dramatic without the government's ration system, which ensured equitable food access and avoided famine (Rosset and Benjamin, 1994; PNAN, 1994; Wright 2005).

TABLE 10.3
Comparison of Nutritional Levels per Capita per Day in 1987 and 1993

Nutrient	Nutritional Needs[a]	Percentage Satisfaction of Recognized Needs	
		1987	1993
Calories	2,972 kcal	97.5	62.7
Protein	86.3 g	89.7	53.0
Fat	92.5 g	95.0	28.0
Iron	16 mg	112.0	68.8
Calcium	1,123 mg	77.4	62.9
Vitamin A	991 mg	100.9	28.8
Vitamin C	224.5 mg	52.2	25.8

Source: PNAN (1994), Pérez Marín and Muñoz (1991).

[a] The nutritional needs for the Cuban population (Porrata et al., 1996) were defined by the FAO standards (FAO/WHO/UNU, 1985).

Despite the economic difficulties, the government continued to reinforce social programs. For example, the infant mortality rate during the first year of life was reduced by almost half during this time—from 11.1 per 1,000 in 1989 to 6.4 at the close of 1999 (*Granma*, 2000). During the early 1990s, severe economic actions were necessary in order to maintain the main social guarantees while reconstructing the Cuban economy. This phase was officially called the "special period." In order to deal with the crisis, the Cuban government implemented measures of austerity and changed the strategies to reduce negative impacts on the national economy.

In response to the precarious food situation, the Cuban National Program of Action for Nutrition (PNAN) was instigated, as a result of commitments made by the International Nutrition Conference in Rome in 1992. Its overall objective was to buffer the consequences of the crisis using the following basic strategies (PNAN, 1994):

- Strengthen agrarian policy through widespread decentralization of land holdings and management, diversification of agricultural production, and the transformation of land tenure of state lands.
- Encourage the population to participate in agricultural activities for their own nutritional improvement.
- Encourage the creation of *autoconsumos* or on-site farms/gardens to supply the dining halls of residential and educational establishments.
- Promote sustainable development compatible with the environment.
- Reduce postharvest losses through improved methods, such as direct sales of food from producers to consumers in the cities (e.g., urban agriculture).
- Incorporate nutritional objectives in programs and plans of agricultural development.

Many of these measures taken by the state were key factors in the proliferation of a more sustainable Cuban agriculture. However, the success of these strategies has been muted by a variety of factors, including the difficulty of adapting specialized large-scale agriculture to new practices, a lack of monetary resources and materials to enact these solutions, and a small workforce in the countryside.

10.6 CHANGES IN AGRARIAN PRODUCTIVE STRUCTURES

In general, certain technical and organizational measures were taken to reduce the impact of the crisis on agriculture. Decentralization and reduction in the scale of big state enterprises was a necessity due to their inefficiency. In 1993, the government created Basic Units of Cooperative Production (UBPCs). This effective measure gave usufruct rights (land use free and for an "indefinite" time) to farmers who were previously workers of state farm enterprises. Other forms of land distribution were also developed that provided interested urban dwellers the opportunity to return to the countryside. Eventually, 10 distinct forms of organization in Cuban agriculture were created; these coexist within three sectors: the state sector, the nonstate sector, and the mixed sector (Table 10.4).

These changes in the agrarian structure of the country were characterized by transfers of land from the state to the other sectors. By January 1995 the state had

TABLE 10.4

Organization of Cuban Agriculture

State sector		State farms
		New-type state farms (GENT)
		Revolutionary Armed Forces (FAR) farms, including farms of the Young Workers' Army (EJT) and the Ministry of Interior (MININT)
		Self-provisioning farms at workplaces and public institutions
Nonstate sector	Collective production	Basic Unit of Cooperative Production (UBPC)
		Agricultural Production Cooperatives (CPA)
	Individual production	Credit and Service Cooperatives (CCS)
		Individual farmers, in usufruct
		Individual farmers, private property
Mixed sector		Joint ventures between the state and foreign capital

Source: Martín (2002).

granted usufruct rights to 58% of the arable land it had controlled at the beginning of 1990 (which had constituted, at that time, 83% of total arable land). This shift in land ownership is informally called the silent third Cuban agrarian reform. During a five-year period, about 150,000 workers were incorporated into the UBPCs (Pérez Rojas et al., 1999). A chronological analysis of the percentage of national agricultural area shows that the UBPCs quickly predominated (Table 10.5). The private, *campesino* sector also increased its land area in the distribution process, an acknowledgment of its management capacity and increasing role in food production. Compared to state enterprises, the UBPC is a more decentralized form of production (Villegas, 1999).

With the creation of the UBPCs, the state was able to both better manage production and save on scarce resources. The size of large mixed crop enterprises was reduced 10-fold, while the size of livestock enterprises was reduced on average 20-fold, reaching a size similar to that of the Agricultural Production Cooperatives

TABLE 10.5

Percentage of Arable Land in Cuba by Form of Land Ownership, 1989–2008

	1989–1992	1993	2000	2008
State	83	47.5	33.1	23.2
Other state sector organizations		9		
UBPC	—	26.5	40.6	39.8
CPA	12	7	26.3	37
Private		10		

Source: PNAN (1994), Pérez Rojas et al. (1999), ONE (2004, 2008).

TABLE 10.6
Average Size of State Enterprises, UBPCs, and CPAs

Principle Activity	State Enterprises (ha), 1989	Average Size UBPCs (ha), 1994	Average Size CPAs (ha), 1994
Various crops[a]	4,300	416	483
Citrus and fruit	17,400	101	577
Coffee	—	429	470
Tobacco	3,100	232	510
Rice	27,200	5,040	—
Cattle	28,000	1,597	631

Source: Data from PNAN (1994).

[a] Tubers, roots, vegetables, plantain, grains, and seeds (beans, corn, soybean, sunflower, sesame, etc.).

(CPAs) that had existed for more than 20 years with reasonable levels of production and efficiency (Table 10.6). The strategy of dividing land into smaller plots within the UBPCs was based on recognition of the greater efficiency of production at a smaller scale. (However, even with these reductions, the average sizes of UBPCs were still large for most of the principal agricultural activities, and the lack of resources made many of them almost unmanageable.)

Following the principle of linking people to the land (i.e., allowing farmers to live on the farm), thousands of families became based on the UBPCs, which had been previously uninhabited and controlled by state enterprises. For example, more than 50 families moved to the 1,000 ha that is now the UBPC "26 de Julio" in Bacuranao, Havana—a tract of land occupied some 15 years ago by only two families—after housing was created to attract people knowledgeable about working in agriculture. (Today this UBPC is highly self-sufficient in food production, generates extra production for commercialization, and achieves its commitment of milk production for sale to the state.) The repopulation of rural areas has been one of the major contributions of the UBPC.

As agricultural enterprises worked and managed by the people who live on them, UBPCs facilitated better natural resource management and local farmer decision making. The reduced scale of the UBPCs, along with their greater diversification and more rational use of inputs, machinery, and infrastructure, allowed increases in efficiency and productivity, and this helped mitigate the losses in external inputs and capital.

However, the UBPC model, as a new form of agriculture in Cuba, is still far from achieving its potential benefits. Many organizational methods employed in the state enterprises were replicated in the UBPCs (Pérez Rojas and Echevarria, 2000). The lack of a sense of ownership, the persistant dependency on external inputs, and limited decision making affect the functioning of UBPCs. In summary, even though the UBPCs in their essence have continued to form part of a structure that operates under the direction of the state enterprises, this form of production has created mechanisms favoring the transition to decentralized production that tends to imitate the values, efficiency, and potential of traditional *campesino* (small farmer) production.

10.7 CONTRIBUTION OF THE SMALL FARMER SECTOR

In Cuba, private farming (carried out by *campesinos* at mostly small and middle scales) can be undertaken individually or in groups under two types of cooperative production—CPA and CCS. The first type, the CPA, is composed of farmers who have given their land to the cooperative so that it can be transformed into social or collective property. The second type is composed of farmers who form a cooperative in which they continue to own land and equipment on an individual basis, buy inputs from the state, and receive credit and services (Alvarez, 2002). Both types of producers sell to the state based on agreements over their production potential, and also cultivate crops and raise animals for self-provisioning. They may also sell agricultural products directly in the local market or to middlemen.

Compared to state farms, private farmers have greater experience and a longer tradition with Cuban agriculture, and unsurprisingly, their agricultural systems proved to be more resilient in the face of the crisis. While the state agricultural enterprises were strongly impacted by the loss in inputs and funding, and delayed adapting to change, the *campesino* sector was able to buffer the scarcity of material resources. At the end of the 1980s, the private sector in Cuban agriculture accounted for 18% of the country's arable land; 10 years later it occupied 25% of the agricultural area and participated significantly in production for both internal consumption and export. The relatively high percentage contribution of *campesino* production to total sales in the national agricultural sector during the years of crisis (Table 10.7) demonstrates how efficient is its use of land. It also shows the capacity of small farmers' methods of production and organization to contribute to the national food balance, even with scarce external inputs.

Abolished at the end of the 1980s, the *Mercado libre campesino* (farmers' free market) was reopened at the beginning of 1994 as the *Mercado Agropecuario*

TABLE 10.7
Percentage Contribution of *Campesino* Production to Total Sales to the State for Various Products in Cuba

Product	Percent of Sales to the State	Product	Percent of Sales to the State
Roots, tubers, and vegetables	43	Milk	32
Sugar cane	18.4	Rice	17
Tobacco	85	Fruit	59
Coffee	55	Citrus	10
Cocoa	61	Pork	42.6
Beans	74	Fish	53
Corn	64	Honey	55

Source: Lugo Fonte (2000).

TABLE 10.8

Structure of Livestock Production in Cuba, 2006

Type of Production	Land Area (ha)	Percent of Land Area	Owners	Head (×10³)	Percent of National Herd	Head/Owner
State enterprises[a]	1,221.6	48.3	4,569	1,082.5	27.3	236.9
UBPC	780.1	30.8	2,470	969.6	24.4	392.5
CPA	201.7	8.0	1,063	191.8	4.8	180.5
CCS + individuals	325.8	12.9	236,088[b]	1,728.4	43.5	7.3
Total	2,529.3	100		3,972.3	100	

Source: Adapted from MINAG statistic bulletins and González et al. (2004).

[a] Included are livestock and crop enterprises dedicated to livestock rearing.

[b] Included are individual owners or in CCS and farmers with or without land.

(agricultural market). Despite the new name, it was in essence the same institution. This agricultural market functioned under the law of supply and demand and became an important distribution channel for agricultural products. In 1996, some 70.7% of the total agricultural direct sales to the population were by individual or cooperative farmers (Martín, 2002).

The small farmer sector was particularly successful with livestock. From 1995 to 2000, the number of livestock animals under private sector management increased, as did the production of livestock products, while during the same period state and UBPC livestock production showed no signs of recovery (González et al., 2004). In 2006, the small farmer sector, with only 13% of the grazing land, owned more than 43% of Cuba's livestock (Table 10.8), a fact that demonstrates the efficiency of *campesino* management. Although cattle production at the national level has been depressed by the scarcity of imported feed and adverse climatic conditions, such as prolonged drought, hurricanes, and other natural events, *campesino* production has developed ways of working around these conditions. Consequently, the small farm sector has, for many, served as a model for restructuring Cuban agriculture (Álvarez, 2002).

The Cuban *campesino* is a key link in the preservation of traditional crop and livestock varieties, which are indispensible to genetic improvement and sustainable agriculture from a local perspective (Ríos, 2004; Wright, 2005). Within the National Association of Small Farmers (Asociación Nacional de Agricultores Pequeños, ANAP), the Agroecological Farmer to Farmer Movement (Movimiento Agroecológico Campesino a Campesino, MACAC) has systematized much traditional agricultural experience and reinforced sustainable principles in Cuban agriculture. This movement is represented in 155 municipalities (i.e., 85% of total) at the national level, and at the end of 2004 employed 3,052 facilitators and 9,211 promoters (Perera, 2004). In a parallel effort, more than 4,000 farmers are involved in the Local Agriculture Innovation Programme of the National Institute for Agricultural Sciences (INCA), which is based on participatory grassroots processes (Ríos, 2006).

However, the positive impact of the *campesino* sector in the transformation of Cuban agriculture has not yet been sufficiently addressed. Many *campesino* agroecological experiences throughout the country are still undocumented despite the fact that they are undoubtedly the main resource necessary for the implementation of a sustainable and agroecological approach at a national scale.

10.8   URBAN AGRICULTURE AND FOOD SECURITY

10.8.1   FOUNDATION, STRUCTURE, AND OBJECTIVES

A major new initiative for the promotion of food self-sufficiency has been urban agriculture. This form of agriculture was almost neglected in Cuba when food was affordable. However, urban gardening was the first reaction of the population to overcoming food shortages (Murphy, 1999). By growing within and around cities, people could make use of local resources and not have to pay transportation costs for either inputs or products (Cruz and Sánchez, 2001). At the beginning of the crisis, people organized themselves to cultivate vacant lots, backyards, and rooftops in the cities. Animals were even reared inside houses in order to ensure families' food supply. At first a matter of subsistence production, urban agriculture by the mid-1990s had been transformed into a practice that also included commercial activities and made a significant contribution to the country's food security.

As urban agriculture became more widespread, it also became more organized and began to receive government support. The "horticultural club" formed in the Havana suburb of Santa Fe in 1992–1993 was the first to organize urbanites for the purpose of providing them with technical assistance and creating a framework for urban production. This movement grew very fast in Havana city and subsequently spread around the whole country.

By 1995, there were already 1,613 organoponics (i.e., small plots of abandoned land in the cities where beds of soil and sources of organic matter are used to produce fresh vegetables), 429 intensive gardens, and 26,604 community gardens. In 1997, a network of municipal enterprises and state institutions (the National System of Urban Agriculture) was created to organize the people already involved in urban agriculture. Spatially, this system covers a radius of 10 km from the center of the capital city of each province, a radius of 5 km from the center of municipal capitals, a radius of 2 km around population centers of more than 10,000 residents, and local production for settlements of less than 1,000 people. The government still plays an important role in the promotion and support of this massive movement toward food security.

The principal objective of the Cuban urban agriculture movement is to increase the daily consumption of vegetables to 300 g per citizen, the amount recommended by UN FAO. The following basic principles of urban agriculture in Cuba define its objectives and organization (Companioni et al., 2002).

- A fresh supply of good quality products offered directly to the population, guaranteeing a balanced production of not less than 300 g of vegetables daily per capita and an adequate variety of animal protein.

- Uniform distribution throughout the country (i.e., in every area of the country with an urban population, urban agriculture should be developed).
- Local consumption by the urban population of local production in each region.
- Crop–animal integration with maximum synergy (i.e., internal cycling of nutrients) to boost production.
- Intensive use of organic matter to increase and conserve soil fertility.
- Use of biological pest controls.
- Use of all available land to produce food, guaranteeing intensive but not import-dependent high yields of crops and livestock.
- Multidisciplinary integration and intensive application of science and technology.
- Maximum use of food production potential, including available labor as well as wastes and by-products for plant nutrients and animal feed.

The urban agriculture program is composed of 28 subprograms, each related to a type or aspect of animal or plant production. These subprograms form the organizational and administrative base of the program (GNAU, 2004). They include, for example, management and conservation of soils, use of organic matter, seed production, vegetables and fresh herbs and spices, fruit trees, grassroots or *arroz popular* production of rice, grains, animal feed, apiculture, livestock, aquaculture, marketing, and small agroindustries (Companioni et al., 2002). Taken together, Cuban urban agriculture has the components to achieve a systems approach; however, each program is supervised separately, responding to its specific factors and providing specialized technical assistance.

10.8.2 *Arroz Popular*: Example of a Successful Subprogram

Central to the Cuban diet, rice is consumed together with beans, meat, vegetables, and even fruits. Its per-capita consumption exceeds 44 kg annually, or 265 g per day (Socorro et al., 2002). Rice production in Cuba was developed for many years in large state farms, and it was also one of the prioritized crops at the beginning of the special period, when it appeared "irrefutable" that conventional, high-input methods were the only possible way to supply enough rice to meet the populations' needs (León, 1996). However, even during the 1980s, when unlimited inputs were available, the national demand was not met and it was necessary to import 40% of the rice consumed. High-input rice production proved to be unsustainable at the onset of the crisis of the 1990s. The new "popular rice" program demonstrated that self-organized, low-input agriculture could have a positive impact on national food self-sufficiency (García, 2003).

The popular production of rice (*arroz popular*) was originally, like urban agriculture in general, a grassroots movement toward self-provisioning. People started to cultivate this cereal in abandoned areas, in small plots between sugar cane fields, in road ditches, etc. This movement grew rapidly and achieved unforeseen levels of production and efficiency. In 1997, while the severely affected Union of Rice Enterprises (Unión de Empresas del Arroz) produced 150,000 tons of rice, popular rice production achieved 140,600 tons, involving 73,500 small producers yielding, as a national average, 2.82 tons per hectare without the use of costly inputs (*Granma*,

1998a). This yield compared favorably to that of conventional rice production during 1980s, which achieved a national average yield of between 2 and 3 tons/ha (ANPP, 1991). In 2001, *arroz popular* was responsible for more than 50% of total domestic rice production (García, 2003).

10.8.3 RECENT SUCCESS AND THE FUTURE

In the year 2000, urban agriculture produced more than 1.64 million tons of vegetables and employed 201,000 workers. Two years later, 326,000 people were linked with the program of urban backyard production. In 2005 production was 4.1 million tons, and in 2006 it had risen to 4.2 million tons, employing 354,000 people (*Granma*, 2001, 2003a, 2006a) (Figure 10.1). The reported production of 20 kg/m² achieved by urban agriculture exceeded 300 g of vegetables per citizen per day.

The urban agriculture movement has also contributed to the establishment of a network of 1,270 points of sale of agricultural products in the cities and 932 agricultural markets (*Granma*, 2003b). The products distributed via this network significantly contribute to food security, although the prices are still high considering the average buying capacity of the population.

The quantity of people dedicated to agricultural labor in the city periphery continues to increase. However, Cruz and Sánchez (2001) consider that this type of agriculture, emerging as a solution to food scarcity and unemployment in the cities, ought to look for a more integrated approach that goes beyond a temporary solution to the crisis and toward goals other than food security—such as preservation of urban environments, the permanent management of resources in urban settings, avoidance of air and water pollution, and creating a culture of nature conservation.

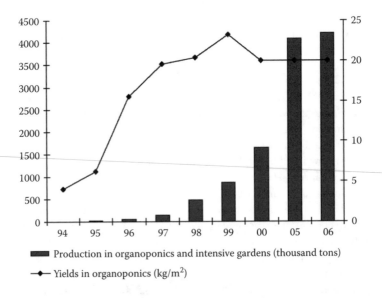

■ Production in organoponics and intensive gardens (thousand tons)

◆ Yields in organoponics (kg/m²)

FIGURE 10.1 Vegetable production from organoponics and intensive gardens.

Although cities became productive in terms of food, urban agriculture still satisfied a small part of the country's overall needs. Thus, it was necessary to develop participatory, low-input rural food production at the onset of the 1990s. An alternative model to the prevailing conventional agriculture paradigm—that of input substitution—was established at a national level, not only in state enterprises and the UBPCs, but also in private individual and cooperative production.

10.9 THE INPUT SUBSTITUTION STRATEGY

Gliessman (2001, 2007) describes three levels or stages in the process of converting from conventional to sustainable agroecosystems. At level 1 farmers "increase the efficiency of conventional practices," and at level 2 they "substitute conventional inputs and practices with alternative practices." Input-substituted systems at the second level, though demonstrably more sustainable than conventional systems, may nevertheless have many of the same problems that occur in conventional systems (e.g., the use of monoculture). These problems will persist until changes in agroecosystem design (i.e., on the basis of a new set of ecological processes) take place at level 3. This conversion process has been widely analyzed by Altieri (1987), who attributes the main cause of ecological disorders in conventional agriculture to monocultural patterns.

During the 1980s, a certain amount of research in Cuba focused on aspects of input substitution—reducing the use of fertilizers, pesticides, and concentrated feed for livestock. These investigations were applied to the most economically important and largest-scale agricultural activities (Funes, 2002). Although the main objective was the reduction of production costs in commercial agriculture through the substitution of biological inputs for agrochemical, these studies—underpinned by ecological principles—formed the basis for scaling up the application of ecological practices when no alternatives were available. As a result, input substitution in Cuba reached a scale never previously attempted in any other country, and its effectiveness and positive impact were significant (Rosset and Benjamin, 1994; Funes et al., 2002).

10.9.1 ALTERNATIVES FOR THE ECOLOGICAL MANAGEMENT OF SOIL

Many microbiological preparations had first been developed for a range of crops as part of general research on nitrogen fixation and solubilization of phosphorus. In the search for input substitution, a wide range of these biofertilizers have been successfully developed and applied on a commercial, main-crop scale, substituting for a significant percentage of chemical fertilizers (Table 10.9).

Research results confirmed the effectivness of using green manures and cover crops in commercial crop production. These studies included the use of sesbania (*Sesbania rostrata*) in rice production (Cabello et al., 1989) and the use of crotalaria (*Crotalaria juncea*), jack bean (*Canavalia ensiformis*), velvet bean (*Mucuna pruriens*), and dolichos lablab bean (*Lablab purpureus*) in other commercial crops (García and Treto, 1997). The inclusion of these plants in local systems was found to fulfill most nutrient needs of the crops. These green manures were able to substitute for high levels of nitrogen fertilization (i.e., the equivalent of 67 to 255 kg/

TABLE 10.9
Principal Uses of Biofertilizers in Cuba

Biofertilizers	Crops	Substitution Achieved
Rhizobium	Beans, peanuts, and cowpeas	75–80% of the N fertilizer
Bradyrhizobium	Soybeans and forage legumes	80% of the N fertilizer
Azotobacter	Vegetables, cassava, sweet potato, maize, rice	15–50% of the N fertilizer
Azospirillum	Rice	25% of the N fertilizer
Phosphorus-solubilizing bacteria	Vegetables, cassava, sweet potato, citrus fruits, coffee nurseries	50–100% of the P fertilizer
Mycorrhizae	Coffee nurseries	30% of the N and K fertilizers

Source: Martínez Viera and Hernández (1995), Treto et al. (2002).

ha of N, 7 to 22 kg/ha of P, and 36 to 211 kg/ha of K) and to improve the physical characteristics of the soil (Treto et al., 2002). In commercial tobacco production, chemical applications were reduced through the use of green manures for soil fertility improvement. Other traditional farming practices were also recovered, including the use of oxen teams for cultivation, which avoided soil compaction, conserved physical soil conditions, and eliminated weeds by mechanical means rather than with herbicides.

Worm humus (or vermicompost) and compost production were applied on a large scale. Between 1994 and 1998, national production of these two organic fertilizers together was between 500,000 and 700,000 MT/year. Small-scale compost and worm humus production became popular, especially in urban agriculture, due to the high levels of organic fertilizers demanded by organoponic vegetable production in beds. At the industrial scale, the use of *cachaza* "filter cake" (impurities filtered from cane juice, a by-product from the sugar industry) allowed a considerable reduction or elimination of chemical fertilizer demand in most of the important commercial crops, especially sugar cane, one of the most fertilizer-demanding crops. With an application of 120 to 160 t/ha, this organic fertilizer completely replaced chemical fertilizers over three years in sandy soils, and the same result was achieved with application of 180 to 240 t/ha over five years in soils with a higher clay content (Treto et al., 2002).

10.9.2 Biological Control

After 1990, as a response to the scarcity of pesticides, biological control became a principal strategy for pest control in Cuba. The rapid implementation of this broad strategy at a national scale in the 1990s was possible because of long-term experience in biological control and the existence, beginning in 1960, of five laboratories for its study. Entomophagous and Entomopathogenous Reproduction Centres (CREEs) were created throughout the country for the production of biological control agents to manage the most important agricultural pests. Some 276 CREEs were widely distributed throughout the nation: 54 for sugar cane cultivation areas and 222 for lands producing vegetables, tubers, fruits, and other crops (Pérez and Vázquez, 2002). The actual production of these biocontrol agents (fungus, bacteria, nematodes, and beneficial insects) was small scale and decentralized, and the CREEs provided services to state farms, cooperatives, and private farmers (Fernández-Larrea, 1997). Their use was widespread, covering about 1 million ha in the nonsugar sector in 1999 (Pérez and Vázquez, 2002).

Although Cuba never halted pesticide imports, they were reduced to about one-third of what was previously purchased before the 1990s (Pérez and Vázquez, 2002). Integrated pest management (IPM) programs, combining biological and chemical pest control together with cultural management, were the most common strategy for confronting the pesticide shortage. The effectiveness of biological control strategies, however, has allowed a continuing decrease in the use of pesticides. Pesticide applications on cash crops were reduced 20-fold in a 15-year period, from 20,000 metric tons in 1989 to around 1,000 metric tons in 2004 (*Granma Internacional*, 2004). This indicates not only the effectiveness of the biological practices developed, but also the countrywide need to strengthen sustainable strategies and innovate for nonchemical pest control.

10.9.3 ANIMAL TRACTION

At the end of the 1980s, the number of tractors in Cuba had reached almost 90,000, with imports of 5,000 per year. After 1989, the number of tractors in operation dropped dramatically due to a lack of spare parts, maintenance, and fuel to keep them working. The traditional practice of using oxen for cultivation and transport was revived. About 300,000 oxen teams were trained, conferring a lower fossil fuel dependency to the new production systems. In 1997, 78% of oxen teams were being used in the private sector, this covering only 15% of national agricultural acreage; later the use of oxen was extended to all agricultural sectors (Ríos and Aguerrebere, 1998).

Lowering fossil fuel use was not the only benefit of using oxen for cultivation. Oxen could offer effective mechanical control of weeds, and thus serve as a substitute for herbicides. Substitution of oxen teams for machine power was successful in achieving many agroecological goals; however, the use of oxen is appropriate for traditional small- to mid-size farming systems, less for large-scale monoculture. Thus, changes in land use patterns were necessary to allow the benefits of animal traction to reach their full potential.

The systematic use of oxen in cropping areas required an integration of land for pasture and animal feed production, i.e., mixed use. Many livestock farms that previously specialized in milk or meat production started using oxen to transport cut forages and to plow land that would grow crops for both subsistence and markets. Specialized crop and livestock farms had to adapt their designs to the new conditions. Similarly, many cooperatives previously dedicated to specialized crops such as potatoes, sweet potatoes, vegetables, etc., created "livestock modules" using dual-purpose cattle that produced milk and meat for farmers and could replace oxen teams over time as a source of traction.

10.9.4 Polycropping and Crop Rotation

Crop rotations and polycultures were developed in order to stimulate natural soil fertility, control pests, restore productive capacity, and to obtain higher land equivalency ratios (LERs).* The application of these alternatives—often practiced by traditional farmers—proved to be critical in supporting production levels, and subsequently was expanded through the country, especially in the cooperative sector (Wright, 2005). Both research results and actual production figures showed that polycropping and crop rotation made possible an increase in the yield of the majority of the economically important crops (Casanova et al., 2002). Experiments confirmed, for example, that the use of soybean (*Glycine max*) in rotation with sugar cane increased yields of the latter from 84.4 to 90.6 t/ha with an additional production of 1.7 t/ha of soybean (Leyva and Pohlan, 1995). Polyculture of cassava (*Manihot esculenta*) and beans (*Phaseolus vulgaris*) under different management cropping systems achieved a higher LER than monoculture of cassava or beans (Mojena and Bertolí, 1995). Polyculture of green manures and corn (*Zea mays*) in rotation with potatoes (*Solanum tuberosum*) also increased potato production (Crespo et al., 1997). All these polycropping arrangements made for more efficient land use as well as successful pest control.

10.9.5 Beyond the Input Substitution Strategy

The previous examples of input substitution strategies recognize the positive results of such approaches on national food self-sufficiency and the environment. This model of input substitution prevailed in Cuba during the years of crisis and is considered the first attempt to convert a conventional food system at a national scale (Rosset and Benjamin, 1994). However, these approaches arguably need to evolve if a higher level of agricultural sustainability is desired.

Many farmers in Cuba, lacking an agroecological framework, substitute inputs out of necessity but prefer the use of agrochemicals when they are available, even though they may recognize the negative effects of these inputs on health (Wright, 2005). Along the same lines, most policymakers in Cuba tend to consider the conventional approach as the most viable way to restore soil fertility, control pests, and increase productivity in agriculture. In fact, one present strategy from the state is the "potentiation" of production—increasing imported agrochemical, oil, and feed inputs for use in prioritized cropping or livestock activities. These conventional approaches are again becoming policy, and the lower-yielding systems still receive much less support from the administrative structures than is necessary. Such political trends in Cuban agriculture make it clear that the national input substitution strategy has not yet evolved to an agroecological stage.

---

* "The land equivalent ratio is calculated using the formula

$$LER = \sum \frac{Ypi}{Ymi}$$

where Y_p is the yield of each crop in the intercrop or polyculture, and Y_m is the yield of each crop in the sole crop or monoculture. For each crop (i) a ratio is calculated to determine the partial LER for that crop, then the partial LERs are summed to give the total LER for the intercrop" (Gliessman, 2001, p. 241).

The Cuban alternative model needs to be reinforced with a stronger focus on both a systems approach and an ecological foundation. Only by making more profound changes—considering alternative agricultural systems that are truly regenerative rather than merely input substituted—can long-term sustainability be achieved. The integration of crops and livestock within more diversified production systems—to create what can be called mixed farming systems (MFSs)—is one of these alternatives.

10.10   MIXED FARMING SYSTEMS: AN AGROECOLOGICAL APPROACH TO SUSTAINABILITY

The national input substitution strategy established both infrastructure for and basic knowledge about sustainable farming system management. However, it is necessary to recognize the technological limitations of input substitution to achieve a more integrated and ecologically sound approach. The still prevalent monoculture systems in agriculture, the continued dependence on external inputs, and the restricted degree of internal cycling in agroecosystems are some of these limitations.

10.10.1   CHANGES IN THE STRUCTURE OF LAND USE

The patterns of land use in present Cuban agriculture are of special relevance for more fundamental conversion to an agroecological model at the national scale. During the past 10 years, major structural changes in the agricultural sector have taken place in Cuba that create the preconditions for a nationwide application of a mixed farming strategy.

First, as mentioned previously, the effects of the crisis during the 1990s made necessary the decentralization of state enterprises and the promotion of cooperativization in order to keep the people on the land. Giving usufruct land rights, reducing the scale of production, and diversification were key factors in the agricultural changes.

Second, the deactivation of 110 sugar mills out of the existing 155 during the last five years means that half of the more than 1.4 million ha formerly devoted to the monoculture of sugar cane is available for other agricultural purposes, e.g., crop production, fruits, reforestation, and livestock. In the first stage of this structural change only 71 sugar mills remained working, with their lands covering an area of 700,000 ha. In the year 2002, the Ministry of Sugar (MINAZ) started a restructuring programme (named *Tarea "Alvaro Reynoso"*) in order to use the lands previously belonging to these sugar mills (Rosales del Toro, 2002). This led to further reductions in sugar production; today there are only 45 mills in operation.

Third, about 40% of the 2 million ha covered by pasture (some 900,000 ha) are now invaded by marabú (*Dichrostachys cinerea*) and aroma (*Acacia farnesiana*), two thorny, *fast-growing, woody* leguminous *species*. These plants are difficult to control by hand and expensive to control with machinery. The main causes of this tremendous invasion are the abandonment of areas and inappropriate land use.

The incorporation of mixed farming strategies might be an effective control practice for these weeds where conditions permit. Calculations made by García Trujillo

(1996) have shown that through mixed farming system strategies in the livestock sector, it is possible—even at very low levels of productivity—to fulfill the food requirement of the Cuban population with respect to animal protein and contribute to energy (carbohydrate) needs as well. Under this approach, extensive land use farming systems might be considered a valid strategy for the future of agriculture in Cuba.

Present ecological, economic, and social conditions favor the conversion to agroecological MFSs in the livestock sector. Because of the availability of animals, infrastructure, and long-standing pastureland, there can be immediate positive results when livestock units are converted to manure-fertilized crop and livestock systems (García-Trujillo and Monzote, 1995; Funes-Monzote and Monzote, 2001). In specialized commercial crop production, rotations with an animal component might allow better use of resources such as the fallow biomass, crop residues, or the by-products of food processing.

Although traditional farmers have commonly practiced the integration of crops and livestock at a small scale, the innovative approaches needed for medium-scale mixed farming systems should be researched, implemented, and disseminated. Moreover, strategies need to be developed for overcoming the major constraints to the development of mixed systems. These constraints include the systems' high need for labor in the context of a sparsely populated countryside, the lack of capital, and the priority still given to conventional agriculture and its specialized infrastructure.

Integration of crop and livestock production can be achieved at different scales in time and space. On a large scale (i.e., regional, national) it requires more capital and inputs than at a middle or small scale. For example, long-distance transportation of animal manure, with its high water content, is difficult and costly, and the available machinery makes it difficult to establish polycropping designs in larger areas. The increase in scale will bring decreases in production efficiency as well. In contrast, resource use efficiency is maximized at smaller scales, at the cooperative or farm level, because at these scales interrelationships (e.g., internal nutrient cycling) can be better facilitated. However, at any scale, the priorities, demands, and capacities of producers to carry out such alternatives are key factors in the successful implementation of the MFS model.

10.10.2 GENERAL APPROACH FOR RESEARCHING AND DEVELOPING MFSs

Ultimately, MFSs integrate the specialized knowledge of plant and animal production with the benefits of crop and livestock diversity. Therefore, many individual approaches form part of a more holistic management program. One way to unite these specialized management concepts into a holistic system based on agroecological principles is to apply an approach called DIA systems, which stands for diversified, integrated, and self-sufficient (Monzote et al., 2002).

During the last decade, this approach has been developed and tested at the farm and cooperative levels; its principles seem to have potential application at the regional or national level. Each of the three components of DIA systems has its particular characteristics, but they share several basic principles, including (1) system biodiversification, (2) soil fertility conservation and management, (3) optimization of nutrient and energy cycles and processes, (4) optimal use of natural and local available resources, (5) maintenance of high levels of resilience in terms of systems sustainability and stability, and (6) use of renewable energy (Funes-Monzote and Monzote, 2002). The validity of this approach for the conversion of Cuban agriculture has been assessed by applying ECOFAS (Ecological Framework for the Assessment of Sustainability) methodology to evaluate the process of converting specialized dairy farming systems (DFSs) into MFSs.

ECOFAS consists of a comprehensive three-stage program for evaluating, monitoring, comparing, analyzing, and designing management strategies for converting specialized land use into mixed land use. Each stage is related to a different hierarchical level of analysis. Stage 1 is the experimental assessment of the conversion process. In stage 2, multivariate statistical methods are used to analyze different agroecological variables and indicators of sustainability in a broader array of systems. This second stage, as a scaling up of the results achieved in stage 1, serves as evidence for policymakers. In stage 3, participatory methods of research and action are used to diagnose and characterize farms and monitor their progress toward achieving multiple objectives using a set of agroecological, economic, and social indicators. The potential impact of the application of ECOFAS methodology for improving productivity and achieving the economic, agroecological, and social goals of sustainability is huge.

10.10.3 STUDY OF THE CONVERSION OF SPECIALIZED DAIRY SYSTEMS INTO MIXED FARMS

Seven research teams throughout the country took part in the three stages of this project, designated Designs for Crop-Livestock Integration at Small and Medium Scale by the Ministry of Science, Technology and Environment (CITMA). Using the ECOFAS methodology, all the teams succeeded in identifying locally adapted strategies for mixed farming that have the potential to alleviate barriers to sustainable livestock production in Cuba.

10.10.3.1 Stage 1: The Experimental Scale

To study experimentally the effects of converting specialized "low external input" dairy systems into mixed farming systems, one specialized dairy operation was chosen as the control and two equivalent farms were converted to mixed farm systems with different percentages of their land put into crop production. Data collected over a six-year period demonstrated that productivity, energy efficiency, and economic profitability all improved on the mixed farms, and that these improvements occurred without a decrease in milk production per unit of farm area (Figure 10.2). Greater use of legumes, more intensive crop rotations, diversification of production, and the use of crop residues for animal feed allowed an increase in the stocking rate on the livestock area of mixed farms. The human labor demand was higher at the beginning of the establishment period on the mixed farms, but it decreased by one-third over the six-year period. Energy efficiency, calculated as a ratio of energy output per unit of energy input, was from two to six times higher in the mixed system and increased over time. In economic terms, the mixed farms reached three to five times the net

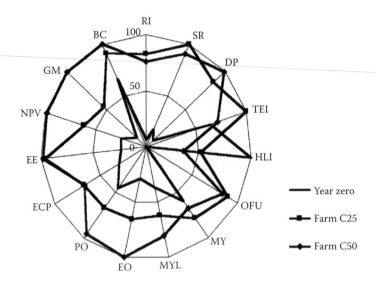

FIGURE 10.2 Agroecological and financial indicators (AEFIs) for the year 0 (specialized dairy farm) and the two experimental mixed farms averaged over the six-year period of the study. Farm C25 had 25% of its land in crop production, and Farm C50 had 50%. Target values for the assessment of each indicator are set at 100%. For Shannon and Margalef index calculation procedures see Gliessman (2007). Indicators: RI—Reforestation (Shannon index); SR—Species richness (Margalef index); DP—Diversity of production (Shannon index); TEI—Total energy inputs (GJ ha^{-1} year^{-1}); HLI—Human labor intensity (hours ha^{-1}day^{-1}); OFU—Organic fertilizers use (t ha^{-1} year^{-1}); MY—Milk yield (t ha^{-1} year^{-1}); MYL—Milk yield per livestock area (t ha^{-1} year^{-1}); EO—Energy output (kg ha^{-1} year^{-1}); PO—Protein output (kg ha^{-1} year^{-1}); ECP—Energy cost of protein production (MJ kg^{-1}); EE—Energy efficiency (GJ produced/GJ input); NPV—Net production value (CUP ha^{-1} year^{-1}); GM—Gross margin (CUP ha^{-1} year^{-1}); B/C—Benefit/cost ratio.

economic value of the original specialized dairy farm, mainly due to high market prices for crop products. In general, among the mixed farms studied, the one with 50% of land in crop proportion (C50) performed better for most of the agroecological and economic indicators than the one with 25% of land in crop proportion (C25). They both had much better performance than the original dairy system in year 0 of the conversion (Figure 10.2).

10.10.3.2 Stage 2: Scaling-Up Experimental Results

Experimental scale results were confirmed by a broader survey of 93 farms covering various soil and climatic conditions in the three main regions of Cuba. The farms under study were classified using multivariate canonical discriminant analysis. Diversity of production, species richness, energy efficiency, and human labor intensity were the primary factors influencing farming systems classification (Funes-Monzote et al., 2004). According to these indicators, integrated crop-livestock farms were more productive and more energy efficient than specialized systems.

In these studies, it was demonstrated that the inclusion of crops into livestock areas enhanced the energy and protein production capacity. This was possible due to the greater energy value of crops, the increase of milk yields achieved in the mixed farms, and the more efficient use of land, capital, and labor at the systems level. A nondetrimental or even positive effect on milk production, through the inclusion of crops in livestock areas, challenged the belief that milk production is reduced when crops are established in pasture-based areas.

10.10.3.3 Stage 3: Application and Conclusions

In stage 3 of the study, project teams characterized mixed and specialized farms in San Antonio de Los Baños municipality as study cases, analyzed their performance by comparing them using a participatorily designed set of agroecological, economic, and social indicators, and then discussed with farmers the possible impact of the results for improving productivity, economic feasibility, and agroecological sustainability of the farms. Application of participatory research methods considered farmers' perspectives in the definition of sustainability goals within strategies for the development of the MFS model at the regional level. The results of the comprehensive farm diagnoses, characterizations, and comparisons provide evidence of the advantages of mixed farming over specialized farming under low-input agriculture conditions.

In summary, implementing mixed crop–livestock designs might solve many problems—relating to adverse environmental effects, productivity, and efficiency—that predominate in specialized dairy systems (Monzote and Funes-Monzote, 1997). Much scientific and practical information demonstrates the advantages of the MFS model; however, more attention should be given to the development of adaptations under a variety of local conditions. A physical description of farming systems and quantification of their ecological flows are commonly found in the literature, but more integrated approaches that document agroecological, economic, and social dimensions are rare.

The application of agroecological approaches through the MFS model can be a further step toward sustainability in Cuban agriculture. Both the technological and practical advantages of MFSs have been scientifically confirmed, and the present economic and social structures of the agrarian sector in Cuba favor this process.

10.11 PRIMARY LESSONS OF THE CONVERSION PROCESS IN CUBAN AGRICULTURE

> The Cuban experiment is the largest attempt at conversion from conventional agriculture to organic or semi-organic farming in human history. We must watch alertly for the lessons we can learn from Cuban successes as well as from Cuban errors.
>
> **—Rosset and Benjamin (1994, p. 82)**

The recent history of Cuban agriculture demonstrates that agrarian reforms will not be effective in the long-term if adaptation to new political situations and ecological perspectives are not taken into account. Therefore, one of the main lessons of the national-scale conversion toward sustainable agriculture in Cuba in the 1990s is that it is necessary to change the prevailing world food production system so that stewardship of natural resources occupies a place as important as socioeconomic or political issues.

The elimination of the *latifundio* in 1959 by itself did not eradicate the many historical problems intrinsic to the Cuban agricultural system. Agrarian reform gave much of the land to those who worked it and reduced the sizes of farms, both of which had positive social impacts. However, the lack of an ecological focus and the concentration of lands by the state as never before in extensive monocultures reinforced the dependency characteristic of the inadequate agricultural development prevailing throughout Cuba's history. Although its intentions were to move toward a more socially just system, the new state agriculture, like that of the *latifundio*, created serious environmental and socioeconomic problems.

The enormous economic, ecological, and social crisis that was unleashed at the beginning of the 1990s was the result of the high level of dependency reached in Cuba's relationship with Eastern Europe and the USSR. Many studies demonstrate the depth of the crisis, and almost all agree with the conclusion that it would have been much worse had there not been the will to change to centralized planning of material resources and to work toward an equitable social structure. Government assistance, together with its encouragement of innovation, the high educational level of the population, and the exchange of resources and knowledge among the people, permitted the creation of a sustainable agriculture movement and its implementation at a national scale.

However, further steps—indeed, profound changes—are necessary in Cuban agriculture. Although innovation has been present in all branches of agriculture and the scientific institutions have tested environmentally sound technologies on a large scale, these efforts have tended to focus on the substitution of inputs, and there remains a disjunction between the biophysical and socioeconomic aspects of agricultural development. If this newest stage in Cuban agriculture, characterized by the emergence of diverse agroecological practices throughout the country, is to progress further, it must be recognized that neither the conventional pattern nor that of input substitution will be versatile enough to cover the technological demands and socioeconomic settings of the country's heterogeneous agriculture. Therefore, it is necessary to develop more integrated, innovative, and locally oriented solutions as opposed to solving specific problems from the top down. The MFS approach, based on agroecological perspectives and participatory methods of dissemination, might aid in reaching a higher stage in the transformation of Cuban agriculture as it moves toward sustainability.

10.12   FINAL REMARKS

Despite the acknowledged successes in the transition toward sustainable agriculture in Cuba, it appears that the impact in terms of national food self-sufficiency is still limited. The country at the moment imports about 50% of its food and only half of the suitable land is cultivated; thus, dependence on external food sources is high and food security is tenuous. Cuban agriculture is responding to this situation with emphases on diversification, decentralization, and greater food self-sufficiency. However, these developments must be systematically supported by science and policy if they are to overcome the food security challenge and allow the agricultural sector to contribute to a viable economy. If the need for economic recovery is used as an argument to return to intensive, industrialized agriculture, sustainability and resource conservation will be threatened. Changes in Cuban agriculture, once driven by the dire necessity for input substitution, must now be guided by more conscientious and scientifically driven policies that aim at development of an agricultural sector that combines production and conservation objectives.

The soaring prices of oil and food on the world market during the last few years emphasize the need for an agricultural reorientation that makes the substitution of food imports with homegrown food products a national priority (Castro, 2008; MINAG, 2008). Mixed crop–livestock farming systems have much to contribute to this goal and to the development of a sustainable agricultural model for Cuba. It is a positive sign that multistakeholder platforms of farmers, scientists, and policymakers have been involved, at various locations in the country, in the design and implementation of these systems in the period since the early 1990s. Rural development strategies are being identified at the local level, technologies adapted to location-specific conditions, and traditional and scientific knowledge integrated to arrive at more sustainable agricultural practices and best uses of available resources.

ACKNOWLEDGMENTS

I am very grateful to Martha Rosemayer at Evergreen College, Washington, for the detailed revisions of the manuscript, and to Julia Wright at Henry Doubleday Research Association (HDRA), England, who made very useful comments based on her experience with Cuban agriculture. I appreciate very much the help of my mother, and tutor of my professional life, Martha Monzote, who made a substantial contribution to this paper. Thanks to my brother, Reinaldo Funes Monzote, for helping with the historical section, and to my father, who revised previous drafts of this article. Without their help, the points of view and opinions in this article (which of course are my own responsibility, as are the mistakes) would not have been as well expressed.

REFERENCES

Alisov, B.P., Paltaraus, B.V. 1974. *Climatología*. Moscow, USSR: Univeridad Estatal de Moscú.

Altieri, M.A. 1987. *Agroecology: The science of sustainable agriculture*. Boulder, CO: Westview Press.

Altieri, M.A. 1993. The implications of Cuba's agricultural conversion for the general Latin American agroecological movement. *Agriculture and Human Values* 3:91–92.

Álvarez, M.D. 2002. Social organization and sustainability of small farm agriculture in Cuba. In *Sustainable agriculture and resistance: Transforming food production in Cuba*, ed. F. Funes, L. García, M. Bourque, N. Pérez, P. Rosset, 72–89. Oakland, CA: Food First Books.

Anon. 1960. *La Reforma Agraria, obra magna de la Revolución en Cuba republicana*. Havana: Oficina del Historiador de la Ciudad.

Anon. 1987. *La Revolución en la Agricultura*. Havana: José Martí.

ANPP (Asamblea Nacional del Poder Popular). 1991. *El programa alimentario*. Havana: José Martí.

Arnaud, J., Fleites-Mestre, P., Chassagne, M., Verdura, T., Garcia Garcia, I., Hernandez-Fernandez, T., Gautier, H., Favier, A., Pérez-Cristiá, R., Barnouin, J., Neuropathy Epidemic Group. 2001. Vitamin B intake and status in healthy Havanan men, 2 years after the Cuban neuropathy epidemic. *British Journal of Nutrition* 85:741–48.

Cabello, R., Rivero, L., Castillo, D., Peña, J.L. 1989. *Informe sobre el estudio de la Sesbania rostrata y S. emerus como abonos verdes en el mejoramiento y conservación de los suelos arroceros con baja fertilidad.* Havana: Instituto de Investigaciones del Arroz.

Casanova, A., Hernández, A., Quintero, P.L. 2002. Intercropping in Cuba. In *Sustainable agriculture and resistance: Transforming food production in Cuba,* ed. F. Funes, L. García, M. Bourque, N. Pérez, P. Rosset, 144–54. Oakland, CA: Food First Books.

Castro, F. 1992. *Discurso pronunciado en la Conferencia de Naciones Unidas sobre Medio Ambiente y Desarrollo.* Rio de Janeiro: Editora Política, Havana, p. 10.

Castro, R. 2008. Mientras mayores sean las dificultades, más exigencia, disciplina y unidad se requieren. Speech at the conclusion of the Constitutive Session of the National Assembly of the Cuban Parliament. *Granma,* February 25, 2008.

Censo Agrícola Nacional. 1951. *Memorias del Censo Agrícola Nacional—1946.* Havana: P. Fernández y Cia.

CITMA (Ministerio de Ciencia, Tecnología y Medio Ambiente). 1997. *Estrategia Nacional Ambiental.* Havana: Ministerio de Ciencia Tecnología y Medioambiente de la República de Cuba.

CITMA. 1998. *Programa de Acción Nacional de Lucha Contra la Desertificación y la Sequía en la República de Cuba.* Havana.

Companioni, N., Ojeda, Y., Páez, E. 2002. The growth of urban agriculture. In *Sustainable agriculture and resistance: Transforming food production in Cuba,* ed. F. Funes, L. García, M. Bourque, N. Pérez, P. Rosset, 220–36. Oakland, CA: Food First Books.

Crespo, G., Fraga, S., Gil, J.I. 1997. *Abonos verdes intercalados con maíz y su efecto en la producción de papa.* Villa Clara. Resúmenes Tercer Encuentro Nacional de Agricultura Orgánica.

Cruz, M.C., Sánchez, R. 2001. *Agricultura y ciudad, una clave para la sustentabilidad.* Bogotá: Linotipia Bolívar.

De la Sagra, R. 1831. *Historia física, económico—política, intelectual y moral de la Isla de Cuba, Relación del último viaje del autor.* París: Librería de L. Hachette y Cª.

Deere, C.D. 1997. Reforming Cuban agriculture. *Development and Change* 28:649–69.

Espinosa, E. 1992. La alimentación en Cuba. Su dimensión social. Tesis presentada para la obtención del grado de Dr. en Ciencias Económicas, Universidad de La Habana.

FAO/WHO/UNU. 1985. *Energy and protein requirements.* Technical Report Series 742, report of a FAO/WHO/UNU Joint Expert Consultation, Geneva.

FAOSTAT. 2004. Database results. Last updated February 2004.

Fernández-Larrea, O. 1997. *Microorganismos en el control fitosanitario en Cuba. Tecnologías de producción.* Villa Clara: Conferencias Tercer Encuentro Nacional de Agricultura Orgánica.

Funes, F. 2002. The organic farming movement in Cuba. In *Sustainable agriculture and resistance: Transforming food production in Cuba,* ed. F. Funes, L. García, M. Bourque, N. Pérez, P. Rosset, 1–26. Oakland, CA: Food First Books.

Funes, F., García, L., Bourque, M., Pérez, N., Rosset, P. 2002. *Sustainable agriculture and resistance: Transforming food production in Cuba.* Oakland, CA: Food First Books.

Funes-Monzote, F.R., Monzote, M. 2001. Integrated agroecological systems as a way forward for Cuban agriculture. *Livestock for Rural Research Development Journal* 13:1. http://www.cipav.org.co/lrrd/lrrd13/1/fune131.htm.

Funes-Monzote, F.R., Monzote, M. 2002. The Cuban experience in integrated crop-livestock-tree farming. *LEISA Newsletter* 18:1, 20–21. http://www.ileia.org/2/18-1/20-21.PDF.

Funes-Monzote, F.R., Monzote, M., van Bruggen, A., Eladio, J. 2004. Indicadores agroecológicos para la clasificación de sistemas de producción integrados y especializados. In *Proceedings II Simposio Internacional sobre Ganadería Agroecológica* (SIGA 2004), Havanna, Cuba, pp. 123–24.

Funes-Monzote, R. 2004. *De bosque a sabana. Azúcar, deforestación y medioambiente en Cuba, 1492–1926*. México: Siglo XXI Editores.

Gaceta Oficial. 1959. Primera Ley de Reforma Agraria, 3 de Junio de 1959, Havana.

Gaceta Oficial. 1997. Ley No. 81 del Medio Ambiente, 11 de Julio de 1997, Havana.

García, A. 2003. Sustitución de importaciones de alimentos en Cuba: necesidad vs. posibilidad. In *Proceedings 8vo Seminario Anual de Economía Cubana*, Havana, pp. 154–204.

García, M., Treto, E. 1997. Contribución al estudio y utilización de los abonos verdes en cultivos y utilización de los abonos verdes en cultivos económicos desarrollados sobre suelos ferralíticos rojos en las condiciones de Cuba. In *Resúmenes del I Taller Nacional de Producción Agroecológica de Cultivos Alimenticios en Condiciones Tropicales*, Instituto de Investigaciones Hortícolas "Liliana Dimitrova," Havana, p. 74.

García Trujillo, R. 1996. *Los animales en los sistemas agroecológicos*. Havana: ACAO.

García Trujillo, R., Monzote, M. 1995. La ganadería cubana en una concepción agroecológica. In *Conferencias Segundo Encuentro Nacional de Agricultura Orgánica*, Havana, p. 60.

Gliessman, S. 2001. *Agroecology: Ecological processes in sustainable agriculture*. Boca Raton, FL: CRC Lewis Publishers.

Gliessman, S. 2007. *Agroecology: The ecology of sustainable food systems*. Boca Raton, FL: CRC Lewis Publishers.

GNAU (Grupo Nacional de Agricultura Urbana). 2004. *Lineamientos para los subprogramas de la Agricultura Urbana (2005–2007)*. Leaflet. Habana: Grupo Nacional de Agricultura Urbana, MINAG.

González, A., Fernández, P., Bu, A., Polanco, C., Aguilar, R., Dresdner, J., Tancini, R. 2004. *La ganadería en Cuba: desempeño y desafíos*. Havana: Instituto Nacional de Investigaciones Económicas.

Granma. 1998. Sobrepasa las 140,000 toneladas el movimiento popular de arroz. 18 de marzo del 1998.

Granma. 2000. Mortalidad infantil en 1999. 4 de enero del 2000.

Granma. 2001. Felicitación de Raúl por los resultados de la agricultura urbana. 30 de enero del 2001.

Granma. 2003a. Crece impacto social de la Agricultura Urbana. 9 de diciembre del 2003.

Granma. 2003b. Fortalecen red estatal de mercados agropecuarios. 19 de diciembre del 2003.

Granma. 2006a. Agrucultura Urbana. Alternativa que más resiste las inclemencias. 5 de enero del 2006.

Granma. 2006b. Consideraciones del Ministerio de la Agricultura sobre la producción y comercialización de productos agropecuarios. 3 de Julio del 2006.

Granma Internacional. 2004. En 15 años: La Isla redujo 20 veces uso de plaguicidas. 26 de noviembre del 2004.

Gutiérrez Domenech, R., Rivero Glean, M. 1999. *Regiones naturales de la isla de Cuba*. Havana: Editorial Científico Técnica.

Koppen, W. 1907. *Climatología*. México: Ed. Fondo de Cultura Económica.

Lage, C. 1992. Interview on Cuban television, November 6.

Lecha, L.B., Paz, L.R., Lapinel, B. 1994. *El Clima de Cuba*. Havana: Ed. Academia.

León, J.J. 1996. *The greening of Cuba*. Video film produced by the Institute for Food and Development Policy, Food First, 38 minutes.

Le Riverend, J. 1970. *Historia económica de Cuba*. Havana: Instituto Cubano del Libro.

Le Riverend, J. 1992. *Problemas de la formación agraria de Cuba*. Havana: Siglos XVI–XVII, Ciencias Sociales.

Leyva, A., Pohlan, J. 1995. Utilización de los principios de la agricultura sostenible en una finca. 1. Biodiversidad. In *Resúmenes Segundo II Encuentro Nacional de Agricultura Orgánica*, Havana, p. 92.

Lugo Fonte, O. 2000. Nuestro deber patriótico es producir para el pueblo. Entrevista a Orlando Lugo Fonte, presidente de la Asociación Nacional de Agricultores Pequeños (ANAP). *Granma*, 17 de mayo del 2000.

Marrero, L. 1974–1984. *Cuba, Economía y Sociedad,* 15 Vols. Madrid, Spain: Playor.

Martin, L. 2002. Transforming the Cuban countryside: Property, markets, and technological change. In *Sustainable agriculture and resistance: Transforming food production in Cuba*, ed. F. Funes, L. García, M. Bourque, N. Pérez, P. Rosset, 57–71. Oakland, CA: Food First Books.

Martínez Viera, R., Hernández, G. 1995. Los biofertilizantes en la agricultura cubana. Paper presented at II Encuentro Nacional de Agricultura Orgánica, Conferencias y Mesas redondas, Havana, May 17–19, 1995.

MINAG (Ministerio de la Agricultura). 2008, May 14. *Informe del Ministerio de la Agricultura a la Comisión Agroalimentaria de la Asamblea Nacional*. Havana: MINAG.

Mojena, M., Bertolí, M.P. 1995. *Asociación yuca/frijol; una forma de aumentar la eficiencia en el uso de la tierra*. Havana: Resúmenes Segundo II Encuentro Nacional de Agricultura Orgánica.

Monzote, M., Funes-Monzote, F. 1997. Integración ganadería—agricultura. Una necesidad presente y futura. *Revista Agricultura Orgánica* 3:1–7.

Monzote, M., Muñoz, E., Funes-Monzote, F. 2002. The integration of crops and livestock. In *Sustainable agriculture and resistance: Transforming food production in Cuba*, ed. F. Funes, L. García, M. Bourque, N. Pérez, P. Rosset, 190–211. Oakland, CA: Food First Books.

Moreno Fraginals, M. 1978. *El Ingenio. Complejo económico social cubano del azúcar.* Havana: Ciencias Sociales.

Murphy, C. 1999. *Cultivating Havana: Urban agriculture and food security in the years of crisis*. Food First Development Report 12. Oakland, CA: Institute for Food and Development Policy.

ONE (Oficina Nacional de Estadísticas). 2004. *Anuario estadístico de Cuba 2004*. Havana.

PAHO (Pan American Health Organization). 2002. *Health in the Americas*, 198–212. 2002 ed., Vol. II.

Perera, J. 2004. Programa Campesino a Campesino en Cuba: un movimiento agroecológico a escala nacional. In *Proceedings II Simposio Internacional sobre Ganadería Agroecológica* (SIGA 2004), Cuba, pp. 176–79.

Pérez, N., Vázquez. 2002. Ecological pest management. In *Sustainable agriculture and resistance: Transforming food production in Cuba*, ed. F. Funes, L. García, M. Bourque, N. Pérez, P. Rosset, 109–43. Oakland, CA: Food First Books.

Pérez Marín, E., Muñoz Baños, E. 1991. *Agricultura y alimentación en Cuba*. Havana: Editorial Ciencias Sociales.

Pérez Rojas, N., Echevarria, D. 2000. Participación y autonomía de gestión en las UBPC. Estudios de casos. In *La última reforma del siglo*, ed. H. Bucrchardt, 71–102. Caracas: Nueva Sociedad.

Pérez Rojas, N., Echeverría, D., González, E., García, M. 1999. *Cambios tecnológicos, sustentabilidad y participación*. Havana: Universidad de La Habana.

PNAN (Plan Nacional de Acción para la Nutrición). 1994. *Plan Nacional de Acción para la Nutrición*. Havana: República de Cuba.

Porrata, C., Hernández, M., Argueyes, J.M. 1996. *Recomendaciones nutricionales y guías de alimentación para la población cubana*. Havana: Instituto de Nutrición e Higiene de los Alimentos.

Reynoso, A. 1862. *Ensayo sobre el cultivo de la caña de azúcar*. Havana: Empresa Consolidada de Artes Gráficas.

Ríos, A., Aguerrebere, S. 1998. *La tracción animal en Cuba*. Havana: Evento Internacional Agroingeniería.

Ríos, H. 2004. Logros en la implementación del fitomejoramiento participativo en Cuba. *Cultivos Tropicales* 24:17–23.

Ríos, H. 2006. Personal communication.

Rivero Glean, M. 1999. *Conozca Cuba. Flora y Fauna*. Havana: Editorial José Martí.

Rivero Glean, M. 2005. Escuela de Altos Estudios de Hotelería y Turismo (EAEHT), Havana. Personal communication, 24 de augusto.

Rosales del Toro, U. 2002. Intervención del ministro del azúcar en el 48 Congreso de la ATAC, 15 de noviembre 2002, Havana.

Rosset, P., Benjamin, M. 1994. *The greening of the revolution: Cuba's experiment with organic agriculture*. Melbourne: Ocean Press.

Sinclair, M., Thompson, M. 2001. Cuba going against the grain: Agricultural crisis and transformation. An Oxfam America Report, Boston, MA: Oxfam America.

Socorro, M., Alemàn, L., Sánchez, S. 2002. "Cultivo Popular": Small-scale rice production. In *Sustainable agriculture and resistance: Transforming food production in Cuba*, ed. F. Funes, L. García, M. Bourque, N. Pérez, P. Rosset, 237–45. Oakland, CA: Food First Books.

Torres-Cuevas, E., Loyola, O. 2001. *Historia de Cuba 1492–1898*. Havana: Editorial Pueblo y Ediucación.

Treto, E., García, M., Martínez Viera, R., Febles, J.M. 2002. Advances in organic soil management. In *Sustainable agriculture and resistance: Transforming food production in Cuba*, ed. F. Funes, L. García, M. Bourque, N. Pérez, P. Rosset, 164–89. Oakland, CA: Food First Books.

Urquiza, N., Gutiérrez, J. 2003. La cuenca hidrográfica como unidad de manejo ambiental: El caso de Cuba. In *Proceedings of the Workshop on Livestock, Environment and Sustainable Development*, Havana, March 10–12, 2003, pp. 137–43.

Valdés, O. 2003. *Historia de la Reforma Agraria en Cuba*. Havana: Editorial Ciencias Sociales.

Villegas, R. 1999. ¿Qué tipo de propiedad representan las UBPC? In *Cambios tecnológicos, sustentabilidad y participación*, ed. N. Pérez Rojas, D. Echevarría, E. González, M. García, 167–83. Havana: Universidad de La Habana.

Wright, J. 2005. Falta Petroleo! Cuba's experiences in the transformation to a more ecological agriculture and impact on food security. PhD thesis, Wageningen University, The Netherlands.

The European Union
Key Roles for
Institutional Support
and Economic Factors

Gloria I. Guzmán and Antonio M. Alonso

CONTENTS

11.1 INTRODUCTION

The process of transition from industrialized agriculture to ecological agriculture (EA) can be studied at many different scales (worldwide, regional, national, local, and farm). In this chapter, we focus on a regional space, the European Union (EU). The individual states of the European Union obviously vary politically, ecologically, culturally, and socioeconomically, but they are tied together economically through the EU framework and share a common legislation in the agrarian sphere, the Common Agricultural Policy (CAP).

Since the early 1990s, ecological agriculture in the EU has grown considerably and now plays a significant role in the region's agricultural economy. The movement toward ecological agriculture has been driven by several factors, the most important of which may be the institutional support EA receives from the member states and the EU as a whole. Institutional influence became notable after the promulgation, in 1991, of Council Regulation No. 2092/91, a standardized set of regulations governing the ecological production system for the entire European Union. This was followed, in 1992, by the adoption of Council Regulation No. 2078/92, which governed the subsidies directed toward ecological agricultural operations. Another major factor has been the growth of the market for organic products in the European Union, both in supermarkets and in direct marketing schemes, which has been driven by a growing awareness among consumers of the degradation of nature and the food insecurity that derive from industrialized agriculture. These factors together have helped to generate positive impacts on the economic balance at the farm level, accelerating the change toward ecological agriculture in the EU.

However, in spite of sharing an agrarian legislation and common tools for the development of EA, the different state members of the EU vary considerably in the degree to which ecologically grown products have penetrated the market and the rate at which changes have occurred. These differences are instructive, and can help us understand what has been effective in stimulating conversion from industrial to ecological agriculture.

Since the growth of EA is the result of the sum of many individual decisions, we must presuppose that conversion occurs first at the farm level. What are the stimuli and obstacles faced by farmers in the EU, and what part have these factors played in initiating conversion on individual farms? It is clear that the answers to these questions vary depending on the country. Some member countries have developed more efficient solutions to the problems faced by farmers and have simultaneously been better able to stimulate the producers to carry out successful conversions.

Because of this dynamic, we will examine the conversion process in the EU at two levels. First, we will look at conversion from the perspective of farmers—what they see as enabling the conversion process and, conversely, what they perceive as limiting or constraining it. Then we will analyze the growth of EA in each country of the EU, and use these data as a context for understanding the impact of the two factors that most strongly affect farmers' decision making: institutional support for EA and the market for ecologically grown products.

11.2 FACTORS STIMULATING AND OBSTRUCTING CONVERSION AT THE FARM LEVEL

At the farm level, the transition process (our preferred term for the conversion process) implies abandoning capital-dependent, polluting technologies (chemical fertilizers, pesticides, veterinary medicines, etc.) and management techniques that degrade the physical structure of the soil (burning of crop residues, excessive cultivation, etc.)

and replacing them with other strategies (organic compost, biological control of pests and diseases, composting of agricultural wastes, crop rotations, use of cover crops, etc.) that are, in general, less demanding of capital and more locally accessible. In the long-term, these "ecological" techniques allow the maintenance of the biological diversity and the productive capacity of the natural resource base.

Ecologically, based on the degree of specialization and intensification that has previously taken place on the farm, greater or smaller efforts will be required to eliminate synthetic chemicals, to rearrange the flows of nutrients and energy, to reincorporate biodiversity, and to diminish the specific influences of individual companies on input use. Economically, past governmental support, the farm's financial situation, the possibility of access to specific subsidies or easy credit, and market strategy can impede or facilitate the change. Socially, the existence of social support structures that actively promote these transformations, such as accessible technical and commercial extension services, consumer organizations, farmer associations, etc., can be very determinant of the types of management changes that can be made.

In the European Union, Council Regulation No. 2092/91 and its later modifications define ecological agriculture from the legal point of view, and establish the normal duration of the transition period as two to three years. This term depends on the type of crops to which the production area is dedicated; it is shorter for annual crops and pastures and longer for perennial crops. At the end of this period, the products obtained from the farm can be commercialized with the endorsement of ecological production. The duration can be modified primarily as a function of previous farm management. The duration may be lengthened if chemical residues associated with conventional techniques of production persist in the soil—a measure intended to ensure that synthetic chemical residues are absent when the food goes to market.

From the ecological point of view, in contrast, the conversion process is defined as the term of time during which the biotic and abiotic components of the soil ecosystem adjust themselves to a new balance (Culik, 1983) (the soil is considered a better indicator than air or water, because it is less exposed to the influence of the practices of adjacent farms and it has a greater buffering capacity). The length of this ecological transition term is affected by a variety of factors, including previous farm management, agroclimatic conditions, and conversion strategy—and it can be remarkably longer than the legally defined period. As Maire et al. (1990) have demonstrated, some soil biology parameters may not have stabilized as long as 20 years after transitioning to ecological agriculture. From the ecological point of view, then, it is possible that most of the ecological farming operations of the EU are still in transition today, because the majority of them have less than 20 years in this new cropping modality. Nevertheless, many ecologically positive impacts are evident almost immediately after adopting agroecological techniques: less soil erosion, greater plant and animal biodiversity, reductions in the use of nonrenewable energy, less carbon dioxide emission, diminution of water contamination from fertilizer leaching and pesticides, and increased biological activity in the soil, among others (Lampkin, 1997; Stolze et al., 2000; Guzmán and Alonso, 2008).

Finally, we can understand the period of transition as the term of time during which the farmer redesigns his farm operation, making the necessary investments and changes that are needed for the new mode of ecological management. Our experience shows us that for farmers who abruptly cease the use of synthetic chemical agents because of the legal imperative, the redesign of agroecosystems with agroecological criteria (Gliessman, 2007) needs a greater period of time, because of the greater economic investment required. The duration of the transition from this point of view could be considered to be intermediate between the two discussed previously.

Whatever definition we adopt, we are convinced that in the EU we are experiencing a very dynamic transition toward EA. Because there are many barriers to conversion at the farm level, the robust growth of EA indicates that countervailing forces exist that allow farmers to overcome these barriers.

11.2.1 FARMERS' RATIONALES FOR INITIATING AGROECOLOGICAL CONVERSION

The reasons that motivate producers to initiate the conversion toward ecological agriculture are varied and can be classified as environmental, economic, and social. In the environmental category, there is a growing awareness among farmers of the environmental problems of modern agriculture. Although there is concern about the loss of agrobiodiversity—the disappearance of traditional, locally adapted plant varieties and animal races/breeds—soil appears to be the resource that farmers consider the most critical. Other farmer motivations are the fear of further deterioration to human health and the search for a better quality of life for themselves and for their animals.

Among the economic reasons for conversion are the desires to reduce variable costs, to obtain the ecological agriculture price premium, to gain access to specific subsidies, and to maximize economic benefit in the context of intensive agriculture's serious economic crisis. More generally, farmers link food security issues and serious food "scandals" (dioxins in chickens, mad cow disease, etc.) with conventional, industrial agriculture.

Finally, farmers' social reasons for converting to EA include the desire for more autonomy from the multinationals that provide seeds and pesticides, and the desire for a more equitable distribution of the resources between the industrialized countries and the third world.

Concerns about human and environmental health were the original drivers of the movement toward ecological agriculture (MacRae et al., 1990), but in more recent years farmers have tended to make the transformation for merely economic considerations. In spite of this, our general perception is that farmers' motivations are modified during the conversion process: although economic viability is basic at the time of deciding to become an ecological producer, over time new attitudes and values arise, generated by farmers' greater satisfaction with their work.

11.2.2 BARRIERS TO CONVERSION

According to different authors (MacRae et al., 1990; Lampkin, 1992; Boisdon et al., 1997; Bellegem and Eijs, 2002; Alonso et al., 2005; González de Molina et al., 2007), the obstacles to the conversion process in the last decades are of three types: technical, social, and economic. We will review the individual barriers in each category as they apply to the EU.

11.2.2.1   Technical Barriers

1. *There is a lack of information about the existence of alternative production methods.* In a recent study made by Alonso et al. (2005) on ecological agriculture in eight protected areas of Andalucía (Spain), more than half of the residents of these zones (55%) said that they did not know (or had a mistaken idea of) what an ecological product is. This percentage was reduced to 37% in the case of conventional farmers. Nevertheless, although many of the farmers and cattle ranchers have heard about ecological agriculture, they usually do not have specific information, due as much to the lack of research applied to their circumstances as to the inadequate training of local agricultural extensionists. These observations suggest that improving the availability of theoretical and practical information about alternative and ecological methods should be a high priority in the effort to support the transition to ecological agriculture.

2. *Technical advising structures, both public and private, are scarce.* Sometimes there are private advisers, but they are rarely contracted by the undercapitalized agrarian sector.

3. *Actual conversion experiences in particular local regions have not been documented and made available to farmers in those regions.* This is a reflection of the overall deficiency of research on the EA approach. Overcoming this deficiency implies developing strategies to incorporate into national and regional research programs the specific funding of ecological agriculture research and development (R&D) projects that also favor the transdisciplinary and systemic character that must be prioritized in agroecological research.

4. *Concrete technical problems, often tied to degradation of the natural resource base, are considered difficult to solve with an agroecological approach.* This situation is most evident in intensive agricultural contexts, such as greenhouse vegetable production and intensive animal production, where it is assumed that fulfilling conventional production norms with ecological production would be very problematic, even though the environmental and commercial advantages of ecological production are recognized. This perception comes from conceiving of the conversion process in terms of the input substitution model and not considering the complete redesign of the system. Again, the key to overcoming this barrier is better research, training, and information dissemination.

5. *It is more difficult to manage the greater number of crops usually entailed by ecological production.* The increased complexity of managing a more diverse system is manifested not only at the production level, but also in storage, processing, and sale. The increased management burden, of course, is balanced by the reduction of the farmer's agronomic and economic risks and other benefits. A good planning process is the key to tipping the balance in the farmer's favor.

6. *"Organic" inputs are often scarce or expensive.* In some countries, ecological cattle ranchers have faced this problem most acutely, because although the number of cattle is usually well adjusted to the feeding

production capacity of the farm, climatological variations (e.g., drought) or other problems may force the rancher to use supplemental feed, and finding feed of ecological origin at reasonable prices can be difficult. Inadequate availability of other organic inputs also exists, such as homeopathic medicines for cattle, biopesticides allowed by law, etc. The reasons for such deficiencies can sometimes be a shortage in the market or high prices. This situation has been improving progressively. Also, the scarcity of quality organic matter in many Mediterranean countries greatly increases the cost of organic fertilization.

7. *Farmers are now obligated to use seeds of ecological origin.* In both the short- and midterm, this requirement, initiated at the end of 2006, can slow the transition process, mainly for vegetable farms, due to the present lack of certified seed in the EU. To this we add the absence of genetic material adapted to ecological production in commercial seed catalogues, especially for traditional local varieties, or varieties selected under ecological management conditions. This situation also can be looked at positively, since it offers opportunities for the rescue and regeneration of agricultural biodiversity.

11.2.2.2 Social Barriers

1. *The farmer faces the period of transition in isolation, without neighboring farmers with whom to share the process.* This situation becomes less and less frequent as the area dedicated to ecological agriculture increases.

2. *Ecological agriculture is generally more labor intensive than conventional production.* The shortage of agrarian labor available in the EU, the aging of the farming population, and the growth of part-time agriculture all contribute to this difficulty. Nevertheless, in the long-term these circumstances could benefit the EU's struggling family-based agricultural sector.

11.2.2.3 Economic Barriers

1. *It can be difficult to bring to market ecologically produced products that are identified and labeled as "ecologically produced," which limits access to the price premium normally associated with these products.* The cattle sector has suffered from these difficulties due to the strict sanitary regulations that require very expensive facilities for the manipulation and packaging of cattle products, mainly milk and meat, and the difficulty of covering these investments given the geographic dispersion of producers. This has complicated the centralization of the collection and storing processes, the transformation and packaging phases, as well as the later distribution of the meat, milk, and other derivatives of ecological origin. For these reasons, producers have been forced to sell products in the conventional market (in the case of meat, prices can actually be lower than those of conventional products) or downsize their operations to small-scale production, as in the case of milk production. Stolze (1997) cites that 94% of the ecological milk produced in the eastern area of Germany during 1994 was sold to dairies at the conventional price. The commercialization of ecologically grown

agricultural products has also been complicated for many farmers by the difficulty of creating efficient distribution channels. This difficulty has been influenced by several factors, including the following: the financial profile of the ecological farmers (until recently, most small and medium operations had very low levels of capitalization); the discontinuity in the availability of basic feedstock, since the crops are highly seasonal in nature; the limited product range; and the farmers' dispersion. All of this has resulted in an important part of the production being commercialized through short market channels. In contrast, there are cases in which supermarket chains have helped finance the conversion period, as is the case of the ASDA chain in the United Kingdom (Moran, 2002).

2. *The nature of the conversion process—the need for capital-intensive investment combined with postponed economic benefits—can create problems with liquidity and negative balances.* System redesign often involves the construction of "ecological infrastructure," such as planting fences, restoring nonproductive spaces like springs, stream margins, and slopes with appropriate vegetation. Technical infrastructure is also needed, such as new construction or adaptation of existing items. In the case of cattle ranching, there is the need for new machinery such as manure spreaders, mowers, etc. This is especially true for specialized cattle ranches and farms. Therefore, intensive systems of cattle ranching must adapt their expensive facilities to more extensive management, which allows the animals the chance to develop all the aspects of their innate behavior. Farms dedicated to monoculture find that their specialized machinery and facilities are insufficient. At the same time, during the years of conversion the economic benefits can diminish due to possible yield decreases and accompanying increases in production costs, without compensation from the price premium obtained. In addition, the legal requirement that producers not maintain the same crop or cattle species in conventional management on one part of the farm while making the conversion to ecological on another part forces producers to convert the entire farm at the same time. This removes the possibility of one part of the farm supporting the other during the transition period. In addition, there are national and regional rules that some producers must also take into account. For example, it is obligatory in the United Kingdom to use green manures to increase soil fertility during first years of transition in stockless farming, which often makes it impossible to plant more profitable crops (Sparkes et al., 2003).

3. *Neighboring conventional operations may contaminate the fields of ecological producers, causing economic losses.* Chemicals applied to the fields of neighboring conventional farmers or discharged into air or water by other nearby economic activities, such as industry, may reach ecologically produced crops; these polluting agents may be detected in the exhaustive analysis required of ecological products, forcing the products to be sent to conventional markets. This leads to an important loss of income for the ecological producer. This situation has been aggravated in the last few years with genetic contamination derived from the presence of transgenic

crops near the ecological crops, leaving ecological farmers in indefensible situations. Ecological corn producers in Catalonia and Aragón (Spain) have had to abandon their crops due to genetic contamination coming from nearby transgenic corn plantations, leading to important social protests in these regions.

4. *The price premium for ecological products declines over the long-term as ecological food production increases and markets are saturated.* This situation has begun to appear in some countries with certain ecological products, including milk in the Netherlands (Baars and van Ham, 1997), milk and meat from cows, goats, and sheep in Austria (Wlcek et al., 2003), and olive oil in Spain. This can present an obstacle in the short-term growth of the area dedicated to EA, especially if the local consumption of these products does not increase sufficiently to compensate for the decline of the price premium.

5. *Support for industrialized agriculture from financial institutions and organizations puts ecological agriculture at a relative disadvantage.* This institutional support is reflected in multiple ways. The most evident form has been the direct subsidy to production, but there are others. Perhaps most importantly, the ecological and social costs of conventional agriculture have historically been externalized. In other words, the producers themselves do not pay for the damage caused to nature or to human health; instead, the costs are assumed by society as a whole. This gives rise to lower production costs on the surface, making conventional products more competitive in the marketplace. Another source of inequality is that the services to agriculture from the public sector—from research and extension to veterinary services for cattle treatment and the catalogues of commercializable varieties—are directed almost exclusively to conventional producers. The banks also have penalized ecological production by imposing greater interest rates on loans dedicated to this activity, since they believe EA carries a greater risk (Bellegem and Eijs, 2002). As a notable exception, some banks, such as Triodos, invest primarily in projects of environmental interest, dedicating a good part of their attention to ecological agriculture.

6. *For small-scale producers, the cost of certification is relatively high.* This is because the cost of certification is usually not strictly proportional to the scale of the operation, since the certifying companies must carry out a series of basic services and analyses that are independent of farm size. In order to resolve this situation, different initiatives are being developed in which local or regional governmental agencies would assume the cost of the certification. Participative certification processes have also been proposed.

The presence of all of these barriers and challenges means that the producer who wishes to farm or raise livestock more ecologically generally needs institutional support to successfully complete the transition phase. This support can take the form of direct subsidies to the producer, but also important are indirect forms of support, which may include the funding of research and training in EA; programs that

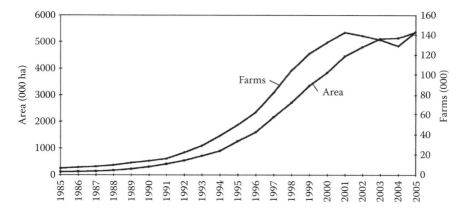

FIGURE 11.1   Growth of ecological agriculture in the European Union. (From: Lampkin, 2003; Eurostat, 2007; Willer and Yussefi, 2007.)

encourage the consumption of ecological products*; investment in more efficient networks of distribution, transformation, and commercialization; and the development of legal structures that benefit ecological production. The institutional support that has been initiated through the Agro-Environmental Measures Program in the EU has had a significant positive impact in the countries in which it has been implemented with more force, as we will see below.

11.3 PRODUCTIVE STRUCTURE OF ECOLOGICAL AGRICULTURE IN THE EU

In the last two decades, ecological agriculture has enjoyed considerable generalized progress in the EU. As can be seen in Figure 11.1, this progress has been particularly rapid since 1993. From a little more than 6,000 farming operations and 100,000 ha in 1985, ecological agriculture in the EU has grown to include, as of 2005, close to 143,000 operations and more than 5.3 million ha in cultivation.†

11.3.1 NATIONAL DIFFERENCES IN GROWTH OF ECOLOGICAL AGRICULTURE

The reasons for the growth in ecological agriculture are multiple and somewhat different for each country (Lampkin and Padel, 1994; Michelsen, 2001; Miele, 2001). Some of the factors that vary among countries are the level of agriculture as an actual

---

* The growth of internal or local consumption requires the planning and coordination of the public and private sectors in a strategy able to inform, interest, and convince the public of the virtues of such consumption. In Austria, for example, public dining halls are a very important channel for these products. Eighty of them are using ecological products to feed up to 15,000 consumers per day. This is in response to a resolution of the Lower Provincial Austrian Government that requires that at least 25% of all the production in volume must be organic, and as a consequence, 27% of the calf meat consumed in these dining halls is ecological (Wlcek, 2003).

† The analysis that is made here uses the configuration of the 15 countries of the European Union (EU) before the last incorporations as its framework.

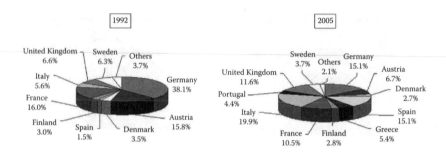

FIGURE 11.2 Percentage share, by country, of the total area in the EU devoted to ecological agriculture area in 1992 and 2005. (From: Eurostat, 2007; Willer and Yussefi, 2007.)

vocation in the country, the level of pressure from social-environmentalist movements, citizen awareness, the organization of the producing sector, the development of agroindustries, and governmental support. Of all of these, governmental support is without a doubt the most important, because it has the capacity to influence all the others. Thus, France and Germany—pioneering countries in the promulgation of legislation supporting ecological agriculture—have been ahead of the others from the beginning. In 1985, of the 100,730 ha devoted to ecological agriculture in the EU, 46% of this area was located in France and 25% in Germany, with only 29% distributed among all the other countries.

Economic subsidies to ecological production began to be implemented in some countries in the latter half of the 1980s: Sweden began in 1985, Denmark in 1987, Germany in 1988, Austria in 1989, and Finland in 1990. Thus, by 1992 ecological agriculture in the EU had become somewhat more evenly distributed. As can be seen in Figure 11.2, Germany (with 38% of the area) had taken the lead from France, Austria had increased its area to 16% of the total, and the hectares devoted to ecological agriculture in several other countries (including Sweden, Denmark, and Finland) had increased significantly, mostly due to institutional support.

Institutional influence became even more significant after the promulgation, in 1991 and 1992, of the sets of standardized regulations governing agricultural production in the EU described above. In addition, numerous countries (Sweden, Denmark, Germany, Austria, Finland, France, Netherlands, some regions of Italy and Wales) began to implement integrated plans for the support of ecological agriculture at the end of the decade of the 1990s; these included increases in subsidies for production, the restoration of economic support for the creation of markets, the promotion of research, and the development of training programs (Willer and Yussefi, 2001). At the beginning of this century, other European regions started up similar plans promoting ecological production, such as the Action Plan to Develop Organic Food and Farming in England-United Kingdom (DEFRA, 2002) and the Ecological Agriculture Promotion Plan in Andalusia-Spain (González de Molina et al., 2007). These plans represent ambitious objectives within a short-term horizon; for example, the Dutch government stated that 10% of the useful agrarian area would have to be dedicated to ecological agriculture by 2010 (MLNV, 2000), and in England the government established the goal of satisfying the internal demand for food with at least 70% ecological products—a large jump from the present 30% quota for the domestic market (DEFRA, 2002). In 2004, the various countries of the EU approved a "European Plan of action for ecological foods and ecological agriculture" that contains 21 action items related to ecological agriculture; this plan aims to develop the ecological foods market and increase consumer confidence by improving quality standards through programs that increase the effectiveness of the standards and the transparency of their development.

An example that illustrates the importance of economic assistance and action plans can be seen in the case of Italy, where from 2001 to 2004 a remarkable reduction in the area (more than 275,000 ha) and in the number of ecological farming operations (more than 19,000) occurred as a result of the retirement of agroenvironmental aid in several regions of the country. This decline in EA was then reversed, beginning in 2004, by the European action plan, which restimulated the development of ecological agriculture in this country. As a consequence, EA in Italy had already recovered more than 112,000 ha and 8,000 farms by 2005.

Other factors, like the draw of new markets and media attention to food scandals (chickens with dioxins, mad cow disease, etc.), have also influenced the development of ecological agriculture (Lampkin and Padel, 1994; Alonso, 2002). Concern about mad cow disease has had the most impact in the central and northern countries of Europe, being particularly notable the case of the United Kingdom (which has moved from 49,000 ha of EA production in 1995 to 620,000 ha in 2005 and now has the fourth largest share of EA area in the EU).

In 2005, EA was distributed among the countries of the EU more evenly than ever before, with no country claiming more than a 20% share. This is the result of significant increases in the number of hectares in ecological production in many countries. The top positions are occupied by Italy (19.9% of the total), followed by Germany and Spain (both with 15.1% of the total). Greece and Portugal have made fast growth in EA in recent years: they have gone from little having relevance to jointly possessing almost 10% of the area of ecological agriculture in the UE. The relative importance of a few countries, however, including Belgium, the Netherlands, Ireland, and Luxembourg, is still very low; together they represent 2.1% of the ecological production area of the European Union.

Table 11.1 charts in more detail the evolution of ecological agriculture in the EU. It shows the annual rates of growth in area and number of farming operations within two periods of time: from 1985 to 1992 and from 1992 to 2004.

Overall in the EU, growth rates declined somewhat in the more recent period; however, these are nevertheless rates of *growth*, and they indicate that ecological agriculture will continue to expand its importance in the future. If the annual rate of growth of land area under ecological production for the EU as a whole continues to average about 20% (slightly higher than the rate for 1992–2005 but lower than the rate for 1985–1992), we can project that by 2013 (the year the new plans of rural development in the EU conclude) there will be 23 million ha in ecological production, which would represent approximately 18% of the useful agricultural area* (UAA) of the European Union.

---

* This includes cultivated land, hay fields, and permanent pasture.

TABLE 11.1

Average Annual Rate of Increase in Two Key Indicators of the Extent of Ecological Agriculture during the Periods 1985–1992 and 1992–2005

	Increase in Area of Land under Ecological Cultivation (%)		Increase in the Number of Ecological Agricultural Operations (%)	
	1985–1992	**1992–2005**	**1985–1992**	**1992–2005**
Germany	34.9%	11.2%	16.7%	10.3%
Austria	44.8%	11.9%	46.2%	9.8%
Belgium	19.1%	22.2%	19.7%	11.1%
Denmark	22.5%	17.1%	26.5%	11.8%
Spain	20.4%	42.8%	12.0%	28.8%
Finland	48.4%	18.7%	55.3%	9.6%
France	9.5%	15.6%	2.5%	10.9%
Greece	—	72.0%	—	50.0%
Netherlands	22.3%	12.9%	12.5%	8.3%
Ireland	26.2%	16.0%	57.8%	13.2%
Italy	29.2%	31.6%	22.6%	24.8%
Luxembourg	5.2%	15.5%	2.6%	14.8%
Portugal	69.4%	44.2%	83.5%	27.1%
United Kingdom	28.7%	24.7%	15.0%	13.8%
Sweden	55.7%	14.8%	38.8%	5.4%
EU Average	26.8%	19.4%	19.6%	15.4%

Source: Lampkin (2003); Willer and Yussefi (2007).

A comparison, in Table 11.1, of second-period growth rates versus first-period growth rates reveals some important differences between countries. In one group of countries that includes Italy, France, and Spain, growth rates increased in the second period relative to the first (in the case of Spain, dramatically so). In another group of countries that includes Germany, Austria, Finland, and Sweden, growth rates have paralleled those of the EU as a whole and have dropped since 1992.

The indicator that normally is used to measure the importance of ecological agriculture in a country is the percentage of UAA devoted to ecological agriculture. As it is possible to appreciate in Figure 11.3, in many countries the area of ecological agriculture already surpasses the 4.2% of UAA that is the average for the EU. Austria, for example, has 11.1% of its UAA in ecological agriculture, and Italy has 8.1%. The Scandinavian countries (Finland, Sweden, and Denmark), which have historically occupied outstanding positions with regard to this indicator, have been joined by Greece and Portugal in the recent years as countries with above-average proportions of land in ecological production. Both of these countries have experienced rapid growth in their ecological production area (especially in olives).

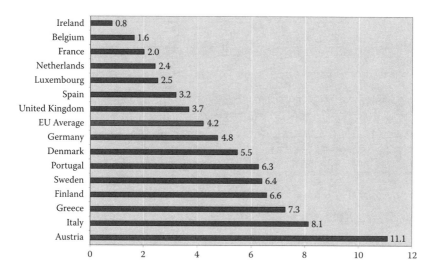

FIGURE 11.3   Land area under ecological production as a percentage of useful agricultural area (UAA) in 2005. (From: Eurostat, 2007; Willer and Yussefi, 2007.)

11.3.2   Farm Size

Looking at both ecological cropping area and the number of farming operations in ecological production allows us to calculate the average size of ecological farming operations. Examining how this figure has changed over time tells us much about the changes that are taking place in ecological agriculture in Europe. At first, most ecological producers had relatively small operations, but the average size of these operations in the EU has been growing steadily as the sector has developed, changing from 16 to more than 25 ha per farm operation between 1985 and 1991. From 1991 to the present time, the European average has continued growing, and is now about 38 ha per farm operation (see Figure 11.4).

There are many reasons for this increase in the average size of ecological farming operations. As ecological markets become more mainstream and the returns decline for conventional agriculture, more medium- and large-scale farmers are enticed into ecological production. Also important is the tendency, on individual farms, for the area in ecological production to increase over time. A farmer tends to initiate the transformation to ecological agriculture on one part of the farm first, so that in successive years more and more surface area comes into ecological management; this occurs as farmers perfect the techniques that allow a gradual improvement in management efficiency. Average farm size increases also as more and more pasture land and land on which crops are grown for cattle feed is incorporated into ecological production.

It is important to note, however, that despite the overall tendency for farm size to increase, a good part of the growth of this production system has come from the addition of many small-scale farms to the sphere of ecological production. Many of these small operations are established by urban people looking for a different kind of lifestyle in rural areas and who develop direct relationships with relatively small markets.

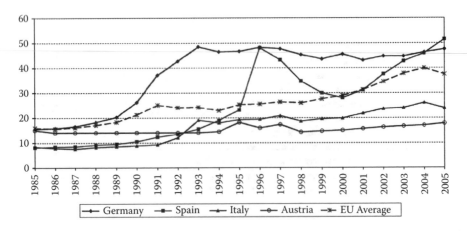

FIGURE 11.4 Change over time in the average size, in hectares, of ecological farming operations in some representative countries. (From: Lampkin, 2003; Eurostat, 2007; Willer and Yussefi, 2007.)

11.3.3 LAND USE AND CROP TYPE

In recent years, one of the most prominent changes in ecological agriculture in the EU has been the growth of ecological cattle ranching. This brings with it the ecological transformation of pasture land and land used to grow animal feedstock, agricultural land uses that generally occur at a more extensive scale than land uses dedicated to other types of food production. Among the factors that explain this growth are the adoption of ecological cattle ranching standards in the EU (primarily based on R(EC)1804/99) and increased market demand for ecologically produced animal products.

The Spanish case serves to illustrate this trend and to highlight the "attractive effect" of the subsidies that are common in EU agriculture. In 1996, when payments for having land in ecological production increased, there was a significant jump in the number of large operations dedicated to ecological production of cereal for livestock feed and pasture in some regions of Spain (Guzmán et al., 2000). When payments decreased later, most of these farmers left the ecological sector; then in 2000 the bullish tendency of the ecological market brought many of these farmers and ranchers back to the ecological sector.

The distribution of ecological production area in the European Union by type of crop in 2005 is shown in Figure 11.5. In general, hay crops and permanent pastures dominate the ecological arena, accounting for an estimated 45% of all ecological area. Only in Denmark, Finland, Italy, and Sweden is the amount of ecological land in permanent pasture surpassed by a different type of agricultural land use. The hay and permanent pasture area is especially high, in absolute terms, in England and Germany.

With respect to cultivated land, it stands out that in all the countries the greatest proportion of land in ecological production is dedicated to extensive crop systems, an important part of which is cereals, leguminous crops, and other forage crops for

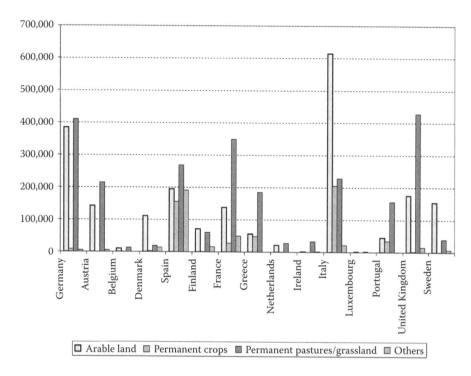

FIGURE 11.5   Ecological agriculture area (in hectares) in the European Union by groups of crops in 2005. (From: Willer and Yussefi, 2007.)

animal feed. For this reason, it would be possible to say that ecological agriculture in most of the countries in the EU is presently dominated by livestock production.

The surface area dedicated to ecologically grown fruit and vegetable products varies depending on market demand; at the moment it is 156,446 ha, most of which is concentrated in five countries: Italy (about 51% of total land in ecological production), France (11.4%), the United Kingdom (9.8%), Germany (9.1%) and Spain (6.3%).

In several countries, symptoms of readjustment between different production systems are being observed. In Austria, for example, producers of ecological milk and milk derivatives experienced a difficult situation several years ago when the demand of the internal market was exceeded and there was strong competition from other countries, such as Denmark. This resulted in a reduction in the number of milk cows. At the same time, however, the production of ecological chickens and eggs grew (Wlcek et al., 2003), reaching around 1.1 million head of poultry by 2005 (Eurostat, 2007). In another example, the government in the United Kingdom, through the Action Plan to Develop Organic Food and Farming, gave special support to cereal grains and fruit and vegetable production because the United Kingdom imported a high percentage of these products (DEFRA, 2002). This differentiated support, together with the relative difficulty of the organic milk market, is modifying the trajectory of ecological production in this country.

Overall, the development of ecological crops and livestock in the EU is characterized by great dynamism. This dynamism is driven primarily by expansion of

consumption of ecological products, the creation of agroindustries dealing in ecologically grown products, and institutional support. The latter factor is the subject of
the next section.

11.4 INFLUENCE OF INSTITUTIONAL SUPPORT ON DEVELOPMENT OF ECOLOGICAL AGRICULTURE

Before the early 1990s, pioneering ecological producers maintained the economic
viability of their farming operations without a legal endorsement by establishing
connections with diverse organizations and groups where they sold their products
(Michelsen et al., 1999; Miele, 2001; Tovey, 1997; Guzmán et al., 2000). The effort to
consolidate and develop ecological agriculture in Europe rested upon their success or
failure. With the arrival of European-wide legislation in the period between 1990 and
1992, however, the situation changed. Together, R(EEC)2092/91, R(EEC)2078/92,
and the reform of the Common Agriculture Policy (CAP) gave ecological food
production a boost by introducing visibly different ecological food products in the
market, recognizing the important role that ecological agriculture could play in sustainable rural development and resource conservation, and creating a framework for
supporting ecological agriculture financially.

In this context, European governments generally have been playing a significant
role in promoting ecological agriculture, although the budgets designated for this
purpose are not large. In 1997, 2001, and 2003, respectively, 259.7, 519.4, and 462.1
million euros* were earmarked. Below, we analyze the influence of these economic
support measures on the ecological sector, using data from these three years.

In the first place, it is possible to say that governments vary in the priority they give
to the support of ecological agriculture. As can be seen in Figure 11.6, the amounts
that the various countries allocate to ecological agriculture, as a percentage of their
total budgets for agroenvironmental measures, are very different. In addition, it is
clear that these amounts vary across time. In 1997, several countries devoted large
portions of their budgets to EA (58.2% for Denmark, 31.7% for Greece, 25.6% for Italy,
and 23.7% for Belgium), while in other countries (including Spain, the Netherlands,
France, Portugal, and the United Kingdom) less than 10% of the agroenvironmental
budget was devoted to EA. In 2001 the situation was rather different. In Spain, the
Netherlands, France, and the United Kingdom, funding for ecological agriculture
rose remarkably relative to 1997, while in Belgium, Finland, and Portugal it was
reduced. Overall in the EU, the percentage of agroenvironmental funds allocated to
ecological agriculture increased from 11.2% in 1997 to 15.8% in 2001.

Although economic support from governments has a positive effect on the growth
of ecological agriculture generally, there is no evidence for a direct relationship
between funding level and the rate of development of the ecological agriculture sector. For example, in 1997 Austria dedicated a mere 12% of its agroenvironmental
funding to EA but the approximately 345,000 ha (see Table 11.2) of ecological area it
had in that year represented a considerable 16% of all the EA land in the EU. In contrast, Greece—a country with a UAA similar to Austria's—devoted almost 38% of

---

* Exchange rate: $1 = €0.74.

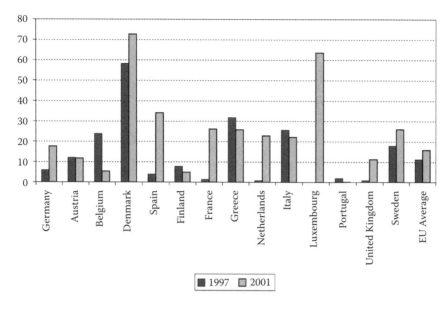

FIGURE 11.6   Percentage of the budget for agroenvironmental measures dedicated to ecological agriculture, by country, in 1997 and 2001. Note: No data exist for Ireland and Luxembourg in 1997. (From: Lampkin et al., 1999; Foster and Lampkin, 2000; Häring et al., 2004.)

its budget for agroenvironmental measures to EA in 1997 but had only 10,000 ha in ecological cultivation, which is less than 0.5% of the EA area of the EU. Something similar could be said of Denmark and Finland for 2001.

Other issues that can be analyzed through the data in Table 11.2 are the evolution of the budget through time, the distribution of the budget, the ecological surface area that receives subsidies with respect to the total, and the average payments made in each country.

With respect to first issue, the evolution of the budget over time, it is possible to say that, in general, there is a tendency to increase the budget for EA in the EU. Within the countries with greater EA area, this increasing tendency is especially clear in Germany, Austria, France, Greece, and Portugal. There are other countries, such as Spain, the United Kingdom, and Sweden, in which the 2003 budget was larger than that for 1997. There are also countries in which the budget has decreased with respect to 1997 (such as Denmark and Finland) and in which it has stayed at the same level (Italy).

The relationships between these budgetary changes and changes in the land area under ecological agriculture vary. Whereas the most common situation is for EA area to increase over time along with budgetary support, there are some exceptions. In Finland the area of EA continues growing despite reductions in support; in Denmark and Italy the area of EA declined along with support between 2001 and 2003. It seems that economic support has had a certain influence on the development of EA in every country, although just how much is not totally clear. Overall, the land area in EA has increased by 2.4 times since 1997, but the budget has increased by only 1.8 times.

TABLE 11.2

Total Ecological Area, Supported Ecological Area, and Economic Investment in Ecological Agriculture in Each Country in 1997, 2001, and 2003

	1997			2001			2003		
	Total Area (000 ha)	Supported Area (000 ha)	Support (10⁶€)	Total Area (000 ha)	Supported Area (000 ha)	Support (10⁶€)	Total Area (000 ha)	Supported Area (000 ha)	Support (10⁶€)
Germany	389.7	229.5	23.5	632.2	533.6	84.5	734	536.8	97.7
Austria	345.4	257.0	65.7	285.5	247.0	67.9	328.8	295.2	85.9
Belgium	6.7	3.4	0.9	22.4	16.6	3.4	24.2	18.9	4.7
Denmark	64.4	50.3	9.4	174.6	158.1	26.5	165.1	110.5	8.7
Spain	152.1	62.9	2.9	485.1	255.1	31.8	725.3	158.2	25.7
Finland	102.3	89.4	20.0	147.9	137.6	16.6	160	142.5	16.9
France	165.4	42.0	4.0	419.8	137.2	26.2	551	207.8	42.2
Greece	10.0	5.7	4.2	31.1	15.5	6.7	244.5	19	7.7
Netherlands	17.0	4.6	0.3	38.0	22.7	4.4	41.9	11	2.5
Ireland	23.6	—	—	30.1	13.7	1.8	28.5	17.7	1.7
Italy	641.1	308.4	100.6	1,230.0	452.2	158.9	1052	297.9	100.3
Luxemburg	0.6	—	—	2.1	2.0	0.3	3	2.3	0.4
Portugal	12.2	10.3	1.2	70.9	27.1	3.7	120.7	27.9	3.9
UK	106.0	29.1	0.7	679.6	408.0	17.5	695.6	249.9	9
Sweden	118.2	205.2	26.3	193.6	430.6	69.0	225.8	407	54.8
Total	2,154.6	1,297.7	259.7	4,442.9	2,857.1	519.4	5,100.4	2,502.6	462.1

Source: Lampkin et al. (1999); Foster and Lampkin (2000); Häring et al. (2004); Stolze and Lampkin (2006).

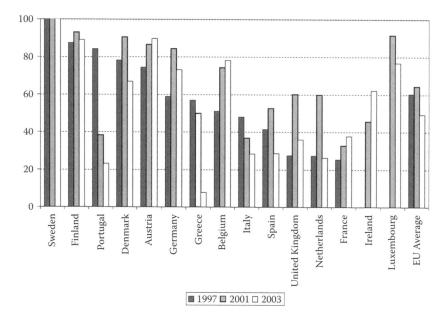

FIGURE 11.7 Percentage of land devoted to ecological agriculture in each country that received subsidies in 1997, 2001, and 2003. (From: Lampkin et al., 1999; Foster and Lampkin, 2000; Häring et al., 2004; Stolze and Lampkin, 2006.)

The relationship between institutional support and EA has also varied over time. In 1997, Spain and France had 7.1 and 7.7% of the ecological area of the EU, respectively, yet they only dedicated 1.1 and 1.5% of their budgets to EA. In 2003, the relation between budget distribution and ecological surface area was more direct.

Figure 11.7 has been constructed with the objective of showing with greater clarity the proportion of ecologically certified agricultural land in each country that received subventions during each of the three years considered. In this figure, the countries have been ordered based on the 1997 data. In the first place, it is necessary to clarify that data for Sweden surpass 100% because in this country there exists an administrative control that does not require that the ecological area be certified to be part of the program. Even so, in the EU the average amount of subsidized ecological agricultural area has declined from 60% in 1997 to 49% in 2003. In 1997 the countries that subsidized a proportion of ecological area above the European average were the Scandinavian countries (Sweden, Finland, and Denmark), Portugal, and Austria. These same countries, except Portugal, have maintained elevated percentages of subsidized area over time (although in the case of Austria the percentage has remained high in part due to a decline in the total number of hectares under EA). Since 1997, a few countries, notably Germany, Belgium, Ireland, and Luxembourg, have significantly increased the percentage of land in EA that is subsidized and in 2003 surpassed the EU average. Showing the opposite tendency are Portugal, Greece, Italy, and Spain.

Finally, to control for large differences in the total amounts of ecologically certified land, we have constructed Figure 11.8. It shows, for each country, the amount of

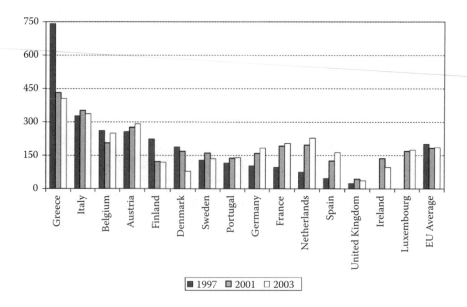

FIGURE 11.8 Subvention per unit of ecologically certified land area in each country in 1997, 2001, and 2003 (€/ha). (From: Lampkin et al., 1999; Häring et al., 2004; Stolze and Lampkin, 2006.)

government subsidy in euros per hectare of ecologically certified land for each of the three years under consideration. As in Figure 11.7, the countries are ordered according to the 1997 data.

As can be seen, Greece in 1997 absolutely outdistanced the rest of countries, with a payment per hectare of around 740 €/ha, which is 2.3 times more than the second country, Italy. Only three other countries (Belgium, Austria, and Finland) were over the average for the EU in 1997. In both Greece and Belgium, it does not seem that these payments promoted the expansion of ecological production, if we consider that the ecological area in Greece was only 10,000 ha, and in Belgium around 6,700 ha. The unit payments have varied very little in the rest of the EU, changing from 200 €/ha in 1997 to nearly 186 €/ha in 2003. Nevertheless, they have changed differently for different countries. There have been countries in which the payment per hectare has been reduced remarkably (Greece, Finland, Denmark), countries in which the payment per hectare has remained stable (Italy, Belgium, United Kingdom), and countries in which the payment per hectare has been increased considerably (e.g., the Netherlands and Spain). In Spain, notably, the per-hectare payment in 2003 was more than three times higher than it was in 1997.

Throughout this account it has been possible to verify the existence of a series of countries (Austria and the Scandinavian countries) that have counted on relatively ample institutional support (as measured by budget devoted to EA, percentage of subsidized ecological area, and payments per hectare). These countries exemplify the importance of economic support in the transition toward the EA, if we consider that these are the countries that show leadership in the relative importance they assign to ecological agriculture (as measured by the percentage of useful agricultural area under EA).

Nevertheless, there are other countries that reveal that the payment of subsidies is not a sufficient condition for the development of EA. In order to exemplify this circumstance, we can pay attention to the cases of Greece, Spain, and the United Kingdom. In Greece, in spite of generous subsidies per hectare, growth in EA was small between 1997 and 2001, with this country occupying the last position in terms of both the relative and absolute importance of EA. In Spain and the United Kingdom, on the other hand, despite historically low institutional support, the development of EA has been remarkable, with more than 1.4 million ha being cultivated ecologically between both countries in 2003.

These "anomalies" can be explained on the basis of other factors—in particular the economic aspects of ecological production in these countries. As we will see next, these economic factors are important and will help clarify the theoretical relationship between direct institutional support and the development of ecological agriculture.

11.5 ECOLOGICAL VERSUS CONVENTIONAL AGRICULTURE: THE ECONOMIC CONTEXT

Ecological agriculture is allowing the generation of positive socioeconomic impacts in the new framework of European rural development (Alonso et al., 2001; Ploeg et al., 2002; Marsden, 2003; O'Connor et al., 2006). It is increasing the profitability of agriculture for many producers and creating more jobs (compared to conventional agriculture) through the production and commercialization of quality products. In this section, we analyze a series of parameters that directly affect the economic viability of these farming operations. On the sales and income side are yields and prices; on the side of costs are production costs, particularly the cost of manual labor. The analysis of these parameters will allow a general comparison between ecological and conventional production.

11.5.1 YIELDS

The yields of crop and livestock production are influenced by numerous factors (climate, topography, soil, technology, etc.). Generally, it is accepted that ecological production yields are lower than conventional ones. This difference may be particularly evident during the transition process (Guzmán et al., 2000), during which there is a reduction in the application of technological inputs (mainly synthetic chemicals) and a slow recovery of the agroecosystem as fertility increases and the soil is decontaminated. Nevertheless, the yield of systems under ecological production (even during the transition process) is not always lower. The time under ecological management, previous production intensity, and the suitability of new management practices are determining factors in the behavior of the yield parameter (Guzmán et al., 2000; Lampkin and Padel, 1994).

Figure 11.9 shows the yields (kilograms or liters per hectare) of diverse ecological products as a percentage of the yields of comparative conventional products. The bar

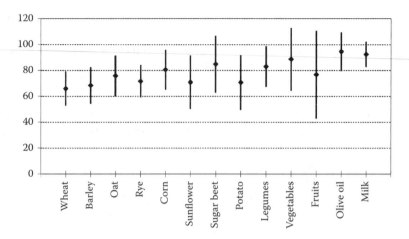

FIGURE 11.9 Yields of ecologically produced products as a percentage of the yields of comparable products produced conventionally. (From: Offermann and Nieberg, 2000; Alonso, 2003; Alonso and Guzmán, 2004.)

for each product represents the average and the standard deviation of that percentage. The data were obtained from several case studies, mostly from Central and Northern Europe, compiled in Offermann and Nieberg (2000), Alonso (2003), and Alonso and Guzmán (2004). One can observe that the yield for ecological products is generally lower than that for comparable conventional products; an average reduction in yield of around 20% can be estimated. This reduction is greatest in the case of cereals, sunflower, potato, and fruits. To put these data in the appropriate context, it is important to recognize that a majority of the farming operations studied had a remarkable degree of intensification prior to their conversion to ecological production. Also, from the ecological point of view, many of these farms still are in transition, and may not have fully recovered their potential fertility. To this fact it is necessary to add that most of the crops that are compared here come from "improved" plants that have lost a good part of their capacity to adapt to adverse conditions (climatologic, defense against diseases, competition with weed flora, etc.) in favor of a greater production response with increasing doses of technological packages of a synthetic chemical character (fertilization, pesticides, and herbicides).

Yield reduction is less common in less intensified systems. For example, Lacasta and Meco (2000) show that ecological cereal systems in dryland areas of the Mediterranean that employ biennial rotations of barley-sunflower and forage-barley can obtain yields equal or superior to those obtained for conventional monoculture of barley. In olive tree cropping systems, yield differences (positive or negative in favor of one or the other system) are small in comparison with other cropping systems; this is because the rusticity of the system has limited the technological intensification in the conventional version of the system, allowing a relatively simple ecological conversion (Alonso et al., 2001). Smaller differences in yield also occur in the case of bovine milk production in semi-intensive farms (with pasturing and cropping), mainly because there is a greater adjustment (to meet obligatory standards) of the cattle stocking rate, which allows the farmer to adapt and orient the local resources toward milk production.

In general, the multiple factors that affect yields are determined to a great extent simply by the inherent differences between ecological and conventional management. Nevertheless, the yields for ecological production are not always lower, and when they are, they are not the sole determinant of production income, since other elements exist, such as the actual price received by farmers, which they can influence themselves in many ways.

11.5.2 PRICES AND MARKETS

The market for ecological products has grown greatly during the last few years. In the period between 1997 and 2005, the value of the total volume of certified ecological food sales in the world nearly tripled, from about 10,000 million euros in 1997 to approximately 26,000 million euros in 2005 (Willer and Yussefi, 2007). The development of this differentiated market is allowing producers to obtain higher prices for their products.

The prices for most ecological products are higher than those for the comparable conventional products. This price difference is known as the ecological price premium. Figure 11.10 shows the price premium for various ecologically produced foods, expressed as a percentage of the price of the conventional product (this means, for example, that a price premium of 100% is equivalent to a price twice as high and a 50% premium is a price half again as high). As in the previous case, the bar for each product represents the average and the standard deviation; the data were obtained from the same set of case studies. It is possible to observe that generally crop farmers receive a greater price premium than producers of animal products. In the case of wheat and potato, conventional prices are exceptionally low to begin

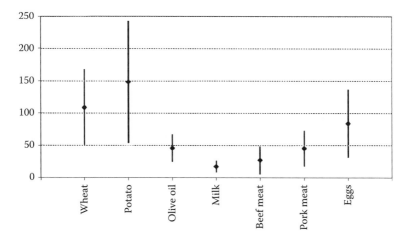

FIGURE 11.10 Differences between prices for ecologically and conventionally produced products, as a percentage of the conventional price. (From: Offermann and Nieberg, 2000; Alonso, 2003; Alonso and Guzmán, 2004.)

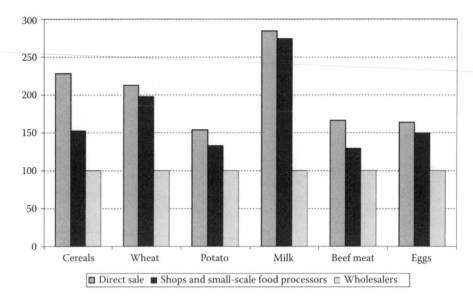

FIGURE 11.11 Average prices received by farmers for ecologically produced products marketed through short channels, as a percentage of prices received through wholesale marketing. (From: Offermann and Nieberg, 2000.)

with, which accentuates the price difference. In the case of olive oil, the surcharge on ecological production is smaller because conventional olive oil is already considered to be a high-quality product and its price reflects this perception.

Within animal products, eggs are the animal product that is relatively better compensated for, because a strong demand exists (with a limited supply) and short channels of distribution are used. For the rest of the products of animal origin, the price premium perceived by ecological producers is lower due to problems coordinating the not very ample supply, the absence of distribution and sales channels, and the tendency of ecological consumers to consume less meat than consumers generally (Offermann and Nieberg, 2000; Lampkin and Padel, 1994; González de Molina et al., 2007).

The prices ecological producers obtain for their products are significantly influenced by the channel of distribution used. The case studies compiled in Offermann and Nieberg (2000) show that ecological producers received much higher prices for their products when they marketed them directly or through small-scale shops or food processors than when they sold them to wholesalers or other intermediaries. The graph in Figure 11.11 shows the data from these compiled case studies. In this graph, the average price obtained through wholesalers is defined as 100%, and the prices obtained through the shorter distribution channels are expressed relative to this price (a price of 200%, for example, represents a price twice as high as that received through wholesaling). As can be seen in Figure 11.11, ecological producers received about 50% more for their potatoes and about 2.5 times more for their milk when they marketed these products through the shorter channels.

A scenario that has been repeated in numerous countries (Miele, 2001) is that pioneering ecological farmers, often isolated geographically and usually producing relatively small amounts, had to make special efforts to sell their products. The channels that were developed initially were a reflection of the ruling productive philosophy, which was to reduce to a minimum the use of natural resources (avoiding unnecessary packaging, reducing distances to market, minimizing transport costs, etc.) and to establish links between producer and consumer that allowed the formation of a high degree of mutual trust and confidence. Doing this required direct contact with the consumer through different systems of direct sale. These included selling on the farm, selling in local markets, delivering to consumers' homes by means of subscription sales, and selling to associations and cooperatives of consumers to which producers themselves belonged.

As the production and consumer demand grew, the original channels grew, which induced other existing conventional channels to take advantage of the growing market for ecological products. Thus, by the end of the twentieth century, ecological products had a presence in most mega-supermarkets and supermarkets in some countries, such as Sweden, Denmark, Finland, United Kingdom, and Austria; in some of these establishments, sales of ecological products surpassed 50% of the total (Willer and Yussefi, 2001). Although consumers generally pay higher prices for ecological products in these conventional retail establishments than they do when purchasing directly from producers (Offermann and Nieberg, 2000; Guzmán et al., 2000; Alonso et al., 2002a), recent social tendencies—including a trend toward purchasing from these types of establishments and an increasing desire for healthy products—have contributed to mitigating the importance of the price factor in the purchase decision.

Some authors (e.g., Michelsen et al., 1999) have related the importance of sales in mega-supermarkets and supermarkets to the development of markets for ecological products; nevertheless, this trend by itself does not explain the growth of the sector. It is evident that ecological product commercialization through large venues has been growing in numerous countries, as is reflected in the data presented in Table 11.3. Nevertheless, one of the results of food chain sales is that they are controlled by a reduced number of large companies that impose prices that do not consider the social, economic, and environmental costs of food production and do not meet the needs of ecological farmers and animal producers. Latacz-Lohmann and

TABLE 11.3

Percentage Share of Total Annual Sales of Ecological Food Products for Three Different Sales Channels for the Years 1999 and 2003

	United Kingdom		France		Germany		Netherlands	
	1999	2003	1999	2003	1999	2003	1999	2003
Supermarkets	74	80	38	48	26	35	2	46.2
Specialized shops	15	10	46	25.5	46	41	96	40.5
Direct sale	6	10	16	16.5	19	17	1	13.3
Other	5		—	10	9	7	1	

Source: Willer and Yussefi (2001); CBI (2005).

TABLE 11.4

Ecological Food Subscription (EFS) Systems in Germany, the Netherlands, and the United Kingdom in 2003

	Germany	Netherlands	United Kingdom
Number of EFS systems	300	55	300
Weekly deliveries (×1,000)	124	41	82
Sale values (millions €)[a]	328.6	14.8	37
Market total value (millions €)	2,752.8	347.8	1,420.8
Market share (%)	8–13%	3.5–4.5%	2.5–3%

Source: Adapted from Haldy (2004).

[a] Rate of exchange: $1 = €0.74.

Foster (1997) have identified this contradiction between ecological agriculture and mainstream commercialization as a structural incompatibility.

Despite the growth in mainstream commercialization, forms of direct sale have held steady or even increased, as have sales to public institutions (student dining halls, hospitals, etc.), airline companies, health clinics, and other organizations. Table 11.3 shows that direct sales increased from 1999 to 2003 in the Netherlands and remained about the same in the United Kingdom, France, and Germany.

After experiencing a decline in some countries in the 1980s, farmers' markets have recently staged a remarkable resurgence. In the United Kingdom, farmers' markets had practically disappeared at the end of the 1980s, but producers were able to reverse the trend in the 1990s: by 1999 more than 4,400 farmers were operating about 200 periodic markets (Alonso et al., 2002), and by 2004 there were nearly 500 farmers' markets (CBI, 2005). At the end of the last century in the Netherlands, a similar circumstance took place: in a period of six years, the number of regular farmers' markets increased from 15 to 34 (Alonso et al., 2002a).

Other direct sale mechanisms also have been developed as an answer to the incompatibilities between ecological production and large distributors; ecological food subscription (EFS) systems are one example (see Table 11.4).

The details of EFS systems vary. In many systems subscribers pay a fixed amount for a weekly basket of vegetables; in others they can select an assortment of vegetables based on consumption preferences. Some systems include the possibility of extending the order to include nonperishable products that are supplied by associations of small farmers and agroindustries. Orders can be placed directly on the farm, by telephone, or even by electronic mail. The products are either picked up on the farm (in these cases it is usually possible to harvest direct from the field), distributed to subscribers' homes, or delivered to a central point at a consumers' association, consumption cooperative, or other similar organization. In some countries these systems of ecological food subscription have acquired certain relevance. Haldy (2004) estimated that in 2003 sales through these systems were 2.5 to 3% of the total of commercial food sales in the United Kingdom and 8 to 13% of the total in Germany.

In some countries, recognition of the benefits of ecological foods for natural resource preservation and human health has promoted their consumption in institutional centers (schools, daycare centers, elder residences, etc.), strengthening the direct sales of local farmers. One of the most excellent examples is Italy, where pilot projects with this focus have been carried out since the 1980s. Today more than 600,000 children in the metropolitan areas of important cities such as Rome, Bologna, Turin, and Padua, as well as many smaller cities, receive ecological food in their schools, for example, 41 cities participate in Milan province, 34 in Trento, 32 in Udine, and 22 in Modena. Political support for these initiatives has been essential, especially in the initial stages. In the Italian region of Emilia-Romagna, for example, the Green Party promoted in 2002 the passage of Law 29/2002, which imposed the requirement that 100% of children between 3 and 10 years old would have a diet of ecological foods at school, and that ecological food constitute at least 35% of the diet in institutes, universities, and hospitals.*

The Green Party has also been pioneering the establishment of ecological product consumption programs in schools and hospitals in the Andalusia region of Spain. It started in 2005 with four groups of producers distributed in four zones who were connected to each other to complement their respective supplies throughout the year (García et al., 2008). In the first year of operation, the program was carried out in 12 schools and three daycare centers, reaching 5,200 people (mainly children). In its second year, the program extended to hospitals and elder residences and reached more than 7,400 people in 56 centers.

Probably the greatest advantage of direct commercialization for the farmers is that it simultaneously allows them to gain access to markets, to retain a higher proportion of the final price, and to have a greater margin of profit (Miele, 2001; Tovey, 1997). It also provides them greater autonomy, and it enables—and in fact encourages—the planting of an ample variety of crops, which reduces the risk of crop failure and allows them to advance toward a more sustainable agriculture. To commit oneself to direct market chains, however, also implies greater costs (new investments, acquisition of training, hiring additional personnel) and the possibility that the commercial activity reduces the time available for actual farming. The expansion of these sale systems among many ecological farmers suggests, however, that the benefits more than outweigh the disadvantages (Michelsen et al., 1999). In Italy more than 34% of the ecological farming operations use short channels of commercialization; in Germany the proportion is about 65%, and in Spain it is 45% (Domínguez et al., 2006; Ploeg et al., 2002).

In spite of the fact that in some countries of the European Union the chains of direct commercialization are losing market share to mainstream consumption, they still play an important role as they synergize with related rural development objectives. The national values included in the European Project (FAIRCT98-4288)† provide evidence for this.

---

* http://www.organic-europe.net/country_reports/italy/default.asp#market (accessed April 2007).
† "The Socio-Economic Impact of Rural Development Policies: Realities and Potentials" (FAIRCT98-4288). Consult www.rural-impact.net, and some results in Sociologia Ruralis (number 40-4), Journal of Environmental Policy and Planning (2001, number 3-2), Ploeg et al. (2002), Marsden (2003), and O'Connor et al. (2006).

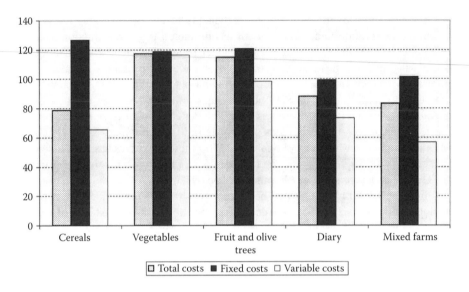

FIGURE 11.12 Per-hectare costs of ecological management as a percentage of the cost of conventional management. (From: Offermann and Nieberg, 2000; Alonso, 2003; Alonso and Guzmán, 2004.)

Overall, then, these short channels of commercialization have many benefits. In addition to generating additional income for farmers, they generate employment (a social benefit), reduce contamination of the environment, and maintain the basic productive structure of smaller towns and those located in marginalized regions.

11.5.3 Production Costs

As is the case with yields, the costs of ecological production depend on multiple factors (type of production orientation, degree of intensification, technology availability, etc.). Figure 11.12 shows the total costs, both fixed and variable, by hectare, for ecological production as a percentage of the management costs of similar conventional systems (these data were obtained from diverse studies).

In general, differences in total costs between the two management types are not substantial, although the costs of ecological management are slightly lower for cereals, bovine milk production, and mixed crop and livestock production, and slightly higher for vegetables and tree fruits. The fixed costs for ecological management tend to be somewhat higher than those for conventional management, fundamentally due to the costs of certification and monitoring, the depreciation of new investments for specific or adapted technology used for the farming conditions (in certain cases new facilities for manipulation and even processing of products are needed), and in some occasions, specialized advising.

In contrast, the variable costs of ecological production are generally lower than those for conventional production, and the ecological standards that prohibit the use of synthetic chemicals—an important component of the variable costs for conventional management—seem to be the main explanation for this circumstance. Seed costs can be greater for ecological operations because in some crops (such as the cereals) higher planting density is usually used to combat weeds, or because there is a greater use of cover crops and legumes (green manure). In some cases as well, ecological seeds have higher prices than conventional equivalents due to limited supply (Lampkin and Padel, 1994; Offermann and Nieberg, 2000). In the case of animal production, the reduction in animal numbers to fit the farm carrying capacity can increase pasturing and reduce the feeding of concentrate, thus reducing feeding costs (Bouilhol et al., 1997).

The reduction in the variable costs of ecological production must be seen in the overall context of the transition process. First, as noted above, the cost savings may be realized only after a period of transition. Second, cost reduction depends on the producer's transition and management strategy. If this strategy entails increasing the system's biological diversity and the recycling of organic matter at the farm level, it is very probable that the farm can take greater advantage of its own resources and use smaller amounts of external inputs, obtaining significant savings in variable costs. If instead the strategy consists of input substitution, it is very probable that the variable costs may actually increase, since, in general, the prices for the replacement organic fertilizer, biopesticides, and equipment allowed by certification and control agencies for use in ecological production are higher than the prices for those used in conventional production.

There is a perception among some people that ecological crop and animal management is essentially a "no management" approach, but nothing could be further from reality. Ecological management requires greater amounts of information, training, and dedication on the producer's part. A farmer must not only consider the external aspects of the farm (new scientific contributions, new technologies, emerging markets, etc.) but also closely follow crop and animal development so as to prevent possible problems before they occur, since the conventional approach of fixing the problem with a synthetic input will not work.

Besides the difficulty of ecological management itself, it must also be accepted that, in general terms, ecological systems require larger inputs of human labor than conventional systems, as is reflected in Figure 11.13. This figure, which uses data collected in many case studies (Offermann and Nieberg, 2000; Alonso, 2003; Alonso and Guzmán, 2004), compares labor use in the two types of systems by showing labor use in ecological crop and animal management systems as a percentage of the labor used in comparative conventional systems. As in previous figures, the bars represent the means and standard deviations.

This additional labor requirement is fundamentally due to the need to control weedy plants (which is done either manually, for horticultural crops, or mechanically, for most other crops) and pests and diseases. Therefore, it is necessary for farmers converting to ecological production to take into account the cost increases that can come about due to additional labor use, and to assess whether there will be adequate availability of labor for carrying out specific seasonal activities. However, it must be noted again that not all ecological operations require greater labor inputs. Some ecological cattle operations, for example, are able to reduce labor needs because of the smaller herd sizes required to match the carrying capacity of the system.

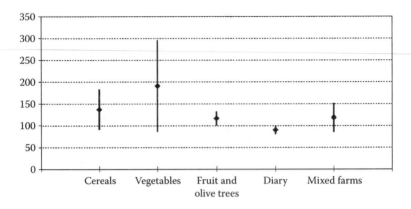

FIGURE 11.13 Labor use per hectare for ecological production systems, as a percentage of labor use for conventional systems. (From: Offermann and Nieberg, 2000; Alonso, 2003; Alonso and Guzmán, 2004.)

11.5.4 ECONOMIC BENEFITS

A compilation of various studies across the European Union shows that ecological systems in general generate larger economic benefits than equivalent conventional systems. Figure 11.14 compares the economic returns for the two types of production by showing the return for various ecological systems as a percentage of the return for comparable conventional systems. As can be seen in the figure, the medians are above 100% for each type of system, but the amplitudes of the standard deviations for some types of systems are rather large, and most drop below the 100% level, indicating that there are some particular ecological operations that obtain inferior economic benefits compared to similar conventional systems (the flip side of this, of

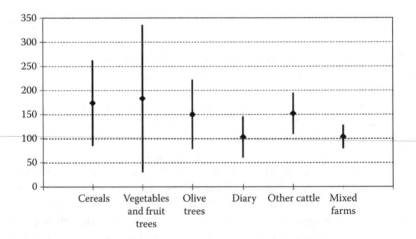

FIGURE 11.14 Economic return per hectare for various ecological systems, as a percentage of the economic return for equivalent conventional systems. (From: Offermann and Nieberg, 2000; Alonso, 2003; Alonso and Guzmán, 2004.)

course, is that there are some particular ecological operations that do exceedingly well economically compared to their conventional counterparts).

Whether a particular ecological operation will obtain superior economic results depends on all the factors that have been discussed above; in other words, multiple factors—including yields, the channels used to bring products to consumers, costs, and the type of production system—determine the final economic result of the farm operation.

Ecological cereal production in general enjoys greater relative benefits, primarily because of the greater price premium and the compensatory payment that many producers receive during the conversion process (Offermann and Nieberg, 2000; Alonso, 2002). The relative economic benefits for ecological animal production systems depend to a large extent on the characteristics of the country in which they operate; in Germany, for example, where the market for ecological meat, milk, and eggs is sufficiently developed, ecological animal operations (especially less intensive ones) face fewer obstacles to obtaining positive economic results than conventional operations. In other countries, such as Spain, where the development of ecological markets and products is still in its early stages, numerous animal operations must sell part of their production in conventional markets without price premium benefits. A similar circumstance occurs for fruit tree and horticulture operations, which can have difficulty selling their products in the relatively undeveloped ecological market.

The greater return from ecological olive groves is attributed to increasing consumer demand in recent years, mainly in Central and Northern European markets, and the resulting higher price premium obtained by farmers. This has spurred a great conversion toward ecological olive oil production in Italy, Spain, Greece, and Portugal (Alonso et al., 2002b).

The growth of the market for ecological products, which allows ecological producers to obtain higher prices and, consequently, realize a greater economic benefit, is acting like a motor of change moving the European Union toward ecological agriculture. In fact, in countries where economic support for the transition to EA has been very small, such as the United Kingdom and Spain, market influence is a key factor in the growth of EA, since it allows ecological farmers and ranchers to obtain the extra economic benefit that they do not obtain through subsidies.

11.6   CONCLUSION

Ecological agriculture in the European Union has enjoyed robust growth in recent years. Although the specifics vary with every country, there are several major reasons for this growth: ecological agriculture is a good match for the agrarian traditions that have been revitalized across Europe; socioenvironmentalist movements have pressured governments to promote EA; awareness of the health, social, and environmental benefits of EA has increased among citizens, and more information about these benefits has become widely available; the farming sector has strengthened its organizational capacity; EA has received significant institutional support; and the market for ecological food products has developed and grown considerably. The two latter factors are most decisive, as has been shown throughout the chapter, since they have very marked influence on the rest of the factors.

Research has affirmed that ecological agriculture, in general terms, offers farmers and society many benefits relative to conventional agriculture. In addition to having positive socioeconomic impacts in rural areas, ecological agriculture helps reduce impacts on natural resources and natural ecosystems relative to conventional agriculture (mainly because its regulated standards eliminate the use of pesticides and synthetic inorganic fertilizers). For these reasons, conversion to ecological agriculture is one of the most promising strategies for the new European rural development framework. However, ecological agriculture still has a long way to go. Consumer demand, the structural organization of the agricultural sector, and governmental support are all going to continue to be key factors determining the future of ecological agriculture in the EU.

REFERENCES

Alonso, A.M. 2002. Desarrollo y situación actual de la agricultura ecológica: elementos de análisis para entender el caso español. *Revista Española de Estudios Agrosociales y Pesqueros* 192:123–59.

Alonso, A. 2003. Análisis de la sostenibilidad de sistemas agrarios: el caso del olivar de Los Pedroches (Córdoba). Tesis doctoral, Universidad de Córdoba, Inédita.

Alonso, A.M., Guzmán, G.I. 2004. *Manual de Olivicultura Ecológica.* Córdoba: ISEC-Universidad de Córdoba.

Alonso, A.M., Knickel, K., Parrott, N. 2002a. Influencia de los canales comerciales en el desarrollo de la agricultura ecológica en Europa. In *Actas del V Congreso de la Sociedad Española de Agricultura Ecológica*, Gijón, Asturias, 16–20 de septiembre, Tomo II, pp. 1409–18.

Alonso, A.M., Jiménez, M., Guzmán, G. 2002b. The production of organic olive oil: The OLIPE co-operative in the Pedroches region. In *Living countrysides. Rural development processes in Europe: The state of the art*, ed. J.D. van der Ploeg, A. Long, J. Banks, 120–27. Doetinchem, Netherlands: Elsevier.

Alonso, A.M., Mudarra, I., Domínguez, M.D., Molero, J., Banda, I. 2005. Productive and institutional multifunctionality: Organic farming in protected areas. In *XXI Congress European Society for Rural Sociology*, Heszthely, Hungría, August 22–27, p. 177.

Alonso, A.M., Sevilla, E., Jiménez, M., Guzmán, G.I. 2001. Rural development and ecological management of endogenous resources: The case of mountain olive tree in Los Pedroches Comarca (Spain). *Journal of Environmental Policy and Planning* 3:163–75.

Baars, T., Ham P.W.M. van. 1997. Future conversion of organic dairy farming in relation to EU regulation. In *Proceedings of the Second ENOF Workshop: Steps in the conversion and development of organic farms*, Barcelona, October 3–4, 1996, pp. 113–21.

Bellegem, T.M. van, Eijs, A. 2002. *Market creation: Organic agriculture in the Netherlands.* ENV/EPOC/GSP/BIO (2001)7/FINAL, Working Group on Economic Aspects of Biodiversity, Organization for Economic Cooperation and Development, Paris.

Boisdon, Y., L'Homme, G., Bouilhol, M. 1997. A farm network for research of references on organic farming, becoming converted or converted farms in the Auvergne area. In *Proceedings of the Second ENOF Workshop*, Barcelona, October 3–4, 1996, pp. 7–11.

Bouilhol, M., Palhieres, H., Passemard, R., Chapput, J.P. 1997. Conversion in upland area sheep farming system (a case study). In *Proceedings of the Second ENOF Workshop*, Barcelona, October 3–4 1996, pp. 69–81.

CBI. 2005. *EU market survey 2005. Organic food products.* Centre for the Promotion of Imports from Developing Countries. http://www.cbi.nl/marketinfo/cbi/index.php?action=downloadFile&doc=639&typ=mid_document (accessed February 2007).

Culik, M.N. 1983. The conversion experiment: Reducing farming cost. *Journal of Soil and Water Conservation* 38:333–35.

DEFRA. 2002. *Action plan to develop organic food and farming in England.* http://www. defra.gov.uk/farm/organic/actionplan/actionplan.pdf (accessed February 2004).

Domínguez, M.D., Alonso, A.M., Simon, X., Mauleón, J.R., Ramos, G., Renting, H. 2006. Catching up with Europe, rural development in Spain. In *Driving rural development: Policy and practice in seven EU countries,* ed. D. O'Connor, H. Renting, M. Gorman, J. Kinsella, 111–44. Assen, The Netherlands: Royal Van Gorcum.

Eurostat. 2007. Statistical information in organic farming. http://www.organic-europe.net/europe_eu/statistics-eurostat.asp#tables (accessed February 2007).

Foster, C., Lampkin, N. 2000. *Organic and in-conversion land area, holdings, livestock and crop production in Europe.* Task 2.1: Technical Deliverable of FAIR3-CT96-1794 project.

García, R., González de Molina, M., Ávila, E., Tobar, E., Alonso, A.M., Gómez, F. 2008. Organic foods for social consumption in Andalusía. In *Proceedings of the Second Scientific Conference of the International Society of Organic Agriculture Research (ISOFAR), Workshop on Organic Public Catering,* 16th IFOAM Organic World Congress, Modena, Italy, June 18–20, CD edition.

Gliessman, S.R. 2007. *Agroecology: Ecology of sustainable food systems.* Boca Raton, FL: CRC/Taylor & Francis.

González de Molina, M., Alonso, A.M., Guzmán, G.I. 2007. La agricultura ecológica en España desde una perspectiva agroecológica. *Revista Española de Estudios Agrosociales y Pesqueros* 214:47–73.

Guzmán, G.I., Alonso, A.M. 2008. A comparison of energy use in conventional and organic olive oil production in Spain. *Agricultural Systems* 98:167–76.

Guzmán, G., González de Molina, M., Sevilla, E., eds. 2000. *Introducción a la agroecología como desarrollo rural sostenible.* Madrid: Mundi-Prensa.

Haldy, H.M. 2004. Organic food subscription schemes in Germany, Denmark, the Netherlands and the United Kingdom. Definitions and patterns of development in an international context. MBA dissertation 2004 at the Aston Business School, Birmingham, United Kingdom. http://www.kmc-consult.com/Organic%20Food%20Subscription%20Schemes%20Dissertation%20by%20HM%20HALDY.pdf (accessed March 2007).

Häring, A.M., Dabbert, S., Aurbacher, J., Bichler, B., Eichert, C., Gambelli, D., Lampkin, N., Offermann, F., Olmos, S., Tuson, J., Zanoli, R. 2004. *Impact of CAP measures on environmentally friendly farming systems: Status quo, analysis and recommendations— The case of organic farming.* Report on the study contract "Environmentally Friendly Farming Systems and the Common Agricultural Policy." http://ec.europa.eu/environment/agriculture/pdf/effscap_report.pdf (accessed April 2004).

Lacasta, C., Meco, R. 2000. Costes energéticos y económicos de agroecosistemas de cereales considerando manejos convencionales y ecológicos. Paper presented at IV Congreso Sociedad Española de Agricultura Ecológica, Córdoba.

Lampkin, N. 1992. *Organic farming.* Ipswich, UK: Farming Press.

Lampkin, N.H. 2003. Statistics of organic farming. GB-SY23 3AL, Welsh Institute of Rural Studies, University of Wales, Aberystwyth. http://www.organic.aber.ac.uk/stats.shtml (accessed March 2003).

Lampkin, N., Foster, C., Padel, S. 1999. *The policy and regulatory environment for organic farming in Europe: Country reports.* Organic Farming in Europe. Economics and Policy, Vol. 2. Stuttgart: University of Hohenheim.

Lampkin, N.H., Padel, S., eds. 1994. *The economics of organic farming. An international perspective.* Wallingford, UK: CAB International.

Latacz-Lohmann, U., Foster, C. 1997. From 'niche' to 'mainstream'—Strategies for marketing organic food in Germany and the UK. *British Food Journal* 99:275–82.

MacRae, R.J., Hill, S.B., Mehuys, G.R., Henning, J. 1990. Farm-scale agronomic and economic conversion from conventional to sustainable agriculture. *Advances in Agronomy* 43:155–98.

Maire, N., Besson, J.M., Suter, H., Hasinger, G., Palasthy, A. 1990. Influence des pratiques culturales sur l'équilibre Physico-chimique et biologique des sols agricoles. *Recherche agronomique en Suisse* 29:61–74.

Marsden, T. 2003. *The condition of rural sustainability.* Assen, The Netherlands: Royal van Gorcum.

Michelsen, J. 2001. Recent development and political acceptance of organic farming in Europe. *Sociologia Ruralis* 41:3–20.

Michelsen, J., Hamm, U., Wynen, E., Roth, E. 1999. *The European market for organic products: Growth and development.* Organic Farming in Europe. Economics and Policy, Vol. 7. Stuttgart: University of Hohenheim.

Miele, M. 2001. Creating sustainability. The social construction of the market for organic produtcs. PhD thesis, Circle for Rural European Studies, Wageningen Universiteit, The Netherlands.

MLNV. 2000. *An organic market to conquer.* Policy Document on Organic Agriculture, 2001–2004, Ministerie van Landbouw, Natuurbeheer en Visserij, Netherlands. http://www.minlnv.nl/international (accessed January 2004).

Moran, D. 2002. *Market creation for biodiversity: The role of organic farming in the EU and US.* ENV/EPOC/GSP/BIO (2001)8/FINAL, Working Group on Economic Aspects of Biodiversity, Organization for Economic Cooperation and Development, Paris.

O'Connor, D., Renting, H., Gorman, M., Kinsella, J., eds. 2006. *Driving rural development: Policy and practice in seven EU countries.* Assen, The Netherlands: Royal Van Gorcum.

Offermann, F., Nieberg, H. 2000. *Economic performance of organic farming in Europe.* Organic Farming in Europe. Economics and Policy, Vol. 5. Stuttgart: University of Hohenheim.

Ploeg, J.D. van der, Long, A., Banks, J., eds. 2002. *Living countrysides. Rural development processes in Europe: The state of the art.* Doetinchem, The Netherlands: Elsevier.

Sparkes, D.L., Wilson, P., Huxham, S.K. 2003. *Organic conversion strategies for stockless farming.* Project Report 307, Home-Grown Cereals Authority, UK.

Stolze, M. 1997. The economic analysis of large-scale organic farms in East Germany. Paper 97103 presented at Proceedings of the Second ENOF Workshop, Barcelona, October 3–4, 1996.

Stolze, M., Lampkin, N. 2006. European organic farming policies: An overview. Paper presented at Joint Organic Congress, Odense, Denmark, May 30–31, 2006. http://orgprints.org/6337/ (accessed May 2007).

Stolze, M., Piorr, A., Häring, A., Dabbert, S. 2000. *Environmental impacts of organic farming in Europe.* Organic Farming in Europe. Economics and Policy, Vol. 6. Stuttgart: University of Hohenheim.

Tovey, H. 1997. Food, environmentalism and rural sociology on the organic farming movements in Ireland. *Sociologia Ruralis* 37:21–37.

Willer, H., Yussefi, M. 2001. *Organic agriculture worldwide. Statistics and future prospect.* Stiftung Ökologie & Landbau (SÖL). http://www.soel.de/inhalte/publikation/s_74_03.pdf (accessed June 2001).

Willer, H., Yussefi, M. 2007. *The world of organic agriculture. Statistics and emerging trends.* Bonn, Germany: International Federations of Organic Agriculture Movements and Research Institute of Organic Agriculture (FiBL).

Wlcek, S., Eder, M., Zollitsch, W. 2003. Organic livestock production and marketing of organic animal products in Austria. Paper presented at Proceedings of the First SAFO Workshop, Florence, Italy, September 5–7, 2003.

12 Japan
Finding Opportunities in the Current Crisis

Joji Muramoto, Kazumasa Hidaka,
and Takuya Mineta

CONTENTS

> I believe it is no exaggeration to say that Japanese agriculture is now faced with crisis that will decide its whole future.
>
> —**Nishida (2003, p. 36)**

12.1 INTRODUCTION

Japanese agriculture is more than 2,000 years old. Today, agriculture represents only 1% of Japan's gross domestic product (GDP). However, the effects of agriculture's external economy on Japanese society are still immeasurable.

Up until the 1940s, Japanese agriculture held many features of sustainability. Most food was grown in integrated, mixed farming systems with closed-loop nutrient cycles. In the postwar era, especially after the 1960s, agricultural modernization changed farming practices and rural life dramatically and many of the sustainable features were lost.

The roots of the current sustainable agriculture movement in Japan can be traced back to the pesticide reduction movement of the late 1960s and the organic movement of the early 1970s. The term *sustainable agriculture* was officially translated in Japanese as *Kankyō Hozengata Nōgyō*, which literally means "environment conservation farming." The Sustainable Agriculture Act, adopted in 1999, signaled a major step forward for the movement. The act discourages pesticide and chemical fertilizer applications and encourages compost application. Passage of the Organic Farming Act in 2006 helped to further promote and extend organic farming.

Although the agricultural workforce and the area of cultivated land keep declining in Japan, we argue that the recent trend of increased awareness among citizens about the interdependence of farming, natural conservation, and food may be a key for developing sustainable agriculture in Japan.

12.1.1 Geography of Japan

Japan is an island country, made up of the more than 3,000 islands of a large stratovolcanic archipelago along the Pacific coast of Asia that stretches from 20°N to 45°N. The four largest islands—Honsyū, Hokkaido, Shikoku, and Kyūsyū—account for 97% of the total land area of 377,900 km^2.

Japan is in a temperate marine climate zone with four distinctive seasons. Due to the north-to-south orientation of the country, however, the climate varies from subtropical in the south to cool temperate in the north. Annual average temperature ranges from 10 to 20°C. Precipitation is relatively high throughout the year, with a rainy season in the early summer. Annual precipitation is 1,000 to 2,000 mm in most parts of the country.

Nine forest ecoregions exist in Japan. They range from subtropical moist broadleaf forests in the southern islands to temperate broadleaf and mixed forests in the mild climate regions of the main islands to temperate coniferous forests in the northern islands (Olson et al., 2001). Net primary production in Japan is relatively high at 10 to 20 tons/ha.

Approximately 73% of the land area of Japan is mountainous; of this mountainous terrain, about 13% is farmed. The scattered plains and intermountain basins, in which the population is concentrated, cover only 25% of the land area. Japan has a population of 128 million people, making it the 10th most populous country in the world. Population density averages as high as 338 people per km^2.

12.1.2 PRIMARY AGROECOSYSTEMS IN JAPAN

Rice (*Oryza sativa* var. *japonica*) has been recognized as the most important crop from economic, political, and cultural perspectives in Japanese agricultural history (Francks, 2000; Ohnuki-Tierney, 1993). Rice is considered one of the oldest domesticated crops in the world; domestication of wild rice may have occurred as early as 8,000 B.C. in China (Chang et al., 2005). Archaeological evidence indicates that intensive wet-paddy rice agriculture was established by 4,000 B.C. in the Lower Yangzte region in China (Fuller and Qin, 2009). Rice is considered to have been introduced to southwestern Japan circa 500 B.C. (Fujiwara, 1998).

Rice plants grow well in Japan, with its abundant rainfall and high summer temperatures. Compared to upland crops, paddy rice has multiple agronomic advantages; it can be grown continuously in the same field, needs less fertilizer, has fewer weed problems, causes almost zero soil erosion, has a relatively high and stable yield, and can be integrated with aquaculture by building irrigation ditches. Because of these facts, rice paddies have been the dominant agroecosystem in Japan, and this remains true to this day. Today, paddy rice occupies more than half of the total cultivated land (Table 12.1).

TABLE 12.1
Area of Land (in thousands of ha) Devoted to Various Agroecosystems in Japan, 1961–2008

	1961	1971	1981	1991	2001	2008
Paddy rice fields	3,388	3,364	3,031	2,825	2,624	2,516
Percentage of 1960 area	100%	99%	89%	83%	77%	74%
Upland fields	2,165	1,409	1,241	1,266	1,179	1,171
Percentage of 1960 area	100%	65%	57%	58%	54%	54%
Orchards	451	616	581	464	349	320
Percentage of 1960 area	100%	137%	129%	103%	77%	71%
Pastures	81	352	589	649	641	621
Percentage of 1960 area	100%	435%	727%	801%	791%	767%
Total	6,085	5,741	5,442	5,204	4,793	4,628
Percentage of 1960 area	100%	94%	89%	86%	79%	76%

Source: MAFF Japan (2008a).

Paddy rice production in Japan generally begins in spring. In May to June, the dried paddy field is flooded and puddled and rice seedlings are transplanted. After the warm and humid summer, the flooded paddy is drained, and rice grains are harvested in August to November. Since a considerable amount of water is required for growing paddy rice, communal irrigation channel systems have been developed in rice growing areas.

Rice paddy fields also provide habitats for aquatic organisms, and in this respect serve as alternative wetlands (Moriyama, 1997) and provide various ecological services, including erosion control and water purification.

Although it has been an important crop since its introduction, rice has never been the staple food for all Japanese people as far as quantity is concerned. In fact, during the medieval (1185–1392) and early modern (1603–1868) periods, rice was a tax taken by governments and was not traded in a free-market economy. Further, the dietary habits of the Japanese in the past varied considerably, depending upon region and class. For example, in northeastern Japan, most people, except warriors and upper-class merchants, ate only millet until the 1960s (Ohnuki-Tierney, 1993, pp. 39–40). Rice has been the major item of food for ritual occasions for most Japanese ever since rice agriculture was introduced.

The long history of paddy rice production has also strongly influenced the Japanese culture. A cosmology based on rice agriculture has retained its importance to this day for Japanese people, expressing itself through religion, folktales, and daily customs (Ohnuki-Tierney, 1993).

Other agroecosystems in Japan include upland systems, orchard systems, and pasture systems. Vegetables, field crops, and industrial crops are grown in upland fields. The top five tree fruits harvested in orchard systems are mandarin oranges (55,000 ha in 2005), apples (43,000 ha), persimmons (25,000 ha), chestnuts (25,000 ha), and grapes (20,000 ha). Due to new pioneer projects, pasture is the only system that has increased in area since 1960. The land area devoted to each system is shown in Table 12.1.

Many vegetables, flowers, and some fruits, such as strawberries, are grown in greenhouses in Japan. The total area of greenhouses in Japan, including plastic greenhouses, large plastic tunnels, small plastic tunnels, and glass greenhouses, is 107,000 ha, the second largest area in the world after China (Peet and Welles, 2005, p. 258).

Not including Hokkaido, the average farm size per farm-household in Japan is 1.36 ha. Hokkaido is the only island with a large deviation from the mean, with an average farm size of 19.3 ha per household (MAFF Japan, 2009a).

According to modern Japanese historian Ann Waswo, Japanese agriculture has three distinctive features relative to agriculture in Western countries: (1) until fairly recently, the total population and total labor force of Japan was made up of a relatively high proportion of farm households; (2) family farming on relatively small holdings persisted throughout the twentieth century; and (3) a single crop, rice, is and has been central in agricultural production (Waswo, 2003, pp. 4–6). Current Japanese agroecosystems in general are characterized as small-scale, family-operated, paddy-centered systems managed intensively by aged, part-time farmers.

TABLE 12.2
Estimated Values of Ecosystem Services
Provided by Agriculture in Japan Annually

Function	Value (trillions of yen)
Flood control	¥3.5
Recreation	¥2.4
River and watershed protection	¥1.5
Landslide protection	¥0.5
Soil erosion protection	¥0.3

Source: Science Council of Japan (2001).

12.1.3 MULTIFUNCTIONALITY OF JAPANESE AGROECOSYSTEMS

Agroecosystems in Japan have many functions besides agricultural production. The Science Council of Japan (2001) listed the following functions:

1. Sustainable food production provides security for citizens into the future.
2. Agricultural activities provide valuable ecosystem services and complement natural cycles, thus contributing to conservation of the environment.
3. Agricultural systems form the foundation of rural communities and maintain their economies.

To this list can be added the functions of serving as the basis of Japan's traditional food culture and being the primary force shaping the Japanese landscape—both of which have crucial roles in Japanese culture.

The value of some of Japanese agriculture's ecosystem-service functions can be estimated in monetary terms. One set of estimated values is given in Table 12.2. The total value of these ecosystem services, 8.2 trillion yen, compares favorably with the total gross output of one year's agricultural production, which was 8.5 trillion yen in 2005.

12.2 PRIMARY FACTORS LIMITING SUSTAINABILITY IN JAPANESE AGROECOSYSTEMS

Japanese agroecosystems were once a model of sustainability. In the early 1900s, an American soil scientist, F.H. King (King, 1911), visited China, Korea, and Japan and observed systems that had maintained their productivity for centuries. According to his reports, these systems employed many of the principles known today to form the basis of sustainable practices:

> In selecting rice as their staple crop; in developing and maintaining their systems of combined irrigation and drainage, notwithstanding they have a large summer rainfall; in their systems of multiple cropping; in their extensive and persistent use of legumes;

in their rotations for green manure to maintain the humus of their soils and for composting; and in the almost religious fidelity with which they have returned to their fields every form of waste which can replace plant food removed by the crops, these nations have demonstrated a grasp of essentials and of fundamental principles which may well cause western nations to pause and reflect. (King, 1911, pp. 274–76)

What King observed in Japan was the product of two historical phases of development that greatly expanded the extent and productivity of rice paddy agriculture without greatly compromising its sustainability. In the Edo era (1600–1867) rice production capacity was seen as the key to political power, and developing new fields and improving rice production capacity was a high-priority policy of governments. Many wetlands and alluvial plains were drained and turned into paddy fields. To obtain higher yield, intensive fertilization with "night soil," animal manure, green manure, oil seed cake, and fish cake was practiced. A variety of rice-based multi-cropping systems and paddy-dryland rotation systems, such as the rice-barley (or wheat) system and rice-cotton-beans system, were also developed (Tokunaga, 1997). In the Meiji era (1868–1912), rice yield experienced another sharp increase, mostly because of the greater use of commercial fertilizers. Rice paddy areas continued to expand and reached about 3 million ha in the late Meiji era (Table 12.3). After that time, the area of paddy rice stayed largely unchanged until early in the 1960s.

In the 1960s, there began a period of agricultural modernization focused on increasing yield even further. It was at this time that Japanese agriculture began to abandon many of the age-old practices that had kept nutrient cycles closed and local. In terms of national policy, the Agricultural Basic Law, passed in 1961, was the centerpiece of this effort. During a time when manufacturing industries experienced rapid growth, the law promoted specialization in mono-cropped, nongrain agricultural commodities, expansion of farm size, and mechanization of farm

TABLE 12.3
Development of Paddy Rice Production in Japan

Period	Rice Paddy Area ($\times 10^6$ ha)	Rice Production ($\times 10^6$ tons)	Yield (tons/ha)	Total Population ($\times 10^6$)
729–806	1.05	1.06	1.01	3.70
1532–1615	1.05–1.20	1.80–1.85	1.50–1.77	22.30
1716–1748	1.63	3.15	1.93	26.50
1830–1843	1.55	3.00	1.94	27.00
1878–1887	2.56	4.77	1.86	37.45
1908–1917	2.99	7.94	2.65	50.98
1938–1942	3.15	9.53	3.02	73.27
1959–1965	3.10	12.38	3.99	93.42
1971–1974	2.62	11.70	4.48	107.09
2001–2004	2.60	8.61	5.11	127.57

Source: Andow, H. (1959), cited by Yoshida (1978); MAFF Japan (2008b).

labor. Use of chemical fertilizers and pesticides was intensified for a higher yield. Ironically, the law and its attendant modernization processes weakened agricultural production in Japan by undermining many of the practices that had kept Japanese agriculture sustainable.

Although rice yields did increase dramatically, total agricultural area decreased from 6.0 million ha in 1960 to 4.7 million ha in 2005. In the same period, the agricultural working population went down from 11.96 million to 2.52 million, and agricultural production's share of gross domestic product (GDP) fell from 9.0% to 1.0%. Although rice paddy fields still occupy approximately 50% of the cultivated land in Japan, the area continues to decrease today.

The changes that have occurred in Japanese agriculture since the 1960s have created an unsustainable food production system that is dependent on fossil fuel inputs, harmful to the environment, responsible for the decline of rural agricultural communities, and unable to satisfy the food needs of a growing population. Below, we discuss with more specificity the unsustainable components of this system.

12.2.1 BROKEN NUTRIENT CYCLES

Nutrient cycling, which involves the movement of energy and material among organisms and the environment, is a basic function of ecosystems. Agroecosystems provide key spaces for nutrient cycles in our societies. Crop fields cycle nutrients through decomposition of organic materials and crop harvest (Smaling et al., 1999), and also by offering spaces for the life cycles and food webs of diverse biotic communities (Brussaard and Ferrera-Cerrato, 1997).

In current Japanese agroecosystems, nutrient cycles are disrupted by various human activities at multiple spatial scales. At a field scale, leaks in nutrient cycles are created by excess application of N or P fertilizers in cultivated fields. N efficiency in 2002–2004 in Japan, measured as the ratio of total N uptake by crops and forage to the total N available from fertilizer, livestock manure, and other N inputs, was 40%—the second lowest ratio among OECD countries. P efficiency in Japan in the same period, calculated in the same manner, was 21%—the lowest ratio among OECD countries (OECD, 2008a).

At the farm scale, nutrient cycling between crop fields and livestock has been displaced. During the 1950s and 1970s, due to the mechanization and the specialization of farming led by the state agricultural policy, the number of mixed farms—farms with crop fields and livestock—decreased substantially. The number of farm households with livestock peaked in the 1950s and early 1960s and has dropped sharply since then. (The number of households with cows and cattle stood at 2,600,000 in 1956 and had declined to 122,000 in 2005; 1,030,000 households had pigs in 1962, whereas only 7,800 households had pigs in 2005; the number of households with hens and chickens declined from 4,510,000 in 1955 to 6,330 in 2005.)

Nutrient cycling between cities and farmlands was also virtually lost. In the Edo era, night soil from cities was a critical fertilizer for crop production in surrounding rural villages. This practice completed the nutrient cycle by returning nutrients back to the farms from which they had come. As chemical fertilizers became accessible, however, night soil disappeared from rural Japan. The sludge from current sewage systems in Japan frequently contains high levels of heavy metals (e.g., zinc,

TABLE 12.4

Average Compost Application Rate in Rice Paddies in Japan

	1960	1965	1970	1975	1980	1985	1990	1995	2000
Compost (tons/ha)	6.30	5.45	4.50	2.67	1.98	1.77	1.76	0.76	0.60

Source: Nakajima (2004b).

cadmium, etc.), making it impractical for application on crop fields (Yūkisei odei no ryokunōchi riyō henshūiinkai, 2003).

At a regional scale, nutrient cycling between livestock farms and crop fields no longer occurs to the extent it once did. The average compost application rate in paddy fields in 2000 was less than 1/10 of the rate in 1960 (Table 12.4).

Lastly, at the country scale, Japan—the largest food importer in the world—imports more nitrogen (N) than the agroecosystems in the country can process. The estimated gross balances for N and P in Japanese agroecosystems during 2002–2004 (+171 kg-N/ha and +51 kg-P/ha, respectively) were among the highest in OECD countries (OECD, 2008a).

Disrupted nutrient cycles result in eutrophication of surface water and coastal waters throughout Japan. Only 53% or fewer closed water bodies (lakes and reservoirs) in the country met safe water quality standards during 1973 and 2005. Four percent of well water across Japan contained nitrate concentrations exceeding drinking water standards (10 mg NO_3-N/L) in 2005 (Ministry of the Environment, Japan, 2007). Thirteen coastal water zones of Japan are experiencing hypoxia due to eutrophication (Selman et al., 2008). Along with nutrients from residential areas, N and P from agroecosystems (non-point sources) and livestock operations (point sources) are considered the major causes of eutrophication in Japan (Ministry of the Environment, Japan, 2007).

12.2.2 Loss of Biodiversity in Agroecosystems

Ecological diversity in an agroecosystem has multiple dimensions: species diversity, genetic diversity, vertical and horizontal spatial diversity, structural diversity, functional diversity, and temporal diversity (Gliessman, 2007, p. 220). The higher the diversity in an agroecosystem, the greater the potential for beneficial interactions.

Because agroecosystems are disturbed ecosystems, their biodiversity (also called agrobiodiversity) is generally lower than that in natural ecosystems (Odum, 1971). However, recent studies indicate that similarly high biodiversity can be found in upland fields in Europe (OECD, 2008b) and rice paddy fields in Asia (Hidaka, 1998), where there are thousands of years of cultivation history. Here we focus on biodiversity in rice paddy agroecosystems.

Most paddy fields in Japan were originally wetlands. After about 2,000 years of rice production, very few natural wetlands remain, forcing many aquatic organisms to depend partly or fully on paddy agroecosystems. Paddy agroecosystems in Japan may support diverse organisms, including birds (60 species), fish (70 species),

TABLE 12.5

Average Pesticide Application Rates for Different Crops in Japan in 1998

	Pesticide	Fungicide	Pesticide/ Fungicide	Herbicide	Total
Vegetables (fields)	6.2 (21)	5.0 (8)	0 (0)	0.4 (0)	11.8 (31)
Vegetables (greenhouses)	8.5 (47)	9.6 (11)	0 (0)	0.2 (0)	18.8 (59)
Tree fruits (fields)	6.3 (34)	7.6 (32)	0 (0)	0.5 (1)	15.2 (81)
Tree fruits (greenhouses)	7.0 (18)	5.2 (23)	0 (0)	0.6 (1)	13.8 (47)
Flowers (fields)	14.0 (17)	10.1 (11)	0 (0)	1.5 (2)	25.8 (31)
Flowers (greenhouses)	15.6 (38)	10.9 (21)	0.1 (0)	0.3 (1)	27.1 (60)
Field crops	3.3 (27)	2.6 (8)	0.1 (0)	0.7 (1)	7.4 (41)
Paddy rice	1.6 (1)	3.1 (2)	1.0 (1)	2.2 (2)	8.0 (6)

Source: MAFF Japan (1999), cited by Kiritani (2004, p. 68).

Note: Number of applications per crop season and amount of active pesticide agent per crop season in kg/ha (in parentheses).

reptiles (12 species), amphibians (20 species), arthropods (>600 species), and weeds (190 species, including natives and nonnatives) (Kiritani, 2000). Since the 1960s, however, many native species inhabiting paddy fields have decreased in abundance, mainly due to pesticide applications and the construction of concrete ditches that deprive paddy-dependent organisms of critical habitat.

Under the monsoon climate, Asian rice has greater insect pest pressures than other rice production areas in the world (Kiritani, 2004, p. 16). However, pesticides are used excessively throughout Japanese rice paddies. Extension specialists all over the country have promoted calendar-based communal pesticide applications without monitoring of pest populations (Une, 2005, pp. 256–259). Table 12.5 shows pesticide application rates and the number of applications per crop season for different crops grown in Japan. Rice has the lowest application rate in kg/ha of the crops listed, but at 6 kg/ha this is still a large amount for one rice season (in comparison, pesticide use for rice in California averaged 0.6 kg/ha in 1998 [California Department of Pesticide Regulation, 1998]).

Due in part to pesticides, a number of species inhabiting paddy fields—two insect species, *Lethocerus deyrollei* (Hemiptera: Lethocerinae) and *Cybister tripunctqatus orientalis* (Coleoptera: Ditiscidae), five bird species, one species of fish, one amphibian species, and three plant species—are listed as endangered (Hidaka, 1998).

12.2.3 DEPENDENCE ON ENERGY-INTENSIVE INPUTS

The mechanization of farming practices initiated in the 1950s relieved farmers from heavy field work and reduced field working hours. From an ecological perspective, however, mechanization of farming practices associated with intensive use of agrochemicals transformed farming from an energy-producing process to an energy-

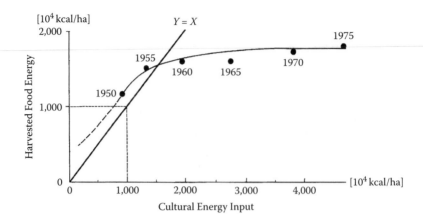

FIGURE 12.1 Energy balance in paddy rice production in Japan from 1950 to 1975. (Modified from Udagawa, 1976.)

consuming process in Japan. Until 1955, in typical rice production in Japan, the ratio of input energy (from nonrenewable energy sources) to output (harvested grain) energy was 1 or less. By 1975, this ratio had become approximately 3, indicating that three times more input energy was needed to produce the same yield of rice (Udagawa, 1976) (Figure 12.1). Because of their smaller size, Japanese farm parcels make use of machinery less efficiently, energy-wise, compared to large-scale farming systems in the United States (Pimentel et al., 1975) and other countries.

12.2.4 AGING OF FARMERS AND DEPOPULATION OF RURAL COMMUNITIES

Since the basic agricultural law in 1961, the population engaged in farming has been declining: from 1961 to 2008, it dropped from 14.5 million to 3.0 million, a decline of more than 79%. The shortage in the working population in farming is probably one of the greatest factors limiting agricultural sustainability in Japan.

At the same time, due to a lack of new farmers and the aging of existing farmers, the depopulation of rural hamlets has progressed at a quickened pace. In 2008, 60% of the population engaged in farming was 65 or more years old. Further, the number of *genkai shūraku* (marginalized farming hamlets) is also increasing. These are defined as farming hamlets in which more than 50% of the population is comprised of people 65 years old or older, in which an increasing number of households are made up of only aged individuals, and in which there are difficulties in conducting daily social functions. In 2006, there were 7,878 *genkai shūraku* in Japan; 423 of these are expected to disappear within 10 years, and another 2,220 are likely to disappear in the near future (MAFF Japan, 2007a).

Rural local governments have been trying to recruit new farmers by providing consulting, networking services, and training programs. Because of these policies, the number of new farmers is gradually increasing. There are also some retired baby boomers showing interest in farming as well as members of younger generations who are dedicated to organic farming (Knight, 2003). During the last decade, there were

60,000 to 80,000 new farmers every year. Nevertheless, during the same period, the working population in agriculture experienced declines of up to 90,000 per year, and no sign of a reversing trend can be seen (MAFF Japan, 2005). Traditionally, Japanese farm households were hereditary, and to acquire agricultural land, permission from the local agricultural commission is necessary; these structures represent system and custom barriers for newcomers.

Meanwhile, there has been an increase in the number of immigrants in rural Japan; these immigrants are either part of the temporary workforce or permanent "foreign brides." The number of the former, officially called trainees for agriculture, reached about 7,500 in 2006. In the case of foreign brides, many come from China, Korea, and other Asian countries, and are seen mostly in northern Japan. In Yamagata prefecture, known to have many cases of such international marriages in rural communities, 1 in 16 marriages are between a Japanese male and a foreign female; this is similar to the rate in Tokyo (Ministry of Health Labour and Welfare, Japan, 2005).

12.2.5 LOW FOOD SELF-SUFFICIENCY

A country's calorie-based food self-sufficiency rate is calculated as the domestically produced calorie supply divided by the total domestic calorie supply, expressed as a percentage. Japan's food self-sufficiency rate dropped from 73% in 1965 to 40% in 2003, the lowest among major industrialized countries (Figure 12.2). Japan is thus the world's largest food importing country. During the 1990s, according to Adachi

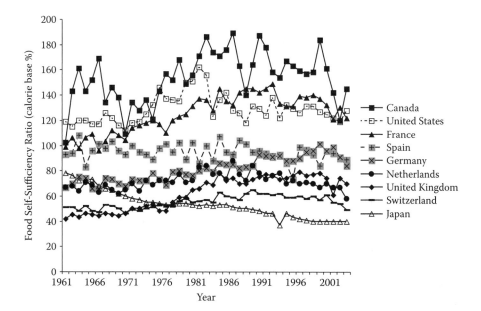

FIGURE 12.2 Food self-sufficiency ratio of some developed countries (calorie bases). (MAFF Japan, 2009b.)

(2003, p. 14), 9.8% of the total amount of produce, fish, and lumber traded globally was imported to Japan, a country with only 2.2% of world's population. In 2006, the food self-sufficiency ratio in Japan was 100% in rice, 13% in wheat, 7% in beans, 79% in vegetables, 39% in fruits, 56% in meat, and 59% in seafood. Although completely self-sufficient in rice, Japan relied almost entirely on imports for wheat and bean supplies.

The causes of low food self-sufficiency in Japan include a decline in rice consumption; an increase in the amount of wheat, meat, and dairy products in the diet; limited farm land per capita; and the feeding of domestic livestock with feed (corn and soybeans) that is almost entirely imported.

The low food self-sufficiency of the country has brought about security concerns among citizens. MAFF is trying to improve the country's food self-sufficiency ratio to 45% by 2015.

12.2.6 INEFFICIENCY OF SMALL-SCALE FARMING

A main goal of the basic agricultural law in 1961 was to narrow the income gap between farm households and factory worker households that expanded during the rapid economic growth of the 1950s. The goal was mostly accomplished by the 1970s. For a majority of farm households, however, the gap was not closed by increasing agricultural income, but by working off farm. Accelerated mechanization of farm tasks did not increase on-farm income, but it did make labor redundant. This was especially true for labor-intensive paddy rice production. Thus, the mechanization provided opportunities for farmers to work off farm. Also, the high capital costs of mechanization often forced farmers to take off-farm jobs to pay the bills. At the same time, the economic growth that continued until the early 1970s provided numerous employment opportunities for rural labor. Consequently, the percentage of part-time farm households increased dramatically; the percentage of class II part-time households (those with primarily nonagricultural income) increased from 27.5% in 1955 to 61.7% in 2005.

A high proportion of full-time farming households grow non-land-intensive crops, such as flowers and vegetables. Many of these families make healthy profits, but their practices may not necessarily be ecologically sustainable.*

Regardless of the government's efforts to expand farm size, a large majority of farms remain small-scale, family-operated, part-time enterprises. This is mainly attributed to policy failures, including a lack of effective policies relating to land zoning, artificially high rice prices that encourage micro-farming households to remain in agriculture, and an acreage reduction policy that has prevented farms from taking full advantage of the merits of scale in rice production (Yamashita, 2008).

The small-scale, part-time farms are inefficient in terms of economic gain and energy consumption per area. There are considerable pressures on these households both globally and domestically to improve efficiency of production. However, when

---

* For example, as seen in Table 12.5, flower and vegetable production tends to use larger amounts of pesticides than other crops.

the noneconomic functions of agriculture are taken into account, the overall merits of this argument become less clear. We discuss this further below.

12.3 CONVERSION TO SUSTAINABLE AGRICULTURE

The previous section reviewed issues and factors limiting agricultural sustainability in Japan that were brought about by the modernization of agriculture. In response, various movements emerged to restore Japanese agriculture in the early 1970s. Among these the organic movement and the pesticide reduction movement. The organic movement represents a grassroots, bottom-up approach to sustainability, whereas the pesticide reduction movement involves both bottom-up and top-down approaches.

12.3.1 THE ORGANIC MOVEMENT IN JAPAN

The organic movement in Japan officially began with the establishment of the Japan Organic Agriculture Association (JOAA) in 1971. Farming without chemical fertilizers and pesticides, however, had been practiced in Japan prior to this time by the "natural farming" group led by Mokichi Okada and Masanobu Fukuoka (Fukuoka, 1978). JOAA was created at a time when rapid economic growth was ending and issues having to do with environmental pollution, fossil fuel energy use, and food safety were emerging. Its founders wanted to create a movement toward "the way agriculture should be" by facilitating the convergence and cooperation of three important groups: farmers who were critical of or suffered from the conventional farming practices that depend heavily on the use of chemical fertilizers and pesticides, consumers who were concerned about food safety and environmental pollution by agriculture, and scientists who were worried about agricultural modernization being promoted primarily from a capitalist viewpoint.

In JOAA's statement of purpose (JOAA, 1971), Teruo Ichiraku, a founding member of the association and the leader of the movement, declared that agriculture should (1) prioritize health and national survival over economic concerns, (2) produce tasty and healthy food, (3) maintain and promote the health of farmers and consumers, (4) cause minimum environmental pollution, and (5) maintain and improve soil fertility. More specifically, in contrast to modern agriculture's mechanization, dependence upon agrochemicals, and large-scale monoculture, the organic movement focused on promoting farming practices in which safe and tasty food is produced by small-scale mixed farms managed upon the natural principle of working with material cycles under given local conditions (JOAA, 1999).

In its early years, the Japanese organic movement realized the importance of "organic relationships between producers and consumers" and developed the *teikei* system. *Teikei* is a trust-based organic food distribution system in which producers and consumers establish face-to-face relationships. The emphasis in *teikei* is on mutual support between individuals rather than on the seller-buyer relationship that normally exists between producers and consumers (Masugata, 2008). In *teikei*, the economic risks of farming organically are shared by producers and consumers. It is a grassroots movement aimed at expanding the self-sufficiency of farmers and connecting them with larger local communities (Adachi, 2003). *Teikei* is considered as one of the origins of community-supported agriculture in the United States and other countries (Henderson and Van En, 2007, pp. 258–266).

The Japanese organic movement was also influenced by Western organic movements from the beginning. *Pay Dirt* by Jerome Rodale was published in Japanese in 1950 and 1974; Sir Albert Howard's *An Agricultural Testament* (1959, 1985, 2003) and *The Soil and Health* (1987) were also translated into Japanese.

Unlike Western organic movements, however, the Japanese organic movement did not establish an organic certification system in its early period, mainly because *teikei* made it seem unnecessary. But the nuclear accident in Chernobyl in 1986 triggered a change; it prompted a surge of consumer demand for organic foods in Japan. In response, there developed a network of specialized distributors, natural food stores, and co-ops where organic produce could be purchased. At the same time, unfortunately, regular markets were filled with "organic" produce with unknown integrity. To deal with this problem, organic labeling guidelines were established in Japan in 1993. Japan moved closer to the next step—establishing organic standards—when a delegation of organic growers and distributors visited in 1995 to promote the importation of "certified" organic foods into the Japanese market. A set of national organic standards (the JAS organic standards) were drafted in 1999 and enacted in 2001.

The introduction of JAS organic standards stimulated an increase in organic foods in regular Japanese markets, but this was not without its drawbacks. Many *teikei*-based organic farmers were forced to make a hard choice between becoming certified or not. Some of them decided against certification because of the extra costs involved; moreover, certification by a third party was not necessary for those who had been maintaining a good relationship with a consumer group through *teikei* (Anzen na tabemono wo tsukutte taberu kai 30 nenshi kankō iinkai, 2005, pp. 247–250). They decided to stop using the word *organic* for their produce even though it went beyond organic standards. Further, it was recognized that JAS organic standards did not provide any support to organic communities (it only burdened them with regulations), and its most notable impact was to increase the amount of imported organic foods in Japanese markets (Nakajima, 2004a).

Meanwhile, the term *organic* became popular in Japanese society and the expanded organic movement, which encompassed organic food distributors, natural food stores, organic certifiers, and *teikei* groups, started to work with legislators to establish an organic farming act. With some scholarly support from the Japanese Society for Organic Agriculture Research (JSOAR) (established in 1999), the act was enacted in December 2006.

The Organic Farming Act was designed to promote and provide public support for organic farming; it therefore went much further than the JAS organic standards, which only established regulations for producing organic commodities (Nakajima, 2008). The act allocated 457 million Japanese yen for FY 2008 for a nationwide organic farming group support program, an organic farming promotion program, and a local organic farming infrastructure development program. It also funded the establishment of 45 "organic model towns" throughout the country.

With the relatively recent introduction of the national organic standards, the market share of organic produce in Japan is still small. According to MAFF, the amount of domestically certified organic produce in 2001 made up only 0.10% of total domestic agricultural produce.* This statistic is lower than the farm area that did not use any chemical pesticides and chemical fertilizers in the same year (0.87%; Table 12.6).

Nevertheless, organic farming in Japan appears to have a good potential. According to a survey conducted by MAFF, over 90% of Japanese consumers have either purchased organic produce (43.8%) or are willing to purchase organic produce with some conditions (54.7%), and 49.7% of farmers who are not currently practicing organic farming were interested in the transition to organic farming (MAFF Japan, 2007b). The full impact of the Organic Farming Act—in terms of stimulating growth in the organic sector—has yet to be seen.

Although organic agriculture is fully integrated into the mainstream economy of Japan and mostly represents conversion at level 2 only, the broader goals of the organic movement are more far-reaching than to merely increase market share. The organic movement in Japan also seeks to transform the food production system. According to Kiichi Nakajima, the president of JSOAR,

> Organic farming practices are neither the technological know-how for farming without pesticides and chemical fertilizers, nor practices to meet JAS organic standards. Organic farming aims to reestablish the way farming should be; it applies nature's ecological principles, fully derives given potentials of crops, produces healthy foods, and creates a unique culture based upon Japanese natural environment. Organic farming practices are to be created through such organic farming processes. They are also a suite of practices that support and help develop organic farming. (Nakajima, 2008, p. 30)

12.3.2 Organic Weed Management Options in Paddy Rice Systems

Although temporary flooding in paddy rice agroecosystems reduces weed germination, weed management is a major challenge in rice production. As seen in Table 12.5, paddy rice systems received the heaviest herbicide applications among the different farming systems in Japan, suggesting that weed management without herbicides in paddies is not an easy task. Since using herbicides for this purpose is a major reason why conventional production of Japan's primary crop is unsustainable, developing and transitioning to nonchemical options for weed management in paddies is an important part of the conversion to sustainable agriculture in Japan. During the last three decades, organic farmers and scientists have developed a suite of such options for paddy weed management. Here, some of the popular options and their potential ecological impacts are described.

12.3.2.1 Aigamo–Paddy Rice Farming

The Aigamo-rice system, or *Aigamo-Suitō Dōji-Saku*, is a mixed farming system in which the farmer simultaneously grows rice and raises hybrid ducks called Aigamo in an enclosed paddy field (Furuno, 2001). The system was developed by Dr. Takao

---

* The number increased to 0.19% in 2007.

TABLE 12.6

Types of Environment Conservation Farming (ECF) in Japan (2001)

Category		Without Compost Applications			With Compost Application(s)			Certification Category
Pesticides	Chemical Fertilizers	Land Area (×1,000 ha)	Percent of ECF Area	Percent of All Cultivated Area	Land Area (×1,000 ha)	% of ECF Area	Percent of All Cultivated Area	
None	None	38	5.4	0.87	29	4.1	0.66	Organic/transitional
None	≥50% reduced	14	1.9	0.31	11	1.5	0.24	
None	<50% reduced	16	2.3	0.37	12	1.7	0.27	
≥50% reduced	None	24	3.4	0.55	16	2.3	0.37	
≥50% reduced	≥50% reduced	149	20.9	3.36	104	14.6	2.34	Specially grown produce
≥50% reduced	<50% reduced	54	7.6	1.22	31	4.4	0.71	
<50% reduced	None	17	2.4	0.39	12	1.7	0.27	
<50% reduced	≥50% reduced	74	10.4	1.67	50	7.0	1.12	
<50% reduced	<50% reduced	325	45.7	7.34	228	32.0	5.14	None
Total		711	100.0	16.1	493	69.3	11.1	

Source: MAFF Japan (2002).

Note: Environment conservation farming is defined by the Sustainable Agriculture Act (1999) as "a farming system that includes at least one of the following practices: 1) reduction of chemical N fertilizer application rate from the local standard, 2) reduction of chemical pesticide application from the local standard, or 3) soil fertility management using compost application."

Furuno, a farmer in Fukuoka, Japan, in the early 1990s. After transplanting rice, two-week-old ducklings are released at a rate of 250 per hectare into flooded rice paddies enclosed by nets or an electric fence. The ducks eat weeds and pests of rice and their swimming activity stimulates rice growth. Besides, their droppings provide nutrients for the rice. Water fern (*Azolla cristata*), an N-fixing aquatic fern, can be added to the paddy to feed the ducks and supply N for the rice. Fish such as loach (*Misgurnus anguillicaudatus*) can also be raised in the same paddy. Approximately 10,000 Japanese farmers have adopted the system, and it has been spreading to rice areas in China, Korea, the Philippines, and Vietnam. As a mixed farming system making use of biotic interactions, Aigamo rice farming is a good example of conversion at level 3.

Due to their aggressive feeding habit, however, Aigamo may reduce biodiversity in rice paddy systems. Yamada et al. (2004) found a lower diversity index in Aigamo-rice paddies compared to non-Aigamo controls. Further, *Azolla cristata* is considered an invasive species in Japan (Ministry of Environment Japan, 2008). Escaped *A. cristata* may cause the extinction of *A. imbricate* and *A. japonica*, both native endangered species.

There are several barriers to wider adoption of Aigamo-rice systems in Japan. These include the costs of ducklings and electric fences, the need to develop the skills required to raise Aigamo, and a limited market for Aigamo.

12.3.2.2 Apple Snail Farming

Apple snail (*Pomacea canaliculata*) is a pest of aquatic crop-plant species, including rice, but it can be used as a biological weed control agent in paddy rice systems. The snail was first introduced to Japan in 1971 for human consumption and was cultured all over Japan. Commercial production failed before long and apple snails escaped, particularly in the Western region of Japan. In 1983, the Japanese MAFF registered apple snail as a harmful animal to paddy rice. Since 1999, more than 7,000 ha of paddy rice, mainly in the Kyūsyū region, have been damaged by apple snails every year (Lowland Crop Rotation Team at National Agricultural Research Center for Kyūshū Okinawa Region, 2009).

Despite these facts, using the apple snail for weed control in rice paddies is a popular practice among farmers in western Japan, making it possible to produce paddy rice with fewer herbicides or none at all. In Ehime, Japan, Hidaka et al. (2007) observed lower weed species numbers in apple snail–invaded fields than in uninvaded fields. For this method to be successful, however, precise water management is a must: shallow flooding has to be maintained once the rice has been transplanted to avoid snail damage on rice plants. The practice is also popular in South Korea, where apple snails are commercially available. Systems using apple snail weed management can be seen as having reached conversion level 2.

Apple snail poses a serious threat to biodiversity in many tropical and temperate aquatic ecosystems in the world. Carlsson et al. (2004) reported on the devastating effects of apple snail on biodiversity and ecosystem functions in tropical wetland ecosystems in Southeast Asia. In regions where the snail has yet to escape, there is concern about its invasion and the serious damage to ecosystems that could result.

12.3.2.3   Living Mulch Farming

A living mulch is a cover crop that is interplanted or undersown with a main crop with the aim of providing the functions of a mulch, which include weed suppression. Chinese milk vetch (*Astragalus sinicus*) and hairy vetch (*Vicia villosa*) are two of the most common cover crops in living mulch rice paddy systems in Japan. The leguminous cover crop also serves as a green manure, providing fixed N to the system. Living mulch suppresses weeds physically through shading, ecologically through competitive pressure, and chemically through allelopathy (Hill et al., 2006) and the release of organic acids during the anaerobic decomposition of the cover crop under flooded water (Yasue, 1993).

Diverse versions of living mulch rice farming systems exist, from no-till to reduced till and from transplanting to direct rice seeding (Cho et al., 2001, 2003). For example, a sequential rice/wheat cropping system developed by Masanobu Fukuoka, the advocator of "nature farming," is a no-till living mulch system using white clover as a cover crop (Fukuoka, 1978). A study of a no-till transplanting system suggests that the greater the cover crop biomass, the higher the weed suppression, though a higher cover crop biomass may also suppress rice growth in its early stages (Takishima and Sakuma, 1961). A vertical harrow is required in no-till living mulch systems and for incorporating living mulch. Because the use of living mulch represents a fundamental change in the dynamics of the system, it is an example of conversion at level 3.

12.3.2.4   Recycled Paper Mulch Farming

The idea of using recycled paper mulch for weed control in rice paddy systems was derived from a traditional weed management practice using grass mulch (such as *Phragmites australis*) in paddy systems (Ueno, 1999). A specially designed tractor transplants rice seedlings as it lays recycled paper mulch on the paddy surface. For about 60 days, the mulch suppresses weeds physically by creating shade and a physical barrier; then it decomposes naturally. This is an example of conversion at level 2.

The main hindrance to adoption of this system is the substantial cost for the rolls of recycled paper and the specially made tractor needed for mulch application, though direct-seeding paper mulch that does not require a special tractor is also available. A negative consequence of paper mulch farming is its possible reduction of the diversity of aquatic biota in the early stages of rice paddy production.

12.3.2.5   Weed Suppression by Plant-Based Organic Fertilizers

Plant-based organic fertilizers such as rice bran and oil seed meal cake supply nutrients to the rice crop, but they can also suppress weeds in rice paddies. At least three mechanisms of weed suppression are hypothesized: (1) the organic acids produced during anaerobic decomposition of the fertilizer are toxic to weed seedlings, (2) anaerobic decomposition of the fertilizer creates a temporary oxygen deficiency in the paddy (Chiba et al., 2001), and (3) the fertilizer promotes the growth of soil microbial biomass—observed by farmers as a thin, organic-rich layer called *toro-toro*—that enhances the growth of worms (*Tubifex tubifex*) that bury weed seeds as they feed.

Organic fertilizer is applied at a rate of 1 to 2 tons/ha when weed seeds germinate in spring. Since applied fertilizer is decomposed in 10 to 20 days, the timing and the application rate determine a good part of the method's success; if it goes wrong, rice plants may be damaged and weeds may not be suppressed. Growers have developed a variety of versions of this system, varying in the types and combinations of organic fertilizers used, and they have integrated it with no-till and reduced-till techniques as well.

12.3.2.6 Winter-Flooded Rice Farming

Since the 1980s winter-flooded rice farming has spread to many parts of Japan. The method aims to suppress spring weeds in fallowed paddies by flooding paddies that would otherwise be dry during the winter. Summer weeds are controlled by deep flooding during the rice growing period. To stimulate biological activity, especially of earthworms in the soil, organic fertilizers are applied after the rice harvest in the fall prior to flooding. Besides suppressing weeds, winter flooding can increase soil N and P fertility, thereby increasing rice yield. Further, it also provides habitat for migratory water birds and is expected to improve biodiversity in rice paddy systems.

The method might not be an option for areas where access to winter irrigation water is limited. Another concern is that expansion of winter-flooded paddies in southern Japan and in Asian countries may also encourage invasive species (e.g., apple snails) or mosquito population growth. Mosquito populations need to be closely monitored because of their potential for spreading human disease (e.g., malaria caused by *Plasmodium* spp.) (Shin et al., 2005; Wu et al., 1991).

12.3.3 THE PESTICIDE REDUCTION MOVEMENT AND ENVIRONMENT CONSERVATION FARMING

In the late 1960s, slightly before the launch of the organic movement, Dr. Keiji Kiritani, an entomologist who worked at the Kochi Prefectural Institute of Agricultural and Forest Science, found that BHC, a popular organic chloride pesticide for rice pests in Japan at that time, destroyed a group of natural enemies of pests, including spiders (Kiritani, 1971; Ladd, 1979). He also demonstrated the first IPM trial aimed at finding the minimum pesticide requirement for pest control (Kiritani et al., 1972). This research marked the beginning of *Gen-Nōyaku Undō* (pesticide reduction movement) in Japan.

Following Kiritani's trial, similar efforts emerged in many parts of Japan, including Akita, Niigata, Miyagi, Fukuoka, and Nara (Kiritani, 2004, pp. 73–79). In Fukuoka, Une found that farmers do not know how to distinguish pests and natural enemies and consequently could not decide by themselves if they should spray or not (Une, 1984). In an attempt to change this situation, he developed the *Mushimi-Ban* (insect observation plate) to assist in insect identification; 150,000 copies of the device have since been sold. Further, Une and his team improved the entire production system based on IPM. In 2001, the group produced 480 tons of no-pesticide rice and 780 tons of reduced-pesticide rice. Currently the movement is active throughout the country and has become active in biodiversity conservation in paddy rice systems and in proposing a menu for a bioindicator-based decoupling policy (Seibutsu tayōsei nōgyō shien senta, 2008). Figure 12.3 is a part of a popular poster developed by Une

FIGURE 12.3 Part of a Japanese poster for improving farmer and consumer awareness of agrobiodiversity in paddy fields. The central figure conveys the message that eating a bowl of rice supports 35 tadpoles in a paddy, suggesting a connection between food, farming, and biodiversity conservation. Numbers of species are averages from the national survey of agrobiodiversity in paddy fields in Japan. Clockwise from the upper left, species listed are *Laccotrephes japonensis, Sympetrum frequens, Rana nigromaculata, Branchinella kugenumaensis, Hyla japonica,* Collembola, Family Lycosidae, *Cynops pyrrhogaster, Hirundo rustica, Ranatra chinensis, Rhabdophis tigrinus,* and *Misgurnus anguillicaudatus.* (Une, Y. 2004. A part of "illustrated poster of rice and living organisms" published by *Nou-to-Shizen-no-Kenkyujo* [the Institute of Farming and Nature]. With permission.)

showing relationships between rice (food), rice production (agriculture), and biodiversity in paddies (environment); it shows that eating a bowl of domestic rice could support not only farming but also local biodiversity in the rice paddies.

The pesticide reduction movement gained official sanction in Japan when the Sustainable Farming Act was enacted in 1999 (as mentioned above, "environment conservation farming" is the literal meaning of the official Japanese term for sustainable agriculture). Like the LISA program (low-input sustainable agriculture) in the United States, this act promotes farming with reduced use of pesticides and chemical fertilizers and increased use of composts. Under this act, farmers may sell their produce as "specially grown crop" based on the reduction rates. In 2001, cropland managed by environment conservation farming practices occupied 16.1% of the total cultivated land in Japan (Table 12.6).

The pesticide reduction movement represents a perfect example of level 1 conversion. Farmers who are not ready to jump into organic farming can join this movement. They can learn how to monitor pests, neutral insects, and natural enemy populations, and make decisions toward pesticide reduction based on what they find. Further, this often leads farmers to level 2 conversion. An important character of this movement is the fact that the door is wide open to any rice farmers.

12.4 PRINCIPLES FOR MOVING FORWARD

In the previous section, we reviewed movements toward sustainable agriculture in Japan, focusing on the organic movement, the development of nonherbicide techniques for weed control in paddy rice systems, and the pesticide reduction movement. Conversion in Japan, as elsewhere, is an ongoing process and there are still many challenges ahead of us. This section discusses principles that, based on what we have learned from the last three decades of conversion processes, are instrumental for moving forward toward greater sustainability in Japanese agriculture.

12.4.1 USE BIOTIC INTERACTIONS IN INNOVATIVE WAYS

Based on recent trends in the relationship between energy input and harvested calories in Japanese paddy systems (Figure 12.1), Hidaka proposed five possible directions, or approaches, for future farming systems (Hidaka, 1990; Hidaka et al., 2008) (Figure 12.4). In this conception, three approaches offer the possibility of changing the energy balance in Japanese agriculture so that it once again can create more energy than it consumes. One approach is to convert to organic farming practices, which involves reducing the energy input of agroecosystems (number 2 in Figure 12.4). Another approach, developing a new system by learning from traditional farming practices, might achieve a similar result energetically (number 5 in Figure 12.4). The organic weed management options discussed above serve as examples of a third

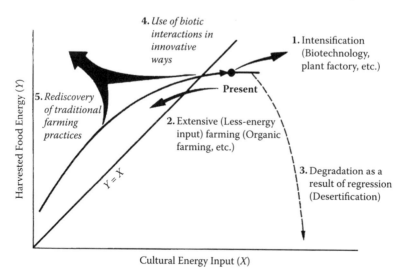

FIGURE 12.4 Five scenarios for future Japanese agroecosystems from an energy balance perspective. (From: Hidaka, 1990; Hidaka et al., 2008.)

approach that may offer the most effective and realistic options—to use biotic interactions in innovative ways (number 4 in Figure 12.4). In this approach, the ecological functions or effects of animals or noncrop plants integrated into agroecosystems take the place of energy-intensive inputs. Examples include the systems that incorporate Aigamo, apple snails, and living mulch.

12.4.2 Carefully Assess Management Options

The introduction of any environment-friendly management option always has the potential for introducing negative effects that may outweigh the positives that caused the farmer to consider the option in the first place. In particular, introduction of innovative approaches that do not take into account long-term biogeographical factors or the technological traditions of the specific site can reduce local biodiversity significantly because of their alteration of the paddy environment. For example, apple snails appear to cause a reduction in the number of vascular plant species numbers in paddy fields (Hidaka et al., 2007); this represents an adverse effect on biodiversity caused by invasive species introduced for the purpose of weed management. A similar negative effect on biodiversity may result from Aigamo duck farming (Yamada et al., 2004; Yamazaki et al., 2004). Rice bran farming can impair water quality and disturb the ecosystem (Koyama and Kidokoro, 2004). When paddies are flooded in the winter, there may be adverse effects on some species such as dragonfly, depending on the region and method employed (Wakasugi and Fujimori, 2005).

Although it is most common to look only at current management options, the biodiversity of a specific paddy ecosystem is determined by multilayered factors under varying spatial and temporal scales. Therefore, when biodiversity restoration in a specific paddy system is planned, it is crucial not only to conduct environmental assessment of the management option to be introduced, but also to make decisions on restoration goals from multilayered evaluation criteria consisting of local crop environmental history, developmental history of the paddy, and biogeographical factors of the location (Hidaka et al., 2008).

12.4.3 Move from IPM to IBM

Although the concept was introduced relatively recently, biodiversity conservation in agroecosystems is becoming a hot topic in Japan. "The First National Strategies on Biodiversity," developed by the Japanese government in 1995, have been revised in 2002 and 2007.

The trend toward biodiversity conservation as a priority can be seen as the integration of two movements: the organic movement, which has sought ecologically based farming practices in paddies, and the natural conservation movement, which has been concerned with biodiversity conservation in agroecosystems (Figure 12.3). According to Yutaka Une, who initiated the Institute for Agriculture and Nature in 2000 and has been active in nationwide biodiversity surveys in paddy ecosystems, "The natural conservation movement initiated from the crisis of organisms, and the organic or pesticide reduction movement met the natural conservation movement when they kicked 'productivity' aside" (Une, 2004, p. 38).

Kiritani has proposed the concept of integrated biodiversity management (IBM), under which integrated pest management (IPM) and conservation are reconciled and made compatible with each other (Kiritani, 2000). Kiritani (2004, pp. 165–166) lists the fundamentals of IBM in paddy ecosystems as follows: (1) recognize a paddy as a place for production and as an alternative wetland, (2) manage the paddy ecosystem to enhance its spatial and temporal heterogeneity, (3) seek farming practices and structures suitable for the goal of conserving species in the area, and (4) protect against the introduction of invasive species as far as possible.

12.4.4   RETHINK THE ECONOMICS OF THE FOOD SYSTEM

Different opinions exist for how to move agriculture in Japan toward sustainability. One of the most provocative ideas comes from Sōichi Yamashita, a conventional citrus farmer and award-winning writer, who has lived through the modernization in Saga, Japan.

Yamashita believes that the agricultural modernization policy made three basic mistakes (Yamashita, 2004):

1. *Modernization made a means the object.* For farmers, agriculture is a means for their livelihood and the livelihood is the object. Modernization policy made the means the object, resulting in the decline of agriculture and fading of rural communities.
2. *Modernization destroyed cycles.* Two types of cycles were destroyed by modernization. First, cycles within agriculture were destroyed by the specialization (monoculture) encouraged by the policy. Second, material and monetary cycles within the community were destroyed, which impoverished farmers and local retailers.
3. *Modernization forced growth.* Nature does not grow nor improve; instead, it cycles. In its huge cycle, there exist agriculture, diet, and the foundation of the livelihood. Forcing agriculture to grow denies its key principle.

Yamashita believes that correcting these three mistakes is the way to reach sustainable agriculture. He also suggests that there are three economies for agriculture and rural communities: the market economy, the local economy, and the self-sufficient economy. Japanese agriculture has been shaken by the market economy, which leads to its destruction. By making a ratio among the three economies of approximately 3:3:4, says Yamashita, we should be able to create a cycle-based society that is much more stable and does not depend on economic growth for its vitality.

12.4.5   REMEMBER THAT AGRICULTURE IS MORE THAN ECONOMICS

Further, Raymond Jussaume, a political sociologist who has studied part-time farming in Japan, argues that part-time farming households, regardless of their small scales and "inefficiency," do not choose to abandon farming because they "combine economic and non-economic goals into their decision making." He goes on to state:

> For example, part-time farming strategies are utilized by some Japanese farm house-holds to augment elderly members' self-worth by providing them with an opportunity to contribute to their household's well being. Thus, as in Europe, the presence of part-time farming is indicative of a desire by many rural households to balance household members' needs and maintain a valued way of life. For these households, farming is not simply a matter of maximizing returns on land, labour and capital resources. (Jussaume, 2003, pp. 217–218)

This is in line with Yamashita's argument that the goal of a farm household is to pursue a livelihood, and to make farming part of that livelihood, not the goal itself. If policymakers want to seek out and promote "advancement for the benefit of local populations" (Jussaume, 1991, p. 166), they would be wise to take this into account. In addition, as the multifunctions of agriculture are recognized, the importance of part-time "lifestyle farming" (as opposed to "industrialized farming") in Japan may be increased (Nakajima, 2004c, pp. 53–93).

12.4.6   BUILD ON EXISTING SUPPORT FOR A CULTURE OF SUSTAINABILITY

Even though certified organic farming itself represents level 2 conversion, the farm-ers and consumers who founded the organic movement in Japan really started from conversion level 4 in the sense that they aimed to create a culture of sustainability on a small community scale. Although they were few in number, the culture of sustain-ability is still alive within the organic movement. In fact, the concepts of *chisan-chisyō* (produce locally and consume locally) and *shindo-fuji* (uniformity of body and soil), used by people to promote the organic movement, have become popular throughout Japanese society (Yamashita, 1998; Oe, 2008).

In this cultural context, organic farmers and eat-local food activists are no longer seen as cranks. The time has come for them to take leadership in strengthening the emerging culture of sustainability. In the future, level 4 conversion in Japan may fur-ther proceed as a citizen's movement as more people become conscious of the inter-dependence of food, agriculture, and environment, and conversion efforts receive appropriate public support and research funding.

12.5   CONCLUSIONS

This chapter examined the factors limiting the sustainability of Japanese agriculture and described the various steps being taken to move the country in a more sustain-able direction. We argue that the increasing awareness among farmers, consumers, policymakers, and scientists of the interdependence of farming, natural conserva-tion, and food may be a key for developing sustainable agriculture in Japan for the future.

Given the steep decline in rural communities, support for new farmers as well as the nonproduction activities of existing farmers (decoupling) should be the priority of public funding. Also, the support systems for rural immigrants warrant more attention and reform.

The conversion to sustainable agriculture is not carried out by farmers alone, but by all citizens, particularly in their roles as consumers. And sustainability cannot be reached without a society-wide consensus about the multifunctionality of agriculture. As noted in this chapter, Japanese agriculture is in serious crisis, but opportunities for change still exist, especially as more people become aware of the simple facts that food can be imported but the environment cannot, and there are no alternatives to agriculture (Yamashita, 2004).

ACKNOWLEDGMENTS

Many thanks to Dr. Eric Engles for his valuable comments and editorial help on a previous version of the manuscript. The work was partially supported by Grants-in-Aid for Scientific Research (No. 20380180) and the Global Center of Excellence (Global COE) Program "Global Eco-Risk Management from Asian Viewpoints" at Yokohama National University, both from the Ministry of Education, Culture, Sports, Science, and Technology, Japan.

REFERENCES

Adachi, K. 2003. *Shoku-nō dōgen* [How we eat determines the quality of food and agriculture]. Tokyo: Commons.

Anzen na tabemono wo tsukutte taberu kai 30 nenshi kankō iinkai. 2005. *Mura to toshi wo musubu miyoshi yasai* [Miyoshi's vegetables connecting the village and cities]. Nagano: Volonte.

Brussaard, L., and Ferrera-Cerrato, R. 1997. *Soil ecology in sustainable agricultural systems.* Boca Raton, FL: CRC/Lewis Publishers.

California Department of Pesticide Regulation. 1998. *Summary of pesticide use report data, 1998.* http://www.cdpr.ca.gov/docs/pur/pur98rep/chmrpt98.pdf.

Carlsson, N.O.L., Broumlnmark, C., and Hansson, L.-A. 2004. Invading herbivory: The golden apple snail alters ecosystem functioning in Asian wetlands. *Ecology* 85:1575–80.

Chang, K.-c., Xu, P., Lu, L., and Allan, S. 2005. *The formation of Chinese civilization: An archaeological perspective.* New Haven, CT: Yale University Press.

Chiba, K., Yoshida, T., Saito, N., and Tachiro, T. 2001. Herbicidal efficacy of rice bran and its effects on the growth and yield of rice plants. *Tohoku Journal of Crop Science* 44:27–30 (in Japanese).

Cho, Y.S., Choe, Z.R., and Ockerby, S.E. 2001. Managing tillage, sowing rate and nitrogen top-dressing level to sustain rice yield in a low-input, direct-sown, rice-vetch cropping system. *Australian Journal of Experimental Agriculture* 41:61–69.

Cho, Y.S., Hidaka, K., and Mineta, T. 2003. Evaluation of white clover and rye grown in rotation with no-tilled rice. *Field Crops Research* 83:237–50.

Francks, P. 2000. Japan and an East Asian model of agriculture's role in industrialization. *Japan Forum* 12:43–52.

Fujiwara, H. 1998. *Inasaku no kigen wo saguru* [Exploring the origin of rice cultivation]. Iwanami Shinsho 554. Tokyo: Iwanami Shoten.

Fukuoka, M. 1978. *The one-straw revolution: An introduction to natural farming.* Emmaus, PA: Rodale Press.

Fuller, D.Q., and Qin, L. 2009. Water management and labour in the origins and dispersal of Asian rice. *World Archaeology* 41:88–111.

Furuno, T. 2001. *The power of duck: Integrated rice and duck farming*. Tasmania: Tagari Publications of the Permaculture Institute.

Gliessman, S.R. 2007. *Agroecology: The ecology of sustainable food systems*. 2nd ed. Boca Raton, FL: CRC Press.

Henderson, E., and Van En, R. 2007. *Sharing the harvest*. (Rev. Ed.). White River Junction, VT: Chelsea Green.

Hidaka, K. 1990. *Shizen-yūki nōhō to gaichū* [Nature/organic farming and insect pests]. Tokyo: Tōju Sha.

Hidaka, K. 1998. Biodiversity conservation in rice paddy fields and environmentally regenerated farming system. *Japanese Journal of Ecology* 48:167–78 (in Japanese).

Hidaka, K., Mineta, T., and Osawa, S. 2008. Integrated considerations of historical establishment of biodiversity in irrigated rice fields and strategy for nature restoration in agricultural rural area. *Journal of Rural Planning Association* 27:20–25 (in Japanese).

Hidaka, K., Mineta, T., and Tokuoka, M. 2007. Impact evaluation of apple snail invasion on vascular plant flora in rice fields. *Journal of Rural Planning Association* 26:233–38 (in Japanese).

Hill, E.C., Ngouajio, M., and Nair, M.G. 2006. Differential response of weeds and vegetable of hairy crops to aqueous extracts vetch and cowpea. *Hortscience* 41:695–700.

JOAA. 1971. Statement of Purpose. http://www.joaa.net/english/index-eng.htm.

JOAA. 1999. *Yûkî rôgyô hando bukky* [Handbook of Organic Agriculture]. Tokyo: Nôsangyoson Bunka Kyôkai.

Jussaume, R. 1991. *Japanese part-time farming: Evolution and impacts*. 1st ed. Ames: Iowa State University Press.

Jussaume, R. 2003. Part-time farming and the structure of agriculture in postwar Japan. In *Farmers and village life in twentieth-century Japan*, ed. A. Waswo and Y. Nishida, 199–220. London: Routledge Curzon.

King, F.H. 1911. *Farmers of forty centuries; or, Permanent agriculture in China, Korea, and Japan*. Emmaus, PA: Rodale Press.

Kiritani, K. 1971. Environmental pollution caused by organic chloride pesticides. *Shikoku Shokubutsu Bōeki Kenkyu* 6:1–44 (in Japanese).

Kiritani, K. 1979. Pest management in rice. *Annual Review of Entomology* 24:279–312.

Kiritani, K. 2000. Integrated biodiversity management in paddy fields: Shift of paradigm from IPM toward IBM. *Integrated Pest Management Reviews* 5:175–83.

Kiritani, K. 2004. *Tadano mushi wo mushishinai nōgyō* [Toward IBM in paddy ecosystem]. Tokyo: Tsukiji Shokan.

Kiritani, K., Inoue, T., Nakasuji, F., Kawahara, S., and Sasaba, T. 1972. An approach to the integrated control of rice pests: Control with selective, low dosage insecticides by reduced number of applications. *Japanese Journal of Applied Entomology and Zoology* 16:94–106 (in Japanese).

Knight, J. 2003. Organic farming settlers in Kumano. In *Farmers and village life in twentieth-century Japan*, ed. A. Waswo and Y. Nishida, 267–84. London: Routledge Curzon.

Koyama, J., and Kidokoro, T. 2004. Influences of water control and application of rice bran on density of Japanese tree frog, *Hyla japonica*, Larvae and spiders in paddy field. *Journal of the Society of Plant Protection of North Japan* 55:173–75 (in Japanese).

Lowland Crop Rotation Team at National Agricultural Research Center for Kyushu Okinawa Region. 2009. *Higai bunpu* [Damaged area by apple snails]. http://ss.knaes.affrc.go.jp/kiban/g_seitai/applesnail/bunpu.html.

MAFF Japan. 2002. *Heisei 13 nendo jizokuteki seisan kankyō ni kansuru jittai chōsa* [Survey on sustainable farming environment 2001]. http://www.maff.go.jp/j/seisan/kankyo/hozen_type/h_torikumi/pdf/h13_cyosa_kekka.pdf.

MAFF Japan. 2005. *Nōgyō kōzō dōtai chōsa hōkokusho* [Report on changes in agricultural structure survey].

MAFF Japan. 2007a. *Kokudo keisei keikaku sakuteinotameno syūrakuno jōkyōnikansuru genkyō haaku chōsa* [A survey on current status of hamlets for developing a national land use plan].

MAFF Japan. 2007b. *Yūkinōgyō wo hajimetosuru kankyō hozengata nōgyō ni kansuru ishiki ikō chōsa kekka* [Result of survey on awareness and inclination regarding organic agriculture and sustainable agriculture]. http://www.maff.go.jp/j/seisan/kankyo/yuuki/y_zyoho/pdf/20071102.pdf.

MAFF Japan. 2008a. *Heisei 20 nendo kōchi menseki* [Cultivated area in 2008]. http://www.maff.go.jp/toukei/sokuhou/data/kouti2008/kouti2008.pdf.

MAFF Japan. 2008b. *Kōchi oyobi sakutsuke menseki tōkei* [Cultivated fields and planting area statistics]. http://www.tdb.maff.go.jp/toukei/a02smenu?TouID=F001.

MAFF Japan. 2009a. *Nōrin suisan kihon deta syu* [Primary statistics of agriculture, forestry and fisheries]. http://www.maff.go.jp/j/tokei/sihyo/index.html.

MAFF Japan. 2009b. *Shuyō senshinkoku no shokuryō jikyū ritsu 1961-2003 (Shisan)* [Food self-sufficiency ratio of major developed countries 1961–2003 (Trial calculation)]. http://www.maff.go.jp/j/zyukyu/fbs/dat/2-5-2-1.xls.

Masugata, T. 2008. *Yūkinōgyō undou to <teikei> no nettowaku* [Network of the organic movement and <teikei>]. Tokyo: Shinyō Sha.

Ministry of the Environment, Japan. 2007. *Annual report on the environment and the sound material-cycle society in Japan 2007.* http://www.env.go.jp/en/wpaper/2007/index.html, Tokyo, Japan.

Ministry of the Environment, Japan. 2008. *List of regulated living organisms under the Invasive Alien Species Act.* http://www.env.go.jp/nature/intro/1outline/files/siteisyu_list_e.pdf.

Ministry of Health Labour and Welfare, Japan. 2005. *Jinkō dōtai chōsa* [Demographic survey].

Moriyama, H. 1997. *Suiden wo mamoru towa dōiukotoka* [Rice-paddy conservation and its implications]. Tokyo: Nōsan-gyoson Bunka Kyokai.

Nakajima, K. 2004a. Challenges under the regulatory system of JAS organic certification. *Journal of Organic Agriculture Research* 4:2–4 (in Japanese).

Nakajima, K. 2004b. Environmental implications of modernizing paddy rice farming practices. *Journal of Organic Agriculture Research* 4:10–28 (in Japanese).

Nakajima, K. 2004c. *Tabemono to nōgyō wa okane dakedewa hakarenai* [Food and agriculture cannot be evaluated solely by monetary values]. Tokyo: Commons.

Nakajima, K. 2008. Organic movement in Japan: Current status and its history. *Kikan At* 12:18–32 (in Japanese).

Nishida, Y. 2003. Dimensions of change in twentieth-century rural Japan. In *Farmers and village life in twentieth-century Japan*, ed. A. Waswo and Y. Nishida, 7–37. London: Routledge Curzon.

Odum, E.P. 1971. *Fundamentals of Ecology.* 3rd ed. Philadelphia: Saunders.

Oe, T. 2008. *Chiiki no chikara* [The power of communities]. Iwanami Shinsho 1115. Tokyo: Iwanami Shoten.

OECD. 2008a. Nutrients. In *Environmental performance of agriculture in OECD countries since 1990*, chap. 1, sect. 1.2. Paris: OECD. www.oecd.org/tad/env/indicators.

OECD. 2008b. Biodiversity. In *Environmental performance of agriculture in OECD countries since 1990*, chap., sect. 1.8. Paris. www.oecd.org/tad/env/indicators.

Ohnuki-Tierney, E. 1993. *Rice as self: Japanese identities through time.* Princeton, NJ: Princeton University Press.

Olson, D.M., Dinerstein, E., Wikramanayake, E.D., Burgess, N.D., Powell, G.V.N., Underwood, E.C., D'amico, J.A., Itoua, I., Strand, H.E., Morrison, J.C., Loucks, C.J., Allnutt, T.F., Ricketts, T.H., Kura, Y., Lamoreux, J.F., Wettengel, W.W., Hedao, P., and Kassem, K.R. 2001. Terrestrial ecoregions of the world: A new map of life on earth. *Bioscience* 51:933–38.

Peet, M.M., and Welles, G. 2005. Greenhouse tomato production. In *Tomatoes*, ed. E. Heuvelink, 257–304. Wallingford, UK: CABI Pub.

Pimentel, D., Dritschilo, W., Krummel, J., and Kutzman, J. 1975. Energy and land constraints in food protein production. *Science* 190:754–761.

Science Council of Japan. 2001. *Chikyū kankyō ningen seikatsu ni kakawaru nōgyō oyobi shinrin no tamentekina kinōnitsuite (tōshin)* [Multifunctions of agriculture and forestry on the earth environment and human livelihood (Recommendations)]. Tokyo. http://www.scj.go.jp/ja/info/kohyo/pdf/shimon-18-1.pdf.

Seibutsu tayōsei nōgyō shien senta. 2008. *Tanbo no megumi 150* [Biodiversity in paddies 150]. Tokyo: Seibutsu tayōsei nōgyō shien senta.

Selman, M., Greenhalgh, S., Diaz, R., and Sugg, Z. 2008. *Eutrophication and hypoxia in coastal areas: A global assessment of the state of knowledge.* Water Quality: Eutrophication and Hypoxia 1. Washington, DC: World Resources Institute.

Shin, E. H., Lee, W. J., Lee, H. I., Lee, D. K., and Klein, T. A. 2005. Seasonal population density and daily survival of anopheline mosquitoes (Diptera: Culicidae) in a malaria endemic area, Republic of Korea. *Journal of Vector Ecology* 30:33–40.

Smaling, E.M.A., Oenema, O., and Fresco, L. 1999. *Nutrient disequilibria in agroecosystems: Concepts and case studies.* Wallingford, UK: CABI Pub.

Takishima, Y., and Sakuma, H. 1961. Studies on organic acid metabolism in rice paddy soil and its inhibitive effect on rice growth. Part 7. Organic acid formation and growth inhibition induced by Chinese milk vetch application. *Japanese Journal of Soil Science and Plant Nutrition* 32:559–64 (in Japanese).

Tokunaga, M. 1997. *Nihon nōhōshi kenkyū* [Studies on Japanese farming practices]. Tokyo: Nōsan-gyoson Bunka Kyōkai.

Udagawa, T. 1976. Estimating energy input in paddy rice production. *Papers on Environmental Information Science* 5:73–79 (in Japanese).

Ueno, H. 1999. *Kami maruchi* [Paper mulch]. In *Josō zai wo tsukawanai ine tsukuri* (Minkan inasaku kenkyū jo ed.), 126–30. Tokyo: Nōsan-gyoson Bunka Kyōkai.

Une, Y. 1984. *Gen nōyaku inasaku no susume* [A guide to pesticide-reduced paddy rice production]. Fukuoka: Nise Hyakusyō Sha.

Une, Y. 2004. Must we pursue higher "productivity" of farming technology in organic agriculture? *Journal of Organic Agriculture Research* 4:29–42 (in Japanese).

Une, Y. 2005. *Nō no tobirano akekata* [How to open the gate of agriculture]. Tokyo: Zenkoku Nōgyō Kairyō Fukyū Shien Kyōkai.

Wakasugi, S., and Fujimori, S. 2005. Impacts of converting wet paddies to dry paddies on dragonfly habitat and its restoration. *Journal of the Japanese Society of Irrigation, Drainage, and Reclamation Engineering* 73:785–88 (in Japanese).

Waswo, A. 2003. Introduction. In *Farmers and village life in twentieth-century Japan*, ed. A. Waswo and Y. Nishida, 1–6. London: Routledge Curzon.

Wu, N., Liao, G.H., Li, D.F., Luo, Y.L., and Zhong, G.M. 1991. The advantages of mosquito biocontrol by stocking edible fish in rice paddies. *Southeast Asian Journal of Tropical Medicine and Public Health* 22:436–42.

Yamada, H., Kawasaki, S., and Yazawa, M., 2004. Effects of Aigamo duck farming on aquatic biota and water quality in a paddy field. *Papers on Environmental Information Science* 18:495–500 (in Japanese).

Yamashita, K. 2008. The perilous decline of Japanese agriculture. The Tokyo Foundation. http://www.tokyofoundation.org/en/articles/2008/the-perilous-decline-of-japanese-agriculture-1.

Yamashita, S. 1998. *Shindo fuji no tankyu* [Exploring "Shindo fuji"]. Tokyo: Sōrin Sha.

Yamashita, S. 2004. *Nō kara mita nihon* [Japan from a farmer's perspective]. Tokyo: Seiryū Shuppan.

Yamazaki, M., Yasuda, N., Yamada, T., Ota, K., and Kimura, M. 2004. Comparison of aquatic organisms communities between paddy fields under rice-duck (Aigamo) farming and paddy fields under conventional farming. *Soil Science and Plant Nutrition* 50:375–83.

Yasue, T. 1993. *Renge zensho* [Chinese milk vetch]. Tokyo: Nōsan-gyoson Bunka Kyōkai.

Yoshida, T. 1978. *Suiden keishi wa nōgyō wo horobosu* [Neglecting paddies will ruin Japanese agriculture]. Tokyo: Nōsan-gyoson Bunka Kyōkai.

Yūkisei odei no ryokunōchi riyō henshūiinkai. 2003. *Yūkisei odei no ryokunōchi riyō* [Field applications of organic sewage wastes]. Tokyo: Hakuyū Sha.

13 The Middle East
Adapting Food Production to Local Biophysical Realities

Alireza Koocheki

CONTENTS

13.1 INTRODUCTION

There is no broad consensus on what countries make up the region usually referred to as the Middle East. Some geographers define the region as Iran, Turkey, Syria, Lebanon, Jordan, Israel, Saudi Arabia, the United Arab Emirates, Bahrain, Oman, Qatar, Kuwait, and Yemen. Others include some of the North African countries, such as Egypt, Libya, Morocco, Tunisia, and Sudan. Still others stretch the boundaries farther in one or more directions, including in various schemes countries such as Afghanistan and Pakistan in the east and Greece and Cyprus in the west.

To complicate the situation even further, some authorities prefer to define the region more broadly and replace the term *Middle East* with *Near East* (e.g., Koohfkan,

2001); in so doing they add countries in Central Asia, such as Turkmenistan and Kyrgyzstan, and others to the west, such as Malta and Mauritania. One of the broadest definitions of the region is the one employed by the United Nation's Food and Agriculture Organization (FAO), which defines the Near East region as the area extending from the Atlantic Ocean (Mauritania and Morocco) in the west to Pakistan and Kyrgyzstan in the east, and from Turkey in the north to Somalia in the south. This region embraces 29 countries with a total area of 18.5 million km^2 (FAO, 1997). In recent years, an even broader classification comprising 32 countries—Central and West Asia and North Africa (CWANA)—has been used.

For the present purposes, we consider the boundaries of the Middle East to extend from Iran in the northeast to Turkey in the northwest, and from the North African countries of Libya, Egypt, Tunisia, Sudan, and Morocco in the southwest to the eastern edge of the Arabian Peninsula. Geographically, this is the region that is referred to as West Asia and North Africa (WANA) or the Middle East and North Africa (MENA) (e.g., Lofgren and Richards, 2003).

The region (considered as the FAO's Near East) covers 14% of the total area of the world and at present hosts 10% of its population. More than 70% of the region is classified as arid and semiarid, with 200 to 400 mm rainfall annually and a growing season of 70 to 150 days. The majority of the region has a Mediterranean climate, with cool to cold winters and hot, dry summers. Most of the rainfall is in the winter, but it is highly erratic in space and time.

Food production has a long history in the region and traditional agriculture has been practiced for centuries. The first clear domestication of plants and animals occurred in this area and long before the historical period the people of the Near East had become completely dependent upon agriculture for their food (Harlan, 1975). Agricultural land in this region, however, is extremely limited. Arable land and permanent crops comprise only about 7% of the total land area; 25% is classified as pasture, mainly in eco-zones with fewer than 200 mm annual rainfall. About 61% of the land area—mostly desert—is identified as "other" and is unsuited to agricultural use (FAO, 1995).

As an arid region with limited water resources and a growing population, the Middle East faces many challenges in achieving agricultural sustainability. At present, the region lags behind many other parts of the world in converting to more sustainable practices. However, as the negative consequences of policies and practices designed mainly to increase yields in the short-term become more apparent, there will be increasing pressure and motivation to move more rapidly toward sustainable practices.

13.2　MAIN AGRICULTURAL SYSTEMS IN THE REGION

Due to the diverse climatic conditions in the region, cropping systems are also diverse. This has been confirmed in a study of the agrobiodiversity of field crops in Iran (Koocheki et al., 2007). Throughout the region, the vast majority of farmers are cultivating holdings of 10 ha or less and often as little as 1 to 2 ha. At the other extreme, farms of 50 ha or more are a small percentage of the total (AOAD, 1983–1984). Three general types of agroecosystems can be recognized in the region: dryland farming or rain-fed agriculture, irrigated farming, and pastoralism.

13.2.1 DRYLAND FARMING

Dryland farming is the most ancient form of arable land agriculture. It is practiced primarily in the more northern and higher-altitude parts of the region, where precipitation totals are greatest and evaporation rates lowest. It covers a wide range of environments, from intermontane basins in Turkey and Iran to undulating plateau and plains in Syria and Iraq (Beaumont and McLachlan, 1985). Dryland farming has been practiced for many centuries irrespective of yield obtained, and little change has taken place through its history in the region (Brengle, 1982). It is estimated that in the region defined as CWANA, 70% of agricultural land is under rain-fed cultivation. In this type of farming, integration of crops and animals is frequently important. There are three general kinds of rain-fed cropping systems. The amount of rainfall, soil characteristics, and sometimes economic factors determine which of these cropping systems is practiced in a particular area.

- *Continuous cropping of cereals*—This system, in which wheat and barley are cultivated every year, is the most widespread in the region. The crop is managed extensively and yield is normally low. In years with below-average rainfall, immature crops are grazed. In some areas with continuous cereal production, wheat is normally produced in wetter areas and barley in drier areas; the transition between the two crops occurs at about the 300 mm annual rainfall isohyet. In the wheat-based system, the major aim is to grow crops, while in the barley-based system it is animal production. With decreasing rainfall farmers rely increasingly on animals—mainly sheep and goats—and cropping becomes a subsidiary practice (Jones, 1990).
- *Cereal-fallow system*—This system is based on the principle of conserving moisture during the fallow period for the next cropping season. However, conserving soil moisture by fallow is effective only if the amount and distribution of rainfall during the fallow period permit moisture penetration to some depth. The moisture that does enter the soil deeply is then conserved by controlling weed growth, which is achieved by leaving straw mulch and cereal residue on the land and using minimum tillage techniques. There are several types of rotations normally practiced with this system: wheat-weedy fallow (with the weedy fallow being grazed), wheat-cultivated fallow, wheat-barley fallow, and wheat-pulses and barley fallow.
- *Cereal cropping with a forage legume fallow (ley farming)*—The term *ley farming* was normally used whenever several years of arable cropping were followed by several years of forage utilized for livestock. In a more recent meaning related specifically to dryland farming, the term refers to the inclusion of annual forage legumes in the crop rotation. This latter concept is based on a farming system commonly used in southern Australia, in which the fallow period is replaced in the rotation by self-regenerating pastures (Robson, 1990). This integrated system of animal husbandry and cereal cropping has the advantage of relatively low inputs. The annual legumes are self-seeding, so little or no tillage is required, and the legumes provide all the nitrogen requirements of the cereal crop. The cereal can be seeded

into legume residues by a chisel drill without additional tillage. The legume and the stubble of cereal crop produce almost year-round grazing. There are considerable benefits to both cereal and animal production in including self-regenerating pasture legumes in the rotation. The legumes used are mainly annual medics such as *Medicago trancatula, M. regosa,* and *M. litoralis,* and also vetches. Despite its advantages, this system is not widely practiced in the region due to difficulties in technology adoption and environmental constraints—mainly in upland areas where the climatic conditions are not similar enough to those of southern Australia. Attempts have been made by international and national research organizations (e.g., ICARDA) in the area to introduce more suitable varieties with high winter hardiness, and to develop management practices better suited to the socioeconomic criteria of the farmers in the region.

13.2.2 IRRIGATED AGRICULTURE

The second type of arable land farming is irrigated agriculture, which originated in the Middle East (Kenyon, 1969). In many countries of the area, irrigation plays a vital role in agricultural production; in some countries virtually all arable land has to be irrigated. The area of irrigated land in the Middle East and North Africa has been reported to be 21 million ha; the ratio of irrigated areas to cultivated areas is 32%, and irrigation efficiency is 50% (FAO and World Bank, 2001). One of the main features of irrigated agriculture in the area is its dependence on the underground supply of water, the extraction of which is highly energy intensive.

The major irrigated field crops in the area are cereals, pulses (chickpea, beans, and lentil), sugar crops (mainly sugar beet), oil crops (sunflower, safflower, and canola), fiber crops (cotton), and forage crops (maize, alfalfa, clovers, and sainfoin). The latter crops are mainly cultivated in rotation with cereals, particularly with wheat. A wheat-based short rotation is a normal practice. Cereals—including wheat, barley, maize, millet, sorghum, and rice—are by far the most important group of crops grown in irrigated systems. Of the cereals, the most important is wheat, followed by barley as a distant second (Boyce et al., 1991; Beaumont and McLachlan, 1985). Traditionally, irrigated crop production in the area was practiced on a subsistence basis, but commercial production dominates today, and governments encourage the production of export-oriented commodities in irrigated systems.

Medicinal and aromatic plants such as cumin and saffron are produced in some areas, particularly in Iran. These crops are grown under limited irrigation with a strong community collaboration and family involvement in production, processing, and marketing (Koocheki, 2003).

Many types of fruits and nuts are also produced in the area, some of them unique, and most originally domesticated in the region. Olive, citrus, dates, grapes, peaches, plums, apricots, pistachio, almond, and walnut are all produced extensively for local market and for export. In Iran, for example, pistachio and dried fruits are two important export items.

In recent decades, traditional systems of fruit production have been replaced by modern types with all the expected advantages and shortcomings. These systems are based on high external inputs and new technologies. In traditional systems, small orchards are managed based on local resources; the modern systems, in contrast, are energy intensive and require a high level of expertise and knowledge. In recent years, vegetable and fruit production under controlled environments has been growing rapidly.

13.2.3 NOMADIC PASTORALISM

Nomadic pastoralism has played a key role in the agriculture of the Middle East for millennia. The nomadic tribes of the Arabian Peninsula and southern Iran are well known throughout the world (Beaumont and McLachlan, 1985). Because of the low precipitation that prevails over much of the area, biological productivity is low. This means that animals cannot be supported on natural vegetation in any one area for very long, and as a consequence, the animals have to be moved from one area to another. Nomadic groups have commonly integrated small animal production with cereal-based farming; this has been one of the main bases of food production historically and still is practiced to some extent.

The great importance of nomadic pastoralism is that it is often the only way in which the existing low-density vegetation can be harvested in an efficient manner. Through the centuries, nomads and their herds have worked as an integrated part of the natural ecosystems of the region (Koocheki and Gliessman, 2005). Rangelands and woodlands are important components of the environments in many areas, and play an important role in the socioeconomics of communities. There is still a strong dependence on rangelands because a good proportion of the small livestock in the area are raised under nomadic systems.

In recent years, however, nomadic pastoralism has declined dramatically, with no proper substitution, and at the same time, stocking rates have increased. The result is rangeland deterioration, which brings pastoralist systems closer to collapse. The rangelands of Iran, for example, are overstocked with more than 8 million cattle and 81 million goats and sheep (the latter the source of wool for the fabled rug industry), and the country faces increasing rangeland deterioration (Brown, 2001b). The fodder needs of livestock in nearly all the countries of the region now exceed the sustainable yield of rangelands and other forage resources. In addition, because productivity is low the animals need supplementary feeding with barley grain, or grazing of green barley or sown forages.

13.3 PRIMARY FACTORS LIMITING SUSTAINABILITY

In the region as a whole, both cropland and water resources are in short supply relative to the population. Total cropland per capita in many of the countries is presently less than 0.1 ha. Overall internal renewable water resources in the area are among the lowest in the world, with an average of 1,577 m^3 per inhabitant per year, as compared to 7,000 m^3 per inhabitant per year worldwide (FAO, 1997). Accelerating rates of population growth put increasing demands on these already stretched land and water resources, which results in several related trends that may be considered the primary factors limiting sustainability of food production in the region: expansion of agriculture to more fragile and marginal lands, growing dependency upon increases in yield per unit of land (Koohfkan, 2001), and a rise in cropping intensities (mainly in the form of a shortening of fallow periods). Land in use as a percentage of its potential is projected to increase from 76% in 1995 to 82% in 2030 (Koohfkan, 2001). By then, fallow land will have almost disappeared and will probably be concentrated in large farms.

Other factors are also affecting the sustainability of agriculture. These include poverty and inequity related to food distribution and consumption, land ownership and tenure issues associated with rangelands and other natural ecosystems, lack of proper technologies for small holdings, and lack of strategic planning for land utilization. These factors and their many effects on land and water resources—such as rangeland deterioration, groundwater depletion, and water pollution—interact in complex ways. Below, these issues of sustainability are examined as they relate to the use of water resources, and then two additional issues of importance—use of fertilizers and the role of government policy—are discussed briefly.

13.3.1 USE OF WATER RESOURCES

In the Middle East region, the most important environmental constraint on agriculture is lack of water. Only in the lowlands bordering the Black Sea and the Caspian Sea does precipitation occur all year round (Beaumont and McLachlan, 1985). The region nevertheless has a very long history of making efficient use of the available scarce water. There is evidence that the birth of agriculture in the area some 10,000 years ago was related to a warming and drying climate (Stevens, 1965). Settlements sprang up in fertile valleys or near large, permanent wells, and road routes were established from oasis to oasis. The so-called hydraulic civilizations evolved here thousands of years ago in the Euphrates, Tigris, and Nile basins by developing ways to feed relatively large populations despite the arid conditions. In ancient times, various types of runoff agriculture and water harvesting were practiced extensively in many arid parts of the Middle East.

Among the unique water collection systems that presumably arose in the region is the chain of wells. This ancient system of supplying water for irrigation and domestic purposes by means of underground infiltration tunnels or "horizontal wells," called *qanats* (Figure 13.1), is believed to have been invented at least 2,500 years ago on the Iranian plateau (Koocheki, 1994). In Iran, some 40,000 old chains of wells formerly supplied 35% of the country's water (National Academy of Science, 1974).

During the last 50 years, however, there has been a relentless increase in the intensity of water use for agiculture and a corresponding extension in the area of irrigated land. Since the twentieth century, water resource management has concentrated on large dam and reservoir schemes on the major rivers to provide increased amounts of water for irrigation and urban/industrial use. Unfortunately, many of these projects have not been as successful as originally envisioned, and they have had unforeseen negative consequences.

The central problem is that as a result of rising demands put on water resources, the Middle East is running out of water. Water shortages now plague almost every country in North Africa and the Middle East. The people who have built their lives

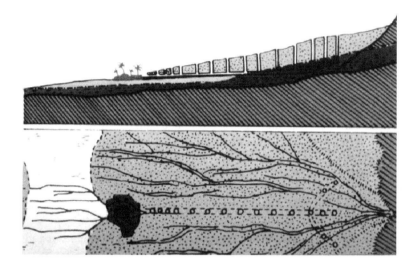

FIGURE 13.1   Chains of wells (*qanats*), an old system of water extraction.

and livelihoods on having reliable sources of fresh water are seeing the shortage of this vital resource impinge on all aspects of the tenuous relations that have developed over the years between nations, between economic sectors, and between individuals and the environment (Wolf, 1996). As the population continues to expand in water-short nations, dependence on imported grain is rising. Algeria, Egypt, Iran, and Morocco are being forced into the world market for 40% or more of their grain supply primarily because they lack sufficient water.

Iran is now facing widespread water shortage. In the northeast, the Chenaran plain—a fertile agricultural region to the east of Mashhad, one of Iran's largest and fastest-growing cities—is quickly losing its water supply (Brown, 2001a). Many wells have gone dry, and some villages have been evacuated. In Iran, more than 80% of available water is allocated to agriculture, but the rapid growth of urban areas means that some of this water must be diverted for local consumption and industrial development. Therefore, more food—mainly wheat—needs to be imported as more land is abandoned. But countries such as Iran that must rely on imported grain may not be able to do so for long. The grain-exporting countries also face rapid population growth and a global shortage of water. Ensuring continuing access to the 1,000 metric tons of "virtual water" needed to grow each 1,000 kg of wheat will not be an easy task for exporting countries (Lofgren and Richards, 2003).

As the demands for water for food production grow, so too do the demands for water for industrial and urban uses. Throughout the region, agriculture is the main user of fresh water, representing over 90% of total water consumption (Koohfkan, 2001). However, it is expected that by 2025 more than 50% of water in the area will be consumed in nonagricultural use. In addition, water quality degradation is quickly joining water scarcity as a major issue in the area. Water pollution from agricultural and urban and industrial wastes, overpumping of groundwater aquifers, and salinization are becoming major problems (Lofgren and Richards, 2003).

13.3.2 FERTILIZER USE AND POLLUTION

In 1997, about 20 million tons of mineral fertilizer was produced in the Middle East region; of this amount, 8.6 million tons was used in the region and the balance was exported to other countries. Fertilizer use per unit area varies from 50 kg nutrient/ha in Sudan to 347 kg/ha in Egypt and to more than 7,000 kg/ha in the protected vegetable production systems in the United Arab Emirates (Koohfkan, 2001). Nitrogen is by far the highest consumed nutrient in the area. A steady increase in N fertilizer consumption in the region has been observed during the past two decades.

The dependence on mineral fertilizer shown by these statistics is unsustainable for many reasons. In addition to the fact that producing the fertilizers requires large inputs of fossil fuels, their use has negative consequences for agroecosystems, the environment, and water resources. Unbalanced fertilization, which is observed in many countries of the region, is causing a serious loss in yields due to declining response to fertilizers (Koohfkan, 2001). Fertilizer use efficiency, as indicated by the ratio of tons of cereal grain produced by one ton of fertilizer nutrient (NPK), showes a steady decline. From an environmental point of view, large amounts of nutrients, especially nitrogen, are lost from agroecosystems through deep percolation, surface runoff, and volatilization, and these can cause serious pollution of surface water and groundwater. Pollution of surface and especially groundwater with nitrate (NO_3-) from nitrogen fertilizers and manure is a problem in many areas (Koohfan, 2001). In areas under high application rates of phosphate fertilizers, there is a potential risk of accumulation of hazardous heavy metals, especially cadmium (Cd) and lead (Pb).

Use of chemical biocides, particularly herbicides, has also been on the increase. This has caused serious problems, including resistance to the chemicals, environmetal pollution, and loss of biodiversity.

13.3.3 GOVERNMENT POLICY

In general, the most important barrier to enhancing the sustainability of the agroecosystems in the area is the lack of concrete support by the respective governments. Their failure to promote more sustainable practices is often due to the misconception that sustainable agriculture would result in lower yields and loss of production, and therefore increase food insecurity.

As shown by Farshad and Zinke (2001) in relation to Iran, many of the underlying causes of unsustainability in agriculture have to do with large-scale socioeconomic, technical, and institutional factors that can be influenced effectively only by governments. Government action is needed to alleviate poverty, which tends to encourage practices that increase production in the short-term but undermine sustainability in the long-term. It is also needed to fully address water scarcity, soil degradation, water quality deterioration, vegetation depletion, land use competition, and inadequate external capital inputs, all of which affect the sustainability of agricultural systems in the country.

FIGURE 13.2 Traditional practice of soil preparation.

13.4 CONVERSION TO SUSTAINABLE PRACTICES

Sustainable agriculture—based on the concept of sustaining the yield for the long-term—has been practiced in the area for centuries in the form of traditional agro-ecosystem management. Based on indigenous knowledge, local communities have developed many environmentally sound technologies (Figure 13.2) for low-input food production in rain-fed agroecosystems, irrigated systems, and pastoral systems (Koocheki and Ghorbani, 2005).

Although traditional technologies, practices, and systems have been replaced by modern and energy-intensive technologies to a great extent, small holdings are still the main basis of farming in the region and sustainable traditional farming is still practiced. At the same time, new concepts for increasing the sustainability of agricultural production are gaining more attention, and there are clear indications that these concepts are increasingly being applied.

13.4.1 INPUT SUBSTITUTION

Although the main focus of farming in the region is to increase yield through agronomic practices such as application of chemical fertilizers, weed control, employment of high-pressure irrigation systems, and so on, there is a great tendency toward making more efficient use of inputs. This is mainly due to increasing cost, environmental concerns, and growing public awareness of the hazards of unhealthy food. Farmers are becoming more concerned about proper utilization of inputs, increasing the energy efficiency of their farming systems, and avoiding environmental pollution.

Practices such as using organic fertilizer, managing pests with nonchemical means, and enhancing the biodiversity of cropping systems are becoming more popular. Other good examples of input substitution are the use of nitrogen-fixing legumes as components of multiple cropping systems, the use of local genotypes and varieties, and the reintroduction of some neglected and underutilized crops.

Iran has a nationwide program to reduce the use of chemical fertilizers and pesticides. As part of this program, integrated pest management, biological controls, and early warning systems have all been encouraged. Chlorophosphorous pesticides have been replaced with biological agents in the control of cotton pests, various kinds of bioinsecticides have been extracted and formulated, and codling moth pheromones have been formulated and mass produced. The government has facilitated biological pest control by providing the agent insects and predators at low cost or free of charge. The result has been a 50% reduction in the use of chemical pesticides since 1990 (Ministry of Agriculture, 2002). In addition, biological fertilizers are being imported or being encouraged to be manufactured in the country.

13.4.2 AGROECOSYSTEM REDESIGN

Diversification of agroecosystems and integration of crops and livestock seem to be the most significant changes at the level of ecosystem redesign. Diversification and integration of animal husbandry with cropping systems is gaining in importance, particularly with small animals such as goats and sheep. Crop residue is used as feed, and manure from the animals is used on the fields. Diversity at the agroecosystem level contributes to greater food security, helps increase employment opportunities, and increases local and national self-reliance by allowing a variety of enterprises, based on products and services, to develop on a national, regional, or community scale (Koocheki and Ghorbani, 2005). A diversity of crop and animal species at the community, farm, or field levels also adds to social and economic stability by reducing reliance on a single enterprise.

13.4.3 BETTER WATER MANAGEMENT

Governments of different political backgrounds in the area are all aware of the water crisis in the region, and they know that the most critical issue they face in the area of resource utilization and sustainability is water management. Indeed, sustainable agriculture in the area is highly dependent on efficient water use. Therefore, water use efficiency and productivity are main topics of interest. One response has been to replace wasteful and inefficient irrigation systems with pressurized irrigation systems (drip and sprinkler irrigation, etc.). Use of these systems is expanding very rapidly; they are, however, energy intensive. Developing water harvesting systems and discovering ways of utilizing salt water, brackish water, and sewage effluent in agriculture are other interesting strategies being pursued, despite the many unsolved problems and unanswered questions associated with these types of resource utilization.

13.4.4 CHANGES IN GOVERNMENT POLICY

Greater awareness of climate change and its potentially serious negative impacts on food production in the region is spurring governments and policymakers to become more concerned about agricultural sustainability. Engaged in developing strategies for mitigating and adapting to the changes, they are seeking ways of utilizing resources more efficiently and reducing dependence on external inputs. Most of the countries in the area are putting an emphasis on adopting holistic approaches that simultaneously attend to conservation, rehabilitation, development, and optimum utilization of the resource base.

Many necessary changes cannot occur unless they are supported or carried out by government policy. Reallocating land used for a particular farming practice to a more sustainable use is one such change. For example, in an attempt to evaluate the suitability of land in Iran for dryland farming, based on sustainability criteria, Koocheki (2002) has concluded that most of the dryland areas, particularly in medium- to low-rainfall areas, should be allocated to rangeland instead of dryland wheat production. A change of this scale would have to be a part of a nationwide strategic plan, and fortunately, such proposals are becoming attractive to policymakers in Iran.

13.4.5 RESEARCH AND EDUCATION

It is promising that research activities in the area of ecological agriculture are accelerating. Across the region, there is a growing interest in research focused on biological pest control, the use of biofertilizers such as mycorrhyzae and rhizobium, integrated weed and pest management, diversification and multiple cropping, and low-input cropping systems.

During the last 10 years in Iran, for example, a growing number of research projects and PhD theses have been focused on different topics of ecological agriculture, and now an MSc program and a minor in ecological agriculture at the Ph.D. level have been introduced in the universities. In addition, the establishment of new institutions dedicated to the fields of environmental studies and ecological agriculture and the publication of related scientific journals offer ample evidence of increasing academic interest in sustainability issues. Academic research will be most effective in creating actual change, however, if it is directed toward on-farm practices and agroecosystem design and management, and if greater emphasis is given to the social aspects of food production systems, from the farm to the consumer level.

13.4.6 ORGANIC AGRICULTURE

Community demands for healthy foods have increased and new international markets for such products have encouraged farmers in the area to produce safe food for export. This trend is expanding under the umbrella of organic agriculture, with its sets of specific rules and regulations. Since this type of farming is more dependent on local resources and biological inputs, and gives more emphasis to the food-related socioeconomic issues that face communities, farmers, and consumers, organic farming can in general be considered a sustainable practice.

In the latest evaluation of organic agriculture worldwide, Willer and Yussefi (2006) state that organic agriculture on the Asian continent is on the upswing, as the area in organic production is increasing. At present, according to these authors, Asia has a total area of 4.1 million ha under organic management (including both fully converted land as well as land in the process of conversion); this compares to 31 million ha under organic management worldwide. For many specific countries, particularly those in the Middle East, no precise figures for organic area or production are available, but it may be assumed that no country has yet put 1% of its agricultural land under organic management.

Small farmers in the Middle East wishing to convert to organic management are faced with a certification and inspection process based on the standards of developed nations, and this can make conversion too expensive and complicated. Therefore, implementation of simple national standards for small farmers, affordable services for inspection and certification, establishment of local certifying organizations, economic incentives, training, and capacity building are all fundamental to expansion of organic farming in the area.

13.4.7 BIODIVERSITY CONSERVATION

As people in the Middle East have become more concerned about what is in the food they eat, they have also increased their awareness of environmental issues such as pollution, destruction of natural habitats, and decline of wild populations of plants and animals. Many nongovernmental organizations (NGOs) are active in protecting the environment and promoting resource and biodiversity conservation, and government policies are increasingly being oriented toward better protection of the environment. One positive sign is an increase in protected areas, national parks, and wildlife refuges.

Case studies on biodiversity protection in different areas of North Africa and the Middle East have been compiled as a part of a book (Lemons et al., 2003). The studies were part of a joint project carried out on conservation of biodiversity in arid regions by the Third World Network of Scientific Organizations (TWNSO) and the Global Environmental Facility (GEF). The case studies document a wide range of biodiversity conservation efforts, from using native trees to sustain biodiversity in Oman to conservation of fig (*Ficus corica* L.) and pomegranate (*Punica granatum* L.) varieties in Tunisia and *in situ* conservation of crop genetic diversity in Morocco.

13.5 CONCLUSIONS

The Middle East has a long way to go to reach agricultural sustainability, and although progress toward this goal is slow, there are many signs that it is accelerating. Population growth and limited water resources are huge barriers to achieving sustainability, but the region as a whole has the advantage of having a long tradition of small-scale, locally adapted, water-efficient, integrated agricultural practices based on rich agrobiodiversity from which it can draw in developing productive and sustainable agroecosystems. In the coming years, the region's progress in moving toward sustainability will be determined in large part by how well traditional agricultural practices are combined with new agroecological principles to create systems that are appropriate to local needs and biophysical conditions.

In addition, it must be born in mind, as Zurayk (2000) has argued, that the prevailing models for the sustainable agriculture approach may lack applicability in many developing countries, especially in the Middle East; therefore, development of appropriate technologies and agroecosystems must be based on the sociocultural background of the communities involved, and the social aspects of sustainability require particular emphasis. It is also important to recognize, as pointed out by Zurayk (2000), that among the major actors involved in sustainable agriculture (i.e., the public sector, the farm sector, the small-scale commercial sector, the industrial sector, and nongovernmental organizations), the public sector has the primary role, as it enables the others to play their roles. Without its participation, sustainable technologies, sustainable agriculture, and sustainable development cannot be nurtured.

REFERENCES

Arab Organization for Agricultural Development (AOAD). 1983–1984. *Yearbook of agricultural statistics.* Vols. 3–4. Khartoum.

Beaumont, P., and K. McLachlan. 1985. *Agricultural development in the Middle East.* Chichester, UK: John Wiley & Sons.

Boyce, K.G., P.G. Tow, and A. Koocheki. 1991. Comparisons of agriculture in countries with Mediterranean type climates. In *Dryland farming: A systems approach*, ed. V. Squires and P. Tow, 250–60. Sydney: Sydney University Press.

Brengle, K.G. 1982. *Principles and practices of dryland farming.* Boulder, CO: Colorado Associated University Press.

Brown, L.R. 2001a. *Economic growth losing momentum*, 108–11. London: Earth Policy Institute.

Brown, L.R. 2001b. *Eco-economy: Building an economy for the earth*, 50–73. London: Earth Policy Institute.

FAO. 1995. *Production yearbook.* Vol. 9. Rome.

FAO. 1997. *Irrigation in the Near East region in figures.* Water Reports 9. Rome.

FAO and World Bank. 2001. *Farming systems and poverty. Improving farmers' livelihoods in a changing world.*

Farshad, A., and J.A. Zinke. 2001. Assessing agricultural sustainability using the six-pillar medel: Iran as a case study. In *Agro-ecosystem sustainability: Developing practical strategies*, pp. 137–151, ed. S.R. Gliessman. Boca Raton, FL: CRC Press.

Harlan, J.R. 1975. *Crops and man*, pp. 179–188. Madison, WI: American Society of Agronomy.

Jones, M.A. 1990. The role of forage legumes in rotation with cereals in Mediterranean areas. In *The role of legumes in the farming systems of Mediterranean areas*, pp. 195–204, ed. A.E. Osman, M.H. Ibrahim, and M.A. Jones. Dordrecht, The Netherlands: Kluwer Academic Publishers.

Kenyon, K. 1969. The origins of the Neolithic. *Advancement of Science* (London) 25:144–60.

Koocheki, A. 1994. Sustainable aspects of traditional land management in Iran. In *Proceedings of International Conference on Land and Water Resources Management in the Mediterranean Region*, Instituto Agronomico Mediterraneo, Valenzano, Bari, Italy, September 4–8, 1994, pp. 559–72.

Koocheki, A. 2002. Evaluation of land suitability for dry land farming in Khorasan province using long-term climatic data. In *Proceedings of an International Conference on Environmentally Sustainable Agriculture for Dry Areas in the 3rd Millennium*, Chinese Academy of Science, Shigiazhuang, China, September 16–19, 2002, p. 14.

Koocheki, A. 2003. Role of indigenous knowledge in agriculture with particular reference to saffron production in Iran. Paper presented at First International Symposium on Saffron Biology and Biotechnology, Albacete, Spain, October.

Koocheki, A., and S. Gliessman. 2005. Pastoral nomadism, a sustainable system for grazing land management in arid areas. *Journal of Sustainable Agriculture* 25:113–131.

Koocheki, A., and R. Ghorbani. 2005. Traditional agriculture in Iran and development challenge for organic agriculture. *International Journal of Biodiversity Science and Management* 11:1–7.

Koocheki, A., M. Nassiri, S. Gliessman, and A. Zarea. 2008. Agrobiodiversity of field crops: A case study for Iran. *Journal of Sustainable Agriculture* 32(1):95–122.

Koohfkan, P. 2001. Food, security, and sustainable development in desert communities. In *Sustainable development of desert communities: A regional symposium*, 169–181. Technical Paper 2, United Nation Development Programme in Iran, Tehran.

Lemons, J., R. Victor, and D. Schaffer. 2003. *Conserving biodiversity in arid regions*, 395–461. Dordrecht, The Netherlands: Kluwer Academic Publishers.

Lofgren, H., and A. Richards. 2003. Food security, poverty, and economic policy in the Middle East and North Africa. In *Food, agriculture, and economic policy in the Middle East and North Africa*, ed. H. Lofgren, 1–33. Vol. 5. Research in Middle East Economics, Amsterdam: JAI.

Ministry of Agriculture. 2002. *Iranian agriculture: Capabilities for development*, 100. Agricultural Planning and Economic Research Institute, Ministry of Jehad-e-Agriculture, Tehran, Iran.

National Academy of Science. 1974. *More water for arid lands: Promising technologies and research opportunities*. Washington, DC.

Robson, A.D. 1990. The role of self-regenerating pasture in rotation with cereals in Mediterranean areas. In *The role of legumes in the farming systems of the Mediterranean area*, pp. 217–236, ed. A.E. Osman, M.H. Ibrahim, and M.A. Jones. Dordrecht, The Netherlands: Kluwer Academic Publishers.

Stevens, G. 1965. *Jordan River partition*. Stanford, CA: The Hoover Institution.

Willer, H., and M. Yussefi. 2006. *Organic agriculture worldwide: Statistics and emerging trends*. http://www.organic-world.net/former.asp.

Wolf, A. T. 1996. *Middle East water conflicts and directions for conflict resolution*. Discussion Paper 12, International Food Policy Research Institute, Food, Agriculture, and the Environment.

Zurayk, R. 2000. Sustainable agriculture in the Middle Eastern context: Why prevailing models won't work. *Culture and Agriculture* 22:37–42.

14 Australia
Farmers Responding to the Need for Conversion

David Dumaresq and Saan Ecker

CONTENTS

14.1 INTRODUCTION

Australia has some 467,000 km^2 of arable land available for agriculture, representing about 6% of the country's land area. By comparison, some 20% of the land area is available for agriculture in the United States. With a population density of about two people per square kilometer and with relatively low fertilizer and agricultural chemical use, there would seem to be little pressure on Australia's agricultural land resource. However, Australia's farms annually feed well over 50 million people elsewhere in the world, providing cereal staples of wheat and rice and animal protein as well as one-third of the world's wool. Adding to this pressure from world food trade are the environmental conditions within the continent. Australia is the world's driest inhabited continent. Two-thirds of the continent is classified as semiarid or arid. This is combined with very high climate variability. Soils have very low nutrient content, generally requiring fertilizer inputs to maintain agricultural productivity (Australian Government, 2006). These conditions produce considerable challenges for agricultural sustainability.

14.2 STATE OF PLAY IN AUSTRALIAN AGRICULTURE

Gray and Lawrence (2001), in their book *A Future for Regional Australia: Escaping Global Misfortune*, maintain that regional Australians are experiencing the effects of global misfortune, which continues to create social, economic, and environmental disadvantage for our rural populations. The three elements that comprise this misfortune are identified as the Australian colonial legacy, the use of inappropriate farming practices in a fragile ecosystem, and the vulnerability of family-based farming systems. To address these problems, concepts relating to farming in Australia are currently evolving to adopt the criteria of the triple bottom line—ideally creating farming systems that are ecologically sustainable, profitable, and socially acceptable. Rickert (2004) rightly identifies that farmers not only provide food and fiber, but also act as stewards of land that provides ecosystem services for the wider community. To remain globally competitive, Australian farmers need to transfer from a yield and paddock focus to a "whole farm" approach embracing key performance indicators across the full range of profit drivers, including water efficiency, farm inputs, machinery, labor, and financing costs.

Australian farm produce is seen as high quality, low in contaminants, free of disease, and about the cheapest in the world. Out of all developed countries, Australians pay virtually the lowest level of effective subsidy to their farmers. The Australian Farm Institute reported that across all OECD countries, governments provide to farmers an average of around 32% of gross farm receipts; in Switzerland, Korea, and Japan this figure is as high as 60%, while in Australia and New Zealand it is around 2% (Australian Farm Institute, 2005).

14.2.1 KEY THEMES

Love (2005) suggests that three themes consistently appear as the major influences on Australian agriculture: (1) global demand from consumers and the supply chain for "clean and green" food and fiber production, (2) increasing government regulation and global and national standards for land stewardship and duty of care, and (3) the need for farmers to manage profitable farming enterprises. Consistent with these themes, *The Trends in Agriculture Report* published by the Australian Productivity Commission in July 2005 identifies three of the key drivers for agriculture as shifts in consumer demand, changes in government policies, and emerging environmental concerns. It adds technological advances and innovation and an unrelenting decline in the sector's terms of trade as two additional influences. While Australia is currently largely protected from market access issues related to perceived or real sustainability issues, there is some concern that inactivity may put Australia's position as a clean and green producer at risk (Australian Government Productivity Commission, 2005).

14.2.2 EMERGING TRENDS

There is ample evidence that a variety of change processes are affecting Australian agriculture. These processes include demographic changes, issues with infrastructure, changes in consumer demand, and shifts in international commodity prices (Productivity Commission, 2005). The number of farms is in a steepening decline and the average farm size is increasing. These trends exist in a context of forecasted climate change that will create recurring drought conditions. Long-term projections show that with no action on climate change there would be a decline in wheat, beef, dairy, and sugar production of 9 to 10% by 2030 and 13 to 19% by 2050. Anticipated agricultural exports would fall 63% by 2030 and 79% by 2050 (ABARE, 2007).

Farmers face a variety of other complex challenges, including new technologies, biosecurity threats, changes to agricultural marketing arrangements (e.g., the recent changes to the Australian wheat board), fluctuating commodity prices in global and domestic markets, new demands for sustainable or low-chemical products, and increasing global competition. Cost increases continue to outstrip profit increases, and the cost of fuel and fertilizer has increased significantly since 2008. The cost of finance is also rising, with farmers carrying some of the highest debt ratios of all businesses, creating extreme vulnerabilities.

Before the current global financial crisis, there were signs that economic growth in agriculture was being boosted by rising commodity prices, despite rising oil prices driving up input costs. While there are potential opportunities, the impact of current and future financial conditions on the agricultural sector is uncertain, adding to existing risk and uncertainty in agricultural industries.

A fall in wool sales is already evident because of the anticipated loss of discretional spending on luxury items, which include wool clothing. A downturn in wool processing in China, where the bulk of Australian wool is processed, is also apparent. According to projections made in 2007, Australia is currently experiencing an 80-year low in national shorn wool production (ABARE, 2007).

Understanding the risk of an increased decline in the terms of trade across different agricultural industries is critical for understanding the vulnerabilities of these industries, but it is also important for understanding the motivations to shift toward organic or sustainable farming. Not only can these farming (and value chain) systems potentially attract a higher-value market, but also they are less dependent on the mainstream agriculture inputs and infrastructures that are currently an area of significant uncertainty. Organic and sustainable farming models, as demonstrated in some of the case studies examined in this chapter, can also increase resilience in the face of climate change, decreasing water availability, and other risks.

14.2.3 RECENT CONCERNS ABOUT SUSTAINABILITY

Beginning in 1991, several studies have been carried out to form a framework for sustainable agriculture in Australia, and to better understand our approaches to sustainability. Studies were conducted by the Standing Committee on Agriculture and Resource Management (SCARM) and its predecessor, the Standing Committee on Agriculture (SCA), federal government scientists, the National Land and Water Resources Audit (NLWRA), the Australian Bureau of Agricultural and Resource Economics (ABARE), and the National Food and Industry Strategy.

In addition to SCARM, Australia currently has a number of processes to measure sustainable development and sustainable agriculture. Table 14.1 provides an overview of these processes.

The National Land and Water Resource Audit summarized studies undertaken between 1991 and 2001 (NLWRA, 2001). This showed the following trends in relation to producers' awareness of the need for change:

- Increasing concern was shown by all farmers about chemical residues in agricultural produce and about the environmental and health effects of agricultural chemicals, but those who are regular users of chemicals, such as cereal or fodder crop producers, were less concerned and showed relatively little change in their use over the period.
- There is increasing awareness that farm practices have impacts beyond the farm boundary, and there are increasingly favorable views nationally toward consideration of the wider public interest in farm decision making.
- There is increasing acceptance of the idea that there will have to be a major transformation of agricultural landscapes if farming is to be sustainable, with just over 46% of respondents agreeing with the proposition that if Australian agriculture is going to have a long-term future, a lot of cleared country will have to be put back to bush and forestry plantations.

Fenton et al. (2000) discuss sustainability as a wider social construct that potentially exerts pressure on farmers to be sustainable as it becomes more of a social standard of behavior and social norm. If sustainability does become more of a social

TABLE 14.1
Broad Sustainability Measures in Australia

Agriculture (Australian Government, 1995)	• Long-term real net income • Natural resource condition • Off-site environmental impacts	• Managerial skills • Socioeconomic impacts
Forests (Montreal Process)	• Biological diversity • Productive capacity • Ecosystem health and vitality • Soil and water resources	• Global carbon cycles • Socioeconomic benefits • An effective legal and institutional framework
Land and water (NLWA 97-01)	• Water availability • Dryland salinity • Vegetation management • Rangeland monitoring	• Agricultural productivity and sustainability • Capacity for change • Ecosystem health
State of environment (Australian Government, 2006)	• Atmosphere • Biodiversity • Coasts and oceans • Human settlements	• Inland waters • Land • Natural and cultural heritage
Fisheries (ESDRA)	• Contribution of the fishery to ecological well-being: (1) retained/nonretained species, (2) general ecosystem • Contribution of the fishery to human well-being: (1) indigenous, (2) local and regional, (3) national social and economic	• Ability to achieve: (1) governance, (2) impact on the environment of the fishery
TBL reporting environmental performance indicators	• Energy • Greenhouse • Water • Materials • Waste: solid and hazardous • Emissions and discharge to air, land, and water	• Biodiversity • Ozone-depleting substances • Suppliers • Products and services • Compliance
Headline indicators (Australian Bureau of Statistics)	• Living standards and economic well-being • Education and skills • Healthy living • Drinking water quality • Air quality • Economic capacity • Industry performance • Economic security • Water management • Forests management	• Fish management • Energy management • Agriculture management • Intragenerational equity • Intergenerational equity • Biodiversity integrity • Climate change • Coastal and marine health • Freshwater health • Land health

norm, then there is also the issue of farmer willingness to conform to this social norm (Fenton et al., 2000).

14.3 RESPONSES

Australia has already made some progress toward reaching the goal of sustainable agricultural production. It has, as an example, the world's largest area of certified organic agricultural land, at more than 11 million ha (Knudsen et al., 2006). Efforts in Australia to support and further promote sustainable agriculture revolve largely around the Landcare programs funded by the federal government and delivered in partnership with regional nongovernment organizations, state governments, community groups, and individuals. The implementation of these programs is dependent on significant volunteer effort. The current program, announced in 2008, is named Caring for Our Country.

14.3.1 Eco-Labeling and Other Forms of Certification

With the exception of a range of organic and biodynamic certification schemes and the environmental management standard ISO 14001 (which has had limited adoption nationally), there are limited formally recognized or internationally observed systems in Australia for endorsing environmentally sustainable production approaches. Australian growers can access international environmental schemes, including EUREPGAP, EMAS, and the European eco-wool certification, but there are no widely recognized national systems designed for Australian growers. While there are a number of regionally recognized sustainable agriculture labeling or branding programs (e.g., the Gippsland Enviro-meat label in Victoria), national coordination to develop a standard or system of endorsement for sustainable agriculture does not yet exist.

The most significant effort leading toward this kind of recognition is represented by the efforts at Environmental Management Systems (EMS) and Environmental Assurance in Australia. The influence of EMS has been particularly strong since 2000, when the federal government provided significant investment into the development of farm EMS programs. This effort was captured in an inventory of related programs and research (Rowland, 2005).

"The most critical issue facing Australian rural industries is how to keep track of everything taking place and how to foster effective and efficient communication channels between these activities in order to promote consistent approaches (to EMS) and prevent both duplication and fragmentation of effort" (Rowland, 2005, p. 11).

Australian governments, industry, and landholders have invested significantly in the area of environmental assurance from 2000 to 2007; the number of land managers involved now reaches into the thousands. Arrangements generally allow for a "flexible, tiered approach allowing producers to 'opt in' at a particular level of verification rigor to suit their enterprise" (Rowland et al., 2005, p. 2). A plethora of resources related to EMS and environmental assurance exist, including workbooks, fact sheets, training kits, and computer software.

Currently sustainable agriculture support at a national scale is available through a range of government programs targeting risk management and drought preparedness (e.g., Farm Ready grants for property management and drought planning) and resource management (e.g., the Improving Farm Practice component of the Caring for Our Country initiative). A number of drivers are leading to policy development in this area.

A number of strategies have been recently developed in the search for more sustainable agricultural practices. It has been suggested that alternative farming practices offer strategies for reducing some of the adverse environmental impacts of agriculture and may be superior to conventional agricultural systems in terms of soil quality and sustainability.

14.3.2 CONVENTIONAL APPROACHES

Without diverging from high-input strategies, some farmers are adopting increasingly sophisticated fertilization strategies and conservation tillage systems to counteract soil nutrient status problems and structural decline. Broadacre cereal cropping systems covering some 24 million ha annually in Australia (ABS, 2006) are increasingly being operated as continuous cropping systems reliant on N and P fertilizers for nutrient replacement and herbicides for weed control. Soil physical integrity is thought to be maintained in these systems by the minimization of tillage and the retention of stubble. Soil biological processes are also thought to be enhanced by such management techniques. The role of soil organic carbon is crucial in these processes (Tisdal and Oades, 1982; Lee and Foster, 1991; Oades and Waters, 1991). In Australia there has been little investigation of alternative approaches to the problems of very large-scale dryland cereal cropping or the investigation of the functioning of these systems as agroecosystems (see Dumaresq et al., 2000).

14.3.3 ALTERNATIVE APPROACHES

Alternative farming systems have gained little attention in Australia until recently and have attracted even less research and development interest. An incomplete summary of alternative farming systems research is to be found in Derrick (1997). A more complete and more recent industry-wide review of Australian organic agriculture commissioned by the federal government can be found in Halpin (2004). Despite the relative lack of research and institutional attention, there is a considerable history in Australia of practitioner-based experimentation with alternative cropping systems, ranging from low-input conventional to organic and biodynamic approaches. Organic farming systems have formed the mainstream of these alternative systems. Broadacre organic cereal cropping systems have been commercially established in Australia since the 1950s (Dumaresq and Greene, 1997; Kondinin, 2000).

As these systems cannot rely on the same high level of off-farm inputs as their conventional counterparts for operations such as nutrient replacement and weed control, they form a contrasting comparison group for the study of the sustainability of mixed cropping systems that comprise the heartland of Australian farming (Rovira, 1993).

14.3.4 EMERGENCE OF "MIDDLE WAY" AGRICULTURE

There is some evidence that there are increasing pressures for recognizing a middle-way approach to sustainable food and fiber production, that is, a system that recognizes efforts toward sustainability as a middle way between the conventional productivist and the certified organic/biodynamic agriculture approaches. This encompasses efforts by conventional growers to create more environmentally and socially benign food and fiber production systems by adopting IPM, eco-efficiency and natural resource protection approaches, and strategies for limiting chemical use.

There are a number of dilemmas for this middle-way approach, which rejects the conventional model of agriculture and yet falls short of or rejects the organic agriculture model as well. These dilemmas include a lack of formal definitions for this type of agriculture and the lack of coordination across this arena. Also, many people, particularly certified organic supply chain participants, are wary of the potential for "greenwashing" through such an approach.

14.4 REGIONAL CASE STUDY: SOUTHEASTERN AUSTRALIA

14.4.1 CASE STUDY OVERVIEW

In 1990, a research project was set up at the Australian National University to investigate the characteristics of broadacre organic systems and make direct, paired comparisons with equivalent conventionally managed farms. The project began with general surveys of the characteristics of broadacre organic stock and crop production enterprises in southeastern Australia. These surveys involved 40 farms, of which half (20) were self-ascribed as organic. This work (Dumaresq, 1992; Dumaresq and Greene, 1997) indicated that in general terms organic farming did not differ greatly from its conventional counterparts in southern Australia. In summary,

- Organic agriculture was not limited to any particular region, climate, rainfall, soil type, or other environmental variable—it was practiced across a wide range of climates and farm types.
- Organic farming was not limited in size of operation—farm size is comparable to conventional farms.
- Organic farming was not limited to any particular set of commodities—all major agricultural commodities produced in Australia have some organic production.

From this work, broadacre organic farming emerges as a clear alternative to the mainstream of conventional agriculture. Early investigation revealed that proponents of organic farming make three central claims:

- Organic farming methods can return yields equivalent to conventional methods.
- Organic methods can provide adequate crop plant nutrition through the management of, and mediation by, soil biological processes.

- Organic farming methods are an environmentally sound form of agriculture compared to their conventional counterparts, yet remain commercially viable.

Investigations of these claims were concentrated into a long-term monitoring project on a commercial farm-scale research site in the eastern Riverina of New South Wales in southeastern Australia (34°22'S, 146°54'W, 200 to 230 m elevation). The site consists of two adjoining organic farms and their conventional neighbors. The organic farms have been operated as single-family enterprises for more than 40 years, with the production converted to organic in 1963. Bordering these two farms are five conventional farms of the same commodity production system with similar long-term histories of development since the land was settled for farming some 90 years ago. Three of these conventional farms have been used in direct comparison trials.

14.4.2 Farming Systems

Farming in this part of Australia is dominated by mixed livestock and cropping systems. The livestock are predominantly sheep, producing both wool and meat, and there are some beef cattle. Cattle production increased in the region over the 10 years of the study as the economic value of sheep for wool production declined.

Pastures are predominantly subterranean clover and ryegrass species and are rotated with crops on a two- to six-year cycle. Crops are winter cereals, predominantly wheat, oats, and barley. Some winter crops of grain legumes (field peas) and oilseeds (canola) are becoming increasingly important. Wheat is the major crop, with cropping intensities generally increasing over the period of the study in the surrounding region as the importance of animal production has declined. Conventional farms in the region have adopted minimum tillage practices with the introduction of herbicides and increased intensities of rotation. Stubble burning is still widely practiced.

14.4.3 Climate

The climate of the area is dominated by long hot summers and cool winters. Winter days average from 4 to 16°C, with summer days ranging from 17 to 29°C. Extremes recorded by researchers during the project were a winter morning temperature of −4°C and an early summer maximum of 45°C during harvest. A mid-summer soil surface temperature of 64°C was also recorded.

Rainfall is distributed fairly evenly throughout the year, with an annual mean of about 450 mm at the site. The growing season for crops and pasture is limited to winter/spring as high evapotranspiration rates in summer and autumn reduce soil moisture content below the threshold for crop production. Soil moisture is thought to be a major limiting factor for plant growth.

14.4.4 Soils

The soils are predominantly red earths (Northcote classification Gn 2.12 to 2.15), with some small areas of yellow earths emerging (Northcote classification Gn 2.42).

The soils are gradational, consisting of silty/sandy loams at the surface, moving to light and medium clays at depth. These soils have very low levels of organic carbon (generally less than 1.4%) and are poorly structured. The soils have low levels of aggregate stability. Surface soils (0 to 50mm) slake slowly but do not disperse. Below 50 mm all soils both slaked and dispersed rapidly; i.e., these soils lose their structural properties when wet, and lose their structural stability rapidly when worked wet. Structural properties decline rapidly at shallow depths, i.e., below 50 mm.

These soils are of general low natural fertility, with low levels of nitrogen (N) and phosphorus (P) usually limiting plant growth. In the farming systems studied N has usually been replaced by legume-based fixation during the pasture phase, although increasingly N fertilizers are being used as rotations shorten. P is added as fertilizer prior to or at crop sowing. Some P fertilization of pastures is done.

14.4.5 FARMING HISTORY

The area was used for extensive sheep grazing as parts of very large pastoral stations established in the 1850s. These stations were subdivided and cropping was introduced in the 1900s. Sheep/wheat farming has been practiced as the dominant form of agriculture since then.

One of the organic farms has been in the control of the same family since 1920, with the other purchased in 1958. Both farms have been under the same organic management since 1963. The conventional farms have been under similar management regimes with ownership periods varying from 1967, 1978, 1986, and 1996. Farming management histories have been established from the late 1940s.

All farms of the research site were under similar conventional management until 1963. Since then the organic farms have been managed without the use of synthetic chemical inputs of herbicides, pesticides, and soluble fertilizers. Full organic certification was achieved in 1987. The conventional farms use a range of such inputs regularly.

14.4.6 CROPPING SYSTEMS

Rotations on the conventional farms have included three years clover/grass pasture followed by three years of cropping, two years pasture with two years cropping, and two years pasture with one year of crop. The cropping sequence was wheat followed by barley, oats, or lupines. All conventional crop areas were prepared for cropping by the application of herbicides to pastures in late spring to control seeding, followed by summer grazing. Primary tillage usually occurred in late autumn one to two months prior to sowing of wheat in mid to late May. One to three cultivations may occur before sowing to help prepare a seedbed and control weeds. Both preemergent and postemergent herbicides were used on some crops.

Conventional crops were sown using fungicide-treated seed sown with fertilizer, usually in the form of di-ammonium phosphate. Some conventional pastures had received fertilizer in the form of superphosphate.

Rotations on the organic farm varied from six to nine years of grass/clover pasture followed by two years of crops. Six years of pasture was seen as usual. Toward the end of the project period, the organic farmer was starting to experiment with two-year pasture rotations. The cropping sequence was wheat followed by oats, rye, wheat, or barley. All pastures were prepared for cropping by grazing prior to primary tillage in spring, usually early October. Cultivation of the fallow continues over summer and autumn, following weed growth from summer rainstorms. This generally results in two to three further cultivations. Fallows are grazed by sheep to further control weed growth.

Organic crops are usually sown using untreated seed retained from the previous year's crop. Fertilizer in the form of reactive phosphate rock is spread on the prepared fallow and worked into the soil surface about three months prior to sowing. In recent times more directed nutrient replacement has been practiced with detailed commercial soil testing determining custom mixes of ground minerals applied annually prior to sowing.

14.4.7 YIELDS

There is considerable dispute about the level of yields achieved by organic systems compared with conventional crops. Wynen (1997) reports in a survey of farmer wheat yield figures that organic and biodynamic yields are equivalent to or exceed their conventional counterparts. In contrast, paired-plot comparison trials done at the Ardlethan site in 1991–1992 show that organic yields are significantly below conventional ones (Dann et al., 1994). These trials indicated that the organic yields were less than half of the conventional yields. Deerida et al. in Western Australia showed similar trends (Deerida et al., 1996). Data on commercial yields for fallow wheat crops (i.e., wheat crops grown immediately after a pasture phase; see Moore and Grace, 1998) were collected from the farms studied in the Ardlethan project. These data indicate that over the study decade conventional yields have exceeded neighboring organic yields.

Long-term organic wheat yields in commercial crops at the research site are 1.9 t/ha with conventional yields being 3.1 t/ha. Australian wheat yields overall averaged 1.91 t/ha for 1998 (ABARE, 2000).

14.4.8 FARMING SYSTEM COMPARISONS

Individual studies were undertaken across some eight years to explore emerging similarities and differences between the conventional and organic farming systems (see, for example, Ryan et al., 1994; Gatehouse, 1995; Newey, 1998; Derrick, 1996; Derrick and Dumaresq, 1999). These culminated in an intensive two-year study across seven wheat crops integrating a wide range of soil, crop, and agronomic measures.

The crops sampled were the routine commercial crops grown by the two farmers using their respective conventional and organic methods. Farmers were not asked to plant crops as a "treatment" for this project; rather, the project sampled those crops grown as the result of the farmers' commercial decisions. Subsequently, a range of combinations of slightly different cropping systems were sampled as the farmers fine-tuned their systems in response to changing markets, seasonal conditions, and costs.

A range of soil and crop measurements were taken. These included soil chemistry (N, P, cation exchange capacity [CEC], pH, cations), soil organic carbon levels, aggregate stability, steady-state infiltration rate, percent root length infected by vesicular-arbuscular mycorrhizae (VAM), soil invertebrate population size, bulk density, and total soil porosity.

Farm and paddock histories were established along with detailed maps of soil types and landscape processes. This enabled direct comparisons in the same cropping year of wheat crops growing under different management but in the same soil type and landscape position, and with as similar cropping and grazing histories as possible. For each crop, detailed agronomic data were collected, including yield, inputs, frequency and timing of each farm operation, and costs. Key similarities and differences for the organic and conventional farming systems are summarized in Table 14.2A,B.

As the previous studies had predicted, the organic cropping system had on average substantially lower yields. These yields were accompanied by significantly lower input costs. Along with the availability of substantial price premiums for organic wheat, this resulted in the organic system achieving substantially higher economic returns for all its crops. The organic farm also achieved consistently better environmental outcomes through its greater reliance on on-farm biological processes and lower reliance on off-farm chemical inputs.

Some quantifiable differences in system outcomes are summarized below as a set of comparative indices (conventional = 100) (Dumaresq and Green, 2001) (Table 4.3).

The southeastern Australian case study described above indicates how Australian broadacre farmers are able to change their in-field management systems, moving toward lower use of external inputs and greater use of in-field ecosystem services to produce commercially viable crops with better environmental outcomes. This study indicates one of several possible key approaches for achieving sustainable farming systems within Australian agriculture. This case study concentrates on on-farm in-field biophysical and agronomic processes and outcomes. The following study looks at how farm enterprises engaging in such change processes focused on sustainability fare in the wider socioeconomic contexts of whole production-to-consumption food systems.

14.5 SOUTHWESTERN AUSTRALIAN REGIONAL CASE STUDY

14.5.1 Case Study Overview

A study completed in 2008 explored sustainability values held within 10 agricultural supply chains. Noncertified supply chains were paired with certified supply chains (e.g., organic and EU eco-wool) to allow comparison between these approaches. The study involved exploring how social and environmental sustainability issues were valued and integrated in 10 production-to-consumption system case studies. The cases represented five different commodities, including wool, dairy, horticulture, grains, and viticulture. The farms sourcing these supply chains were located in the Blackwood Catchment in the southwest of Western Australia. As an established

TABLE 14.2A
Farming System Similarities

Systems Characteristic	Organic	Conventional
Enterprise mix		Same
Tillage practices		Similar
Cropping cycle		Similar
Soil types		Same
Total P		Similar
Total and available N		Similar
Soil infiltration		Similar

TABLE 14.2B
Farming System Differences

Systems Characteristic	Organic	Conventional
Fertilizers	RPR + gypsum, lime	DAP, superphosphate
Herbicides	None	1–4 applications per crop
Rotation length	6–9 years	2–3 years
Fallow length	7–8 months	1–4 months
Yield	Lower*	Higher
Costs[a]	Lower*	Higher
Returns[a]	Higher*	Lower
pH	Higher*	Lower
Cation exchange capacity	Higher*	Lower
Extractable P	Lower*	Higher
Organic carbon	Higher	Lower
Soil invertebrate diversity	Higher*	Lower
VAM presence	Higher*	Lower
Soil porosity	Higher	Lower
Soil macroaggregation	Higher	Lower

[a] Not actual costs and returns—modeled gross margins only. (Derived from Dumaresq and Greene, 2001.)

* $p \leq .05$.

TABLE 14.3

System Outcomes

Systems Characteristic	Organic	Conventional
Yield	65	100
Costs ($/ha)[a]	74	100
Return (organic wheat @ $215/t)[a]	148	100
Return (both systems @ $119/t)[a]	61	100
VAM presence	174	100
Soil faunal diversity (at harvest)	148	100
Soil macroaggregation (at harvest)	105	100
Soil porosity (at harvest)	110	100

[a] Not actual costs and returns—modeled gross margins only.

social catchment, the location provided an important context for the project. A majority of farm case studies were implementing an environmental management system, a voluntary program delivered by the local catchment group called BestFarms. This program assisted farmers in developing environmental management plans for their farms. The information relevant to the farm stage of these supply chain case studies is presented below.

The farmers involved in the 10 case studies all had in common the objective of creating improved food and fiber production systems according to their different values and motivations and interpretations of improved systems. Not all of the farmers interviewed had the objective of converting to alternative food and fiber systems, although some notably did (e.g., biodynamic grain and EU eco-wool case studies). The producers involved in this study, while they may not have set out to challenge current dominant structures in food and fiber supply, nevertheless "contribute to a practical critique of those structures through their actions and discourse" (Holloway et al., 2007, p. 90).

All farmers in this study were considered to be environmental best-practice farmers and were selected on that basis. A number of the case study farmers (many of whom were also company directors of the supply chains) showed enthusiasm about the potential to communicate sustainability values through a potential environmental certification system. However, the certified growers generally believed that the various certification systems (Demeter, NASAA, and EU eco-label) already adequately communicated their sustainability values to consumers and did not support an additional environmental/social assurance system.

The demographics of the case study farming families were varied; the families ranged from young families who had inherited the family farm to city dwellers who had chosen a "farm change" in later life. They all communicated a strong sense of place. Generally, these growers are keen to have their environmental efforts recognized and communicated for a number of reasons. Some growers saw communication of their environmental values as an educational tool, communicating to the wider community what has to happen on a farm to produce, for example, a liter of milk. Others saw the value in improving the general image of farming in the wider community.

14.5.2 FARMING SYSTEMS

Moving toward more sustainable agroecosystems was a common aspiration among the case study farmers. This included systems that utilize water more efficiently, systems that are resilient to variable weather conditions, and systems designed so that the inputs can be turned off as needed during times of reduced resources. The conventional farmers were using a number of techniques commonly associated with organic farming systems.

The use of agroecological principles, including composting, nutrient cycling, natural pest control, and encouraging soil biological activity, were daily toil for the organic and biodynamic growers. However, it appeared that all of the conventional growers were in one way or another moving more toward agroecological systems approaches themselves, including working with agronomists who were more sympathetic to organic systems.

Crop rotations are an important aspect of these systems. The conventional wool case study ensured long rotations between cropping and grazing of up to seven years, significantly longer than the one or two years of neighboring farms. No-till farming, while raising the issue of herbicide resistance, is widely considered an important signal of sustainable farming systems. Both the conventional wool and the conventional grains systems use no-till. They have also found alternatives to burning stubble, still a common practice in the area.

While some movement toward agroecological systems existed among conventional farmers, large transformative changes to production systems were not on the agenda, despite knowledge among conventional farmers about the benefits. Generally, costs were thought to be too high. Susan (biodynamic grains), who with her partner runs a biodynamic grain operation at a scale similar to that of most of her neighbors, observes that this lack of take-up is puzzling: "It seems incredible that we are looking at organic farming resolving salinity, water runoff, and all these issues; I guess there's no blanket solution—but why hasn't there been a greater uptake?"

14.5.3 FINDINGS

14.5.3.1 Alternatives to Chemical Pest Control Methods

Concerns related to the use of chemicals were expressed by the conventional farmers; these were usually raised by the women interviewed. Sarah (conventional grains) commented as follows:

> My opinion is that the whole chemical scenario has come upon us so dramatically. I reckon there is a time space of 15 years since when I first started farming with Robert. There would just be a bit of Round-Up used just as a knockdown for weeds, and they would still conventionally rip up and work back, to what it is now—such a vast array of chemicals used for so many things. And even though there's been some really good things, like with no-tillage, I can see our soils springing up and there's less horsepower

needed. But on the other hand, the use of chemicals has got so big. That's given people lots of different options and the ability to produce greater than they would have been able, but the whole impact I don't really think has been understood or measured and that concerns me, living amongst it.

Peter (conventional milk) made a similar observation about the explosion in the use of chemicals over his more-than-50-year lifetime. Both Sarah (conventional grains) and Nadia (conventional wool) shared their concerns about exposure of family members and other farmworkers to chemicals. Sarah said she herself buys organic food whenever possible (despite the difficulty she has in accessing this in her area), and hence feels somewhat conflicted about the use of chemicals on the farm. Alternatives to chemical pest control such as integrated pest management (IPM) were being explored by some of the conventional farmers in efforts to develop more resilient systems, to save on chemical costs, and to comply with pesticide residue requirements. Alan (conventional strawberries) aims to use minimal pesticides through the use of IPM.

Organic growers were more advanced with system techniques for managing pests and weeds. David and Rebecca (organic wine) applied their impressive scientific problem-solving ability to managing weeds in the vineyard:

> We did a root analysis to find out what was accumulating and it was potassium and calcium. As calcium levels change in the soil a lot of the sorrel disappeared. We use weeds as biomass. We don't disturb the soil structure. If you put in a cover crop you disturb the very nature and microbial activity in the soil. The latest research from CSIRO is you get 7 kg/ha of nitrogen from converting that. We developed a technique of just mulching it and turning it into a mat, so we just turn it into our own humus. We don't slash it, we mulch it.

14.5.3.2 Building the Soil Resource

Organic growers observed that their methods—such as feeding the soil rather than the plant and using whole-systems approaches—improved the general condition of the environment, not just the plant or animal. Improving the soil, is seen as a key philosophical and practical goal and outcome of organic farming. Kurt (biodynamic dairy) notes:

> When we first started, if you walked in there in the summertime, the quartz used to blind you. You can't even see it now because as we've changed the structure of the soils and put some health back into the soil, it's gotten darker and darker. I tried to rip it one day with a single tyne, couldn't rip it, it was that compacted. You can go and dig a hole in it with your hand now. Because we are using BD and the soil starts to work right, everything comes back how it's supposed to be and the soil is working and alive.

Technologies that help farmers maximize productivity according to water availability on their properties are also important. Most farmers in the wheat belt appear to know about the Ron Watkins system, a keyline-like system developed by Western Australian farmer Ron Watkins that uses tree-lined contour banks for water management. It is so named because Ron personally goes to farms and sets it up. However, very few farmers have invested in this community-owned technology.

The Ron Watkins system of water harvesting was used by the conventional wool, conventional grain, and biodynamic grain farmers, all of whom are located in the lower-rainfall areas of the catchment (around 300 to 400 mm/annum). This system ensured maximum water collection and storage.

The biodynamic growers are confident that their farming techniques have increased soil water-holding capacity. They suggest that their property greens up faster and stays greener longer than surrounding properties, essentially creating a longer growing season.

14.5.4 INDIVIDUAL CASES

14.5.4.1 Biodynamic Grains

Susan and Andrew converted the wheat belt farm that had been in Andrew's family to biodynamics the year they got married. They later introduced a postharvest processing stage into their operation because of the lack of profitable markets for biodynamic grain. They established a flour mill at a nearby wheat belt town to guarantee their on-farm grain price. They produce a range of fine gourmet flours, and they also initiated the development of a bakery in the coastal strip of the southwest. Their motivation to farm biodynamically relates to concern about exposing their family to chemicals, but also because they enjoy it immensely. Susan noted a research finding that farmers who do organic farming get the most joy out of farming. The farm has an area of 1,000 ha, and the boundaries follow a catchment, so they are relatively independent from downstream impacts from other farms. Tree planting on the farm started in the 1950s, most of the fences are on contours, and the keyline system has been implemented on some parts of the farm. Production risks were offset against a diversity of products ranging from gourmet flours to stock feed. Because of the available market niche for organic poultry crumble, they were able to use waste creatively.

14.5.4.2 Conventional Grains

Sarah and Robert have a mixed sheep and cropping enterprise located in the southwest wheat belt. They have made significant efforts to balance sustainability and profitability. They are successful farmers and have won primary producer awards. Their environmental priorities are related to maintaining soil health and protecting and enhancing biodiversity. They use a no-till system and minimal sprays. They specialize in growing soft wheat, which is used to make biscuits. Millers want low-protein wheat for making flour designed for use in biscuits and pastry. They also produce higher-protein noodle wheat, which attracts a premium for protein on the richer fertile soils. They have planted extensive biodiversity corridors, with more than 200,000 native seedlings established. The farm is fenced on the contour and uses a system of contour banks to harvest and control surface water and seepage, ensuring that there is adequate stock water even during drought. Waterways are protected and paddocks have been set up to support rotational grazing and efficient movement of stock through a network of laneways. They carefully monitor and manage soil erosion. They use chemicals as little as possible, with a preemergent herbicide applied at seeding and postemergent spraying done only as required. They constantly adapt their system to avoid pesticide resistance and have considered moving to more organic approaches such as integrated pest management.

14.5.4.3 Organic Horticulture

Warren and Olga initiated their organic strawberry operation in 2003. They are certified by National Association for Sustainable Agriculture Australia (NASAA) and produce approximately 20 tons of organic strawberries per annum for Western Australian and interstate markets. They supply to organic retailers as well as supermarket chains. In 2006, with the help of regional development grants, they established a small processing plant on their property with a view to using their seconds to produce their own labeled products. They are also setting up a café and plan to run a farm tour business focusing on production of organic and high-quality food and on the importance of the environment. Partly because of the small scale of this farm (which produces only 5% of the quantity of the paired conventional grower), they are able to attract adequate labor, have markets for their seconds, and receive a premium price for their product. They also have greater latitude in making managerial decisions (such as to focus on taste rather than transportability of fruit).

14.5.4.4 Conventional Horticulture

Alan has been growing strawberries for 26 years. He originally farmed in the southwest, but mainly due to labor shortages he moved his operation to Perth, where he has established a market garden in the outer suburbs. Alan grows the strawberries on constant rotation on his small acreage, producing about 450 tons annually. Alan makes significant efforts to manage his property sustainably, with a key focus on improving water efficiency and using integrated pest management. He is part of the Waterwise program and carefully monitors water use from groundwater supplies. Waterwise is a WA government program aimed at increasing water efficiency in irrigation. Alan's operation is fairly chemical intensive with fumigation and weed control requirements. He is gradually reducing his pesticide use, assisted by an agronomist who checks the crops weekly and advises on targeted pest control rather than broadscale preventative spraying. He is also reducing his fertilizer use through improved soil monitoring programs. He recognizes that alternative methods of leaving the soil fallow and increasing organic matter could decrease his reliance on chemicals; however, he suggests that implementing this would be too expensive.

14.5.4.5 Biodynamic Milk

The dairy is a dryland operation that uses biodynamic principles in combination with a stock nutrition supplement program; the animals are fed *ad lib*, allowed to meet their nutritional requirements through their own "nutritional intelligence." There are a number of significant differences between Kurt's dairy farm and conventional dairies, including no grain feeding and no irrigation of pastures. It is a small operation with approximately 100 cows in milking at any time. The cows are trained to manure outside the dairy, thereby eliminating the need for dairy wash-down and associated effluent issues. There is also less turnover of cows than in most conventional dairies. The heifers are milked for up to 10 lactations, rather than the average of 4 or 5. In the old style, they all have names. They are also allowed significant freedom to range about the property.

14.5.4.6 Conventional Milk

Peter and Elizabeth and their children are dairy farmers in the southwest. The family has a herd of 470 milking cows in two dairies with an equal number in calves. Peter has taken considerable effort in his breeding program, achieving a breeding value in the top 5% in Australia. The science of nutrition and feeding is also critical in their operation. Peter takes considerable effort to manage his farm sustainably. They use minimal sprays and limit medication such as antibiotics for cows. While they meet the best-practice requirements for their two-pond effluent system, Peter suggested that this could be improved. He commented that along with many other dairy effluent ponds in the area, it can overflow in the winter. Peter has taken steps to resolve this through implementing a system that uses effluent as irrigation and is spearheading the way with this in the local area.

14.5.4.7 Organic Wine

David and Rebecca commenced growing vines organically as a later-life venture in the southwest in the early 1990s. The farm is located in a bushland setting, chosen for its aesthetic and wildlife appeal. Their decision to farm organically was influenced by David's involvement in solar energy development in Australia in the 1970s and 1980s. They did not set out to be organic growers but discovered that there was no reason for them to use chemicals. One of the first organic vine growers in the region, they learned through experimentation. Their convictions have influenced not only their farming methods but their whole-farm approach, involving rehabilitation of the creek line and creation of buffer zones and wildlife corridors. They have won environmental awards for their on-farm environmental management. They feel that their NASAA certification adequately communicates their efforts toward sustainability to their buyers. They do not see the need for another environmental certification label on top of the NASAA label, but they do support a system of formal accounting of the environmental impacts of production and think that this would be helpful.

14.5.4.8 Conventional Wine

This syndicate-owned wine company started in 1997, aiming for production for the premium wines market. They produce a range of red and white wines sold in Western Australia, interstate, and overseas, particularly the United States and Asia. Led by company directors Christine and Mark, the key drivers for the establishment of the company were to fuel employment in the southwest and to engage in their passion for winemaking and business and community development. They established the vineyards by purchasing three lots of farmland in different parts of the southwest based on suitable soil types and market availability. They are committed to best practice, use mulching, and keep machinery traffic to a minimum. While they use a regular herbicide and fungicide chemical control program, they have not used insecticides since 1997. They attribute this to the presence of beneficial pests and birdlife and the choice of interrow crops that attract beneficial insects. They are also fortunate that in most seasons the local red gum flowers at the same time as the fruit, so the silver-eye bird pest is attracted to the gum-flower nectar rather than the grapes.

14.5.4.9 Eco-Wool

This case study involves an international company owned by Australian merino fine wool growers. The company specializes in wool marketing and wool supply chain management with a focus on the environmentally conscious outdoor wear market. The company has been in operation since 1998 and has sold eco-wool since 2000. Matthew, a wool grower from southwest Western Australia, was the initiator of the company and maintains the role of production chain manager. Recently incorporated into the standard that has been developed for this product is a farm-scale management system developed with the assistance of the WA Department of Agriculture and Food. This provides growers with best-practice information and benchmarks to assist them in managing their operations with minimal environmental impact. Over and above the chemical residue limit requirements of the eco-label, the company has identified several areas in which it is possible to improve the wool growing environment, including soil acidity, salinity, groundwater levels, erosion, animal welfare, staff training, and natural vegetation management.

14.5.4.10 Conventional Wool

The conventional wool case study is a high-quality merino stud and wool producer located in the wheat belt of the southwest of W. Australia. With the reins recently passed over from the previous generation, farmers Bede and Nadia have taken responsibility for a legacy of environmental sustainability and generations of breeding management. The farm has been in the family since 1905 and covers about 2,500 ha. There are three operations: stud sheep, commercial sheep for wool production, and fat lambs. The merinos bred from this farm have produced some of the most influential bloodlines in the state, with rams often attaining the highest prices at sales. Environmental management was a key focus of Bede's parents and grandparents. Evidence of rising salinity prompted the planting of 100,000 trees in the 1960s. When a geological survey in 1986 showed water tables were still rising, a whole-farm plan approach was implemented. As a result, almost the entire farm has the Ron Watkins system of water harvesting in place. This, along with shelter belt and tree planting throughout the farm, has helped drought-proof the farm. Bede and Nadia also pay particular note to soils and use a soil agronomist who works toward organic systems, advocating composting, lower-input systems, and management of pH and nutrient balance to control weeds. Salinity and wind erosion are also minimized because of the massive tree plantings. Plantings have also been designed to support bird habitat. Bede says that his parents spent hundreds of thousands of dollars on improving the farm environment before funds were available from Landcare. They crop about 35% of the farm compared to the local average of about 60 to 80%. Two years of cropping is rotated with five years of clover-based pastures.

14.5.4.11 Conversion to Sustainable Farming

All of the case study farmers interviewed told stories of conversion or adaptation to more sustainable farming systems, as they described and understood them. Table 14.4

TABLE 14.4
A Summary of Farmer Opinions about the Conversion Processes on Their Farms

Case Study	The Change toward Sustainability	Reason for Change	Strengths	Opportunities Realized
Biodynamic grains	Shifted from conventional to organic Generational change	Concern over chemicals Shift to organic occurred when female partner attained equal farmer status (when they got married)	Cropping system, supply chain relationships, networks	Market development/value adding—development of new supply chain (mill, bakery, and retailer relationships)
Conventional grains	Increasing biodiversity Reducing chemical use Trafficking within conventional system	Understanding the benefits of working with nature	Part of an established and guaranteed commodity market	Market QA pressure driving increased accountability
Organic horticulture	Increasing knowledge leading to practice shifts	The lifestyle ideal included a sustainable farm	Organic farming system	Market development/value adding—development of new supply chain (factory, tearoom, farm stay); able to reduce inputs to adapt to drought
Conventional horticulture	Improved water and fertilizer efficiency	Increasing awareness of hazards, including groundwater contamination from market gardening	In the business 18 years	Addressing environmental regulation—IPM, Waterwise, reduced fertilizer input
Biodynamic milk	Fundamental systems change to biodynamic and low input	Land degradation and production loss Youngest son tries a different tack	Farming system	Low-input farming (but correct inputs)

Continued

TABLE 14.4 (*Continued*)
A Summary of Farmer Opinions about the Conversion Processes on Their Farms

Case Study	The Change toward Sustainability	Reason for Change	Strengths	Opportunities Realized
Conventional milk	Increasing knowledge leading to practice shifts	Realizing that codes of practice (e.g., effluent legislation) not sufficient for environmental protection	Part of an established and guaranteed commodity market	Expansion including better effluent management systems
Organic wine	Sustainable from the start but increasing knowledge leading to practice shifts	Found they didn't need chemicals Influenced by involvement in alternative energy innovations	Problem solving Organic farming system Supply chain relationships	Market development—increasingly targeting environmental market
Conventional wine	Efforts at sustainable from the start but increasing knowledge leading to marketing shifts	Rural decline Need to develop sustainable employment options	Supply chain relationships	Social networks/lifestyle—developing a business that meets socioeconomic needs and sustainability desires
Eco-wool	Shift toward inclusion of environmental, social, and animal welfare accountability systems	Inevitable future market demand Market advantage Future vision	Farming system Bloodlines	Market development—adapting to potential future market demand and a new supply chain (farm plans and ISO 14001)
Conventional wool	Previous generation—shift to high-water-use systems to reduce salinity (keyline) and low-input, long rotations	Previous generation—understanding the extent of salinity Current generation—wanting to be more commercial	Best merino bloodlines in the state Generational commitment Established market	Adaptive farming

summarizes these shifts to organic, biodynamic, or more sustainable farming systems.

14.6 CONCLUSIONS

There are a range of drivers shifting Australian farming toward more sustainable systems. Different stakeholders have different opinions on sustainable farming systems. Signals of sustainable systems innovation and adoption range from the establishment of international and national standards and third-party assessment to individual enterprises developing and selling "trademark" approaches. Sustainable farming systems can be considered those that incorporate recommended management practices, which can range from practices recommended by government extension agencies to local farmer-devised techniques that are known to be beneficial. No one group of individuals, organization, or agency owns sustainable agriculture.

The eastern and western Australia case studies explored here demonstrate that while some farmers undertake full conversion to organically certified systems, others may aim to reduce impacts and optimize inputs and be satisfied with middle-way agricultural systems. Many of those in the latter category are on the way to organic conversion, but their efforts may not always result in certification. It is important to note that the full range of organic farming practices available to organic farmers are also available for adoption by conventional farmers. As the case studies presented in this chapter show, the adoption of these practices is well under way. Conversely, many, but not all, strategies available to conventional farmers are open to organic farmers and can be adopted by them as part of alternative systems innovation.

The exploration of the move toward more sustainable farming systems in the case studies confirmed that the desire to protect the health of people and the environment is a leading factor. This includes the health of the farm family, consumers, and the wider community. Other drivers include wanting to have more resource-efficient systems that work with nature. Evidence from the eastern Australian case studies show that soil health and productivity can be improved through organic systems and costly inputs can be reduced, indicating that the drive to improve the soil resource can be fulfilled without compromising commercial viability.

Sometimes these changes are generational, signaling the desire to do it differently from how it was done in the past or wanting to repair past damage by previous generations through an alternative approach. Often the desire for change arises because the current system of food and fiber production, both on and off farm, is simply not working.

Entrepreneurial drivers are also evident in people's decisions to convert to organic or forms of sustainable farming, with recognition that products from these farming styles can be marketed to attract consumers who share these values. Based on such improved markets and potentially reduced costs for chemicals and other costly inputs, conversion can increase the profitability of farm enterprises.

The case studies also indicate that at the level of the individual farm enterprise to the regional product chain, farmers and value chain participants are successfully addressing many of the long list of problems that government and industry inquiries and reports have indicated beset Australian agriculture.

Australia's current policy directions on sustainable production are *ad hoc* and lack a coordinated focus. There is limited technical support for all forms of alternative production by industry associations and government industry initiatives, especially for organic/biodynamic farmers. This limited support for conversion to organic and sustainable farming, particularly compared to the European Union, has not dented the enthusiasm of farmers such as those considered in this chapter. The organic/biodynamic industry is growing at approximately 20% per annum, with the retail value in Australia growing from $28 million in 1990 to $300 million in 2004 (Biological Farmers of Australia, 2004). This indicates a rapidly growing consumer demand for sustainable products that can continue to drive the development and extension of alternative agricultural systems that enhance agroecological function and maintain food production.

The drivers of conversion to organic and sustainable farming are many and varied, and while generally enterprises must be profitable to continue, the examples considered here demonstrate that a range of social, economic, and environmental factors need to be considered in efforts to understand and potentially enhance conversion to more sustainable farming systems. These examples also demonstrate that Australian agriculture is capable of creating a sustainable future for itself and for the world food trade.

REFERENCES

ABARE. 2000. *Outlook 2000.* Vol. 2. Agriculture and Regional Australia. Canberra: Australian Bureau of Agricultural and Resource Economics.

ABARE. 2007. *2007 commodities: Natural fibres: Outlook for wool and cotton.* Australian Bureau of Agricultural and Resource Economics. http://www.abareconomics. com/ interactive/ac_mar07/htm/fibres.htm (accessed 2007).

Australian Bureau of Statistics. 2006. *Agricultural commodities, Australia, 2006–7.* Australian Bureau of Statistics Catalogue 7121.0, 2008.

Australian Farm Institute. 2005. *Australia's farm-dependent economy: Analysis of the role of agriculture in the Australian economy.* New South Wales: Surry Hills.

Australian Government. 1998. *Sustainable agriculture—Assessing Australia's recent performance.* SCARM Technical Report 70A, report to the Standing Committee on Agriculture and Resource Management (SCARM), CSIRO Publishing, Melbourne.

Australian Government. 2006. *State of the environment: Australia 2006.* Melbourne: CSIRO Publishing.

Australian Government Productivity Commission. 2005. *Trends in Australian agriculture.* Australian Government Productivity Commission Research Paper, July 5.

Biological Farmers of Australia. 2004. *Organic annual report 2004.* http://www.bfa.com. au/_files/Organic_Annual2004_Take2.pdf (accessed 2007).

Dann, P.R., Derrick, J.W., Dumaresq, D.C., and Ryan, M.H. 1996. The response to superphosphate and reactive phosphate rock by organically and conventionally grown wheat. *Australian Journal of Experimental Agriculture* 36:71–78.

Deerida, A.R., Bell, W., and O'Hara, G.W. 1996. Wheat production and soil chemical properties of organic and conventional paired sites in Western Australia. In *Proceedings of the 8th Australian Agronomy Conference,* Toowoomba, pp. 200–2.

Derrick, J.W. 1996. A comparison of agroecosystems: Organic and conventional broadacrefarming in south east Australia. PhD thesis, Geography Department, Australian National University, Canberra.

Derrick, J.W. 1997. Organic agriculture in Australia: On-farm research. In *Organic agriculture in Australia*, ed. D.C. Dumaresq, R.S.B. Greene, and L. van Kerkhoff, 153–72. Rural Industries Research and Development Corporation Research Paper 97/14, RIRDC, Canberra.

Derrick, J.W., and Dumaresq, D.C. 1999. Soil chemical properties under organic and conventional management in southern New South Wales. *Australian Journal of Soil Research* 37:1047–55.

Dumaresq, D.C. 1992. Can alternative farming systems provide the answers? In *Alternative farming systems*. Roseworthy, SA: National Key Centre for Dryland Farming Systems.

Dumaresq, D.C., and Greene, R.S.B. 1997. Overview of the organic industry in Australia. In *Organic agriculture in Australia*, ed. D.C. Dumaresq, R.S.B. Greene, and L. van Kerkhoff, 95–109. Rural Industries Research and Development Corporation Research Paper 97/14, RIRDC, Canberra.

Dumaresq, D.C., and Greene, R.S.B. 2001. *Soil structure, fungi, fauna and phosphorus*. Rural Industries Research and Development Corporation Research Paper 1/130, RIRDC, Canberra.

Dumaresq, D.C., Greene, R.S.B., and van Kerkhoff, L., eds. 1997. *Organic agriculture in Australia*. Rural Industries Research and Development Corporation Research Paper 97/14, RIRDC, Canberra.

Fenton, D., MacGregor, C., and Cary, J. 2000. *Framework for review of capacity and motivation for change to sustainable management practices*. Canberra: Bureau of Rural Sciences.

Gatehouse, R. 1995. Mycorrhizae, soil management and soil structure in neighbouring organic and conventional dryland wheat farms at Ardlethan, NSW. BSc (Hons) thesis, Geography Department, Australian National University, Canberra.

Gray, I., and Lawrence, G. 2001. *A future for regional Australia: Escaping global misfortune*. Cambridge, UK: Cambridge University Press.

Halpin, D. 2004. *The Australian organic industry, a profile*. Australian Government Department of Agriculture, Fisheries and Forestry.

Holloway, L., et al. 2007. Beyond the 'alternative-conventional' divide? Thinking differently about food production-consumption relationships. In *Alternative food geographies. Representation and practice*, pp. 77–94, ed. D. Maye, L. Holloway, and M. Kneafsey. Amsterdam: Elsevier.

Knudsen, M.T., et al. 2006. Global trends in agriculture and food systems. In *Global development of organic agriculture*, pp. 3–48, ed. N. Halberg, H. Alroe, T. Knudsen, and E. Kristensen. Wallingford, UK: CABI Publishing.

Kondinin Group. 2000. *Organic farming in Australia*. RIRDC Publication 00/97.

The Land. 2006. City slickers love it farm fresh. December 14.

Lee, K.E., and Foster, R.C. 1991. Soil fauna and soil structure. *Australian Journal of Soil Research* 29:745–75.

Love C. 2005. Natural resource challenges facing Australian agriculture. Paper presented at Australian Farm Institute Strategic Roundtable Conference Proceedings, Rural Resources Group Pty. Ltd.

Moore, A.D., and Grace, P.R. 1998. Effects of annual pasture composition on subsequent wheat yields in the Waite Permanent Rotation Trial, South Australia. *Australian Journal of Experimental Agriculture* 38:55–59.

National Land and Water Resources Audit. 2001. *Australian agriculture assessment 2001*. Canberra.

Newey, A. 1998. The contributions of VAM fungi and soil fauna to structural properties of a red earth soil under winter wheat. BSc(Hons) thesis, Geography Department, Australian National University, Canberra.

Oades, J.M., and Waters, A.G. 1991. Aggregate hierarchy in soils. *Australian Journal of Soil Research* 29:815–28.

Rickert, K. 2004. *Emerging challenges for farming systems: Lessons from Australian and Dutch agriculture*. Canberra: Rural Industries Research and Development Corporation.

Rovira, A. 1993. Sustainable farming systems in the cereal-livestock areas of the Mediterranean region of Australia. Paper presented at 3rd Wye International Conference on Sustainable Agriculture, Wye College, University of London, September 1993.

Rowland, P. 2005. *Outline of current research and development for environmental management systems*. Canberra: Rural Industries Research and Development Corporation.

Rowland, P., Waller, M., Gorrie, G., and Douglas, B. 2005. *Towards a national approach to certification and information management in Australian agriculture*. Australia 21 Draft Options Paper, Canberra.

Ryan, M.H., Chilvers, G.A., and Dumaresq, D.C. 1994. Colonisation of wheat by VA-mycorrhizal fungi was found to be higher on a farm managed in an organic manner than on a conventional neighbour. *Plant and Soil* 160:33–40.

Tisdale, J.M., and Oades, J.M. 1982. Organic matter and water-stable aggregates in soils. *Journal of Soil Science* 33:141–63.

Wynen, E. 1997. An economic assessment of organic agriculture and implications for future research. In *Organic agriculture in Australia*, ed. D.C. Dumaresq, R.S.B. Greene, and L. van Kerkhoff, 110–15. Rural Industries Research and Development Corporation Research Paper 97/14, RIRDC, Canberra.

Section III

The Way Forward

15 Transforming the Global Food System

Stephen R. Gliessman

CONTENTS

15.1 CONVERSION IN PROCESS

The conversion process in our food systems is under way. The experiences presented in the chapters of this book demonstrate how farmer need and initiative, consumer interest and commitment, and researcher approach and focus are all converging to promote the transformation of our global food system. Members of all parts of the food system are thinking beyond the yield focus of level 1, with many of them making the wholesale substitution of inputs and practices required for level 2. Yet many of these same level 2 farmers and researchers are now confronting a host of limitations that require and promote the move to level 3 action—the redesign of their farming systems. But for level 3 conversion to fully occur, the involvement of all members of the food system is required. From the farm to the table, everyone must participate in the development of level 4 conversion: the thinking, values, and actions that build new and sustainable relationships in the food system. There are some important signs that this is happening (Gliessman, 2007).

15.2 SIGNS OF CHANGE

The USDA agricultural census released in February 2009 shows some encouraging evidence that level 4 change is happening in U.S. agriculture. For the first time in many decades, over the five-year period of the last census (2002–2007) there has been a dramatic increase in the number of farms and farmers (USDA/NASS, 2009). The number of farms grew by 4% and the operators of these farms are more diverse. Although nearly 300,000 new farms began operation since the census of 2002, there was a net increase of only 75,810. But this net gain bucks the trend for the censuses of the past 30 years.

Compared to farms nationally, most of the new farms tend to be more diversified and smaller in size, have smaller gross sales, and be owned by younger farmers who also have off-farm work. More than 36% are classified as residential or lifestyle farms, with annual sales below $250,000, but most have sales below $10,000. Such farms used to be known as hobby farms and were considered to be of insignificant value to agriculture. Another 21% of the new farms are classified as retirement farms, with similarly low total annual income, but operators who report that they are retired from normal-wage labor. Considering the increase in these two types of farms together, it would appear that a major change is happening in the farm sector. Is farming being valued as a lifestyle, not just a way of making a living? Is level 4 thinking providing an incentive for new farmers to diversify livelihood strategies, stay on the farm, and make a contribution to local food systems? Despite the need to maintain nonfarm or off-farm jobs, what we may be seeing with this sector is a different kind of farming where the livelihood strategy that is used is one that integrates activities. Farming is being seen more and more as a way of life, and less a main source of income. But farming in this sense still creates a connection to the land and a value for society. Keeping a farming activity in a family ensures that future generations will have a deeper connection to the land, farming, and sustainability.

Another trend seen in the past five years of data is that U.S. farm operators are becoming more demographically diverse. The 2007 census shows a 30% increase over 2002 in the number of principal farm operators who are women. Hispanic farmers increased 10% during this same time period, and the counts of Asian, American Indian, and African American farmers all went up as well. These changes are cultural expressions of change, and may indicate further conversion at level 4. According to the new secretary of agriculture, Tom Vilsack, the latest census of agriculture "is far more than a tally of numbers. It's a reflection of the people—and their livelihoods—behind those numbers."

Data from the 2007 census also indicate that some of the trends in our food systems that are considered antithetical to sustainability are showing no signs of changing direction. The concentration of farming activity in the large farm sector continues. The number of farms with sales of more than $500,000 grew by 46,000 in the time since the last census. In 2002, 144,000 of the more than 2 million farms produced 75% of the value of U.S. agricultural production. In 2007, the same share of production was concentrated in only 125,000 farms. Another way of looking at concentration is that for 2007, farms with sales of more than $1 million produced 59% of U.S. agricultural production, whereas in 2002 farms in this sales range produced 47% of all production. Farms in the mid-size range between these larger operations and the small farms described above suffered losses in number of operators, number of farms, and value of production. The American Farmland Trust, commenting on the 2007 census, also observed that although the number of farmers may have increased since 2002, the amount of farmland decreased in the United States by 16.2 million acres (AFT, 2009). That's a loss of 2 acres every minute.

Yet there is cause to think that a culture of sustainability is developing nonetheless. In addition to looking at farm numbers, farmer demographics, and the economics of farming, the census also examined other areas, such as organic farming, value-added farming, and specialty farming, all of which showed significant increases, with the greatest increases in the smaller farm sector. Since it is believed that access to information and the capability for rapid and direct communication promote the conversion process, it is interesting that the census also found that 57% of all farmers now have Internet access, an increase from 50% in 2002, with 58% of those with access having high-speed connections. Census questions regarding on-farm income generation and direct marketing raised considerable interest, and even a question about the maintenance of historic barns on farms showed how the farming sector is seeing value in lifestyle aspects that reach beyond production to a culture of sustainability.

The consumer has a crucially important role in the conversion process, and there is good evidence that consumers are beginning to demand food that is grown more sustainably. A recent announcement from the USDA's Agricultural Marketing Service shows the continuation of a recent trend in the growth of direct-to-consumer marketing options in the United States (USDA/AMS, 2008). The number of registered farmers' markets, where farmers are able to sell directly to consumers and avoid the middlemen, had reached a total of 4,685 in August 2008. This is a 6.8% increase from two years earlier, when there were 4,385 farmers' markets in the United States. Community-supported agriculture options, known as CSAs, have enjoyed a similar increase in popularity, expanding from none in 1985 to approximately 1,350 today. CSAs allow consumers to establish long-term relationships with specific farms and farmers through various subscription and direct delivery arrangements that give them fresh, local produce, and the opportunity to know the who, where, and how of the growing of their food. The majority of CSA operations are organically certified or considered more sustainable (Halweil, 2004).

The recent appearance of the concept of "locavorism," with its focus on eating locally, is another emerging trend that offers hope for change in our food systems (Nabhan, 2001). A *locavore* (or localvore) is defined as someone who tries to eat primarily foods from his or her local region, which is often called a foodshed. Consumers who consider themselves to be locavores generally hope to develop a more direct connection with their sources of food and support the local economy, while resisting industrialized and processed food that is shipped long distances and possibly creating a larger carbon footprint. In this sense, the relationship between the eater and grower is more important than the actual distance (somewhere between 100 to 250 miles qualifies as local, depending on the location). Several recent authors have provided engaging examples of how what we eat can create changes in many parts of the food system (Kingsolver, 2007; Pollan, 2008). The growing awareness of the connections between health, food, environment, and sustainability are drivers for change.

15.3 LESSONS LEARNED

With such potential for our food systems to move toward greater sustainability, it is important that we take steps to realize that potential. This can happen only if knowledge about the conversion process—what works and what does not, what barriers exist and how to remove them—is widely disseminated, expanded, and improved upon. A prodigious amount of such knowledge has been presented in the pages of this book. To facilitate its transfer into the minds of researchers and farmers and into the agendas of those individuals and organizations promoting food system change, we offer here a list of concise lessons derived from the foregoing chapters. Each lesson states a principle, generalization, or observation supported by the work contained in more than one of the chapters of this book.*

- Governmental policy that supports and incentivizes the conversion to more sustainable practices can be effective in producing change in the food system (Chapters 2, 7, 10, and 11).
- Changes in consumer preferences are an extremely important driver of bottom-up changes in the food system (Chapters 3, 8, 11, and 12).
- In less developed countries, converting to organic farming methods frequently results in increased production and lower input costs, which contributes to food security and sovereignty (Chapters 2, 9, and 13).
- In many regions of the world, traditional food production systems, threatened by the worldwide switch to more intensive, fossil-fuel-subsidized systems over the last few decades or more, serve as excellent examples, sources, and foundations for locally adapted, resource-use-efficient, sustainable agroecosystems (Chapters 9 and 14).
- Biophysical and environmental constraints—including aridity, easily erodable soils, and a short growing season—present enormous challenges in the development of more sustainable food systems in certain regions (Chapters 4, 5, and 13).
- Among farmers, psychological, social, personal, and community factors often play a more important role in motivating change and perseverance and determining success in conversion than do economic and technical factors (Chapters 2, 4, and 7).
- In many developed countries, trends in farm ownership, farm size, farm production, and division of income into on- and off-farm activities indicate that farming is being valued increasingly as a lifestyle choice, with the values inherent in that lifestyle being consistent with level 4 thinking (Chapters 12 and 15).
- During the transition process (after a farmer has converted the farm system and before it has recovered its fertility and yield potential) support in all its possible forms and sources—financial assistance from government agencies, technical advice from extension agents or researchers, social and psychological support from the community—is an important determinant of the success or failure of the conversion (Chapters 2 and 4).
- Organic production, while a clear improvement over conventional in terms of resource use and consumer health, is easily integrated into the current transnational food system dominated by large corporations driven primarily by profit seeking (Chapters 3 and 15).

---

* The listings of chapters after each statement are not meant to be exhaustive; in each case, the listed chapters are those that we believe contain the most evidence for the statement.

- Organically produced food is becoming increasingly mainstream in developed countries around the world; this trend represents both a danger that change in the global food system will be stalled at level 2 and an opportunity for growing awareness of the unsustainability of the current system to promote change at levels 3 and 4 (Chapters 3 and 14).
- Responsibility for the conversion process is ideally shared between researchers and farmers, with each participating in redesigning food systems as they learn together (Chapters 2, 6, and 14).
- Conversion can be viewed as a stepwise, almost linear process, where lessons learned at one level promote the transition to the next level. But it can also be seen as a set of parallel processes, where lessons learned at one level can drive changes at both lower and higher levels. In particular, the development of level 4 values and priorities can positively impact conversion at all other levels (Chapters 6, 10, and 15).
- The current demand from developed countries for organic products can be a strong driver of conversion in developing countries, but the danger still exists that organic products will only be grown for export in these countries, and not reach local- and national-level consumers (Chapters 8 and 11).

The lessons learned from all of the chapters in this book tell us that there are no "silver bullets" that will create sustainability in our food systems. No one technology, practice, input, or action will solve all issues. Each one must be examined from the interdisciplinary and interactive viewpoint of the complex indicators of sustainability that are emphasized by all chapter authors.

15.4 MOVING TOWARD SUSTAINABILITY

This book is about the opportunities that lie ahead for sustainable food systems. All of its authors operate on the basic premise that a sustainable food system is necessary to meet the urgent challenges immediately ahead. Over the past five or six decades we have built up a highly productive industrial food system, but it is highly subsidized by cheap fossil fuels, abundant land and water resources, and the maximization of cheap, energy-dense but nutrient-deficient calories. Now we face rising energy and food costs, a changing climate, declining water supplies, a growing population, and the paradox of widespread hunger and obesity.

The conversion to organic or ecological agriculture is an important step in the conversion to sustainability. Even though this step is nowhere close to being complete, we must begin thinking in terms of conversion levels 3 and 4. Without changing the design of food systems at the farm level, and without reconnecting the growers of our food with the eaters of our food, sustainability will most likely stay out of reach. A radically different approach is needed for growing food and developing the agriculture of the future. We must build food systems that are organized on a foundation of health: health for our communities, for people, for animals, and for the natural environment. The quality of food, not just its quantity, must guide our agriculture. The ways we grow, distribute, and prepare food should celebrate our cultural diversity and our shared humanity, providing not only sustenance but also justice, beauty, and security.

All participants in the food system, from growers to eaters, have a duty to create regional systems that can provide healthy food for their communities. We all have the duty to honor and respect the workers of the land, without whom we could not survive. As stewards of resources for the next generation, we all have the responsibility to protect the land and water on which we depend from degradation. As demonstrated by the authors of the chapters of this book, many of the needed changes have begun, but we must now accelerate the conversion of all sectors of our food system and make its benefits available to all. Some of the biggest changes in the food system probably need to come at the policy level. Only then will we see greater movement toward a culture of sustainability in all sectors of the food system.

At the national Slow Food Nation event held in July 2008 in San Francisco, an important group of sustainable food system advocates proposed the following set of 12 principles to frame the food and agriculture policy needed to move all agriculture toward sustainability (Food Declaration, 2008).

1. A healthy food and agriculture policy forms the foundation of secure and prosperous societies, healthy communities, and healthy people.
2. A healthy food and agriculture policy provides access to affordable, nutritious food to everyone.
3. A healthy food and agriculture policy prevents the exploitation of farmers, workers, and natural resources; the domination of genomes and markets; and the cruel treatment of animals, by any nation, corporation, or individual.
4. A healthy food and agriculture policy upholds the dignity, safety, and quality of life for all who work to feed us.
5. A healthy food and agriculture policy commits resources to teach children the skills and knowledge essential to food production, preparation, nutrition, and enjoyment.
6. A healthy food and agriculture policy protects the finite resources of productive soils, fresh water, and biological diversity.
7. A healthy food and agriculture policy strives to remove fossil fuel from every link of the food chain and replace it with renewable resources and energy.
8. A healthy food and agriculture policy originates from a biological rather than an industrial framework.
9. A healthy food and agriculture policy fosters diversity in all its relevant forms: diversity of domestic and wild species; diversity of foods, flavors, and traditions; diversity of ownership.
10. A healthy food and agriculture policy requires a national dialogue concerning technologies used in production, and allows regions to adopt their own respective guidelines on such matters.
11. A healthy food and agriculture policy enforces transparency so that citizens know how their food is produced, where it comes from, and what it contains.
12. A healthy food and agriculture policy promotes economic structures and support programs to nurture the development of just and sustainable regional farms and food networks.

As *policy* principles, these statements refer specifically to the steps that governments can take to promote the movement toward agricultural sustainability. It is evident from these policy statements that the specific initiatives, programs, laws, regulations, standards, levies, funding streams, and other policies that would flow from them encompass far more than agriculture; they suggest policies in the areas of education, international trade, commerce, the environment, energy, transportation, health, science funding, and labor. Transforming agriculture in the profound ways required for sustainability requires equally profound change in nearly all of the ways we human beings organize our existence on earth, and governments—to the extent that they represent people and the collective good—have a primary responsibility for engineering and guiding those changes through policy, funding, and governance. Governments, in other words, have an important role in helping to initiate and promote conversion at all its levels, with a particular focus on level 4.

The efforts of the European Union nations to support ecological agriculture (Guzmán and Alonso, this volume) demonstrate the crucial role of government policy in promoting conversion to sustainability. With its organic standards, economic support for farmers converting to ecological agriculture, recognition of the importance of the ecological services provided by agroecosystems, trade policies that protect domestic organic agriculture, and other policies, the EU has engineered a major expansion of ecological agriculture among its member nations. Government support in the EU is responsible, at least in part, for significant increases in the number of ecological farming operations and in the number of hectares of land under organic (ecological) management. In 2002, the EU had five times more hectares of land under organic management than the United States. In terms of the percentage of agricultural land under organic management, the disparity was even greater: the EU's figure is about 10 times that of the United States' (Dimitri and Oberholtzer, 2006).

The United States and other countries would do well to follow the example set by the EU in promoting conversion at level 2. But even this step would be just the beginning, because expanding organic production is merely one of the foundations for the changes that must come at levels 3 and 4. We must begin to institute more fundamental changes in everything related to food production and consumption. There are many policy shifts that might begin to accomplish this goal. It could be very effective, for example, to expand the role of food policy councils, advisory panels of citizens, and other stakeholders to local and regional governments in the United States. There are about 50 food policy councils in the United States at the present time and room for many more. In the area of education, it is time to reimagine home economics as a way of teaching a more environmentally responsible and healthy approach to food consumption. As a way of reorienting the economic basis of agriculture, we must begin to take into account all the environmental and social costs that have heretofore been eliminated (as externalities) from the cost-benefit calculus. Conventional farmers must begin to be penalized for the externalized costs of greenhouse gas emission, pollution of groundwater and waterways, degradation of the soil resource, and exploitation of human labor. These are just a few of the policy avenues that can be followed to promote the conversion to sustainability.

Ultimately, of course, governments are merely the agents of the people; sustainability is our collective responsibility. As we move toward developing more just and sustainable farms and food networks that meet the needs of local communities as well as distant markets, we will need the growth and development of a culture of sustainability in which eating is once again seen as an agricultural act, and level 4 becomes the guiding process for transforming all sectors and corners of our global food system.

REFERENCES

American Farmland Trust. 2009. *Better policy for farms, food, and the environment.* www.farmland.org/programs/farm-bill/9-for-09.asp.

Dimitri, C., and L. Oberholtzer. 2006. *EU and U.S. organic markets face strong demand under different policies.* USDA. http://www.ers.usda.gov/AmberWaves/February06/Features/feature1.htm#sectors.

Food Declaration. 2008. *A declaration for a healthy food and agriculture system.* www.food-declaration.org.

Gliessman, S.R. 2007. *Agroecology: The ecology of sustainable food systems.* 2nd ed. Boca Raton, FL: CRC Press/Taylor & Francis.

Halweil, B. 2004. *Eat here: Reclaiming homegrown pleasures in a global supermarket.* New York: Norton.

Kingsolver, B. 2007. *Animal, vegetable, miracle: A year of food life.* HarperCollins.

Nabhan, G.P. 2001. *Coming home to eat: The pleasures and politics of local foods.* New York: Norton.

Pollan, M. 2008. *In defense of food: An eater's manifesto.* New York: Penguin Press.

U.S. Department of Agriculture, Agricultural Marketing Service. 2008. *Number of farmers markets continues to rise in U.S.* AMS 178-08. www.ams.usda.gov/FarmersMarkets.

U.S. Department of Agriculture, National Agricultural Statistics Service. 2009. *2007 census of agriculture.* www.agcensus.usda.gov.

Index

Page numbers followed by *f* indicate figures.
Page numbers followed by *t* indicate tables.